THE ART OF RANDOMNESS

Randomized Algorithms in the Real World

by Ronald T. Kneusel

no starch press®

San Francisco

Printed in the United States of America

First printing

28 27 26 25 24   1 2 3 4 5

ISBN-13: 978-1-7185-0324-3 (print)
ISBN-13: 978-1-7185-0325-0 (ebook)

 Published by No Starch Press®, Inc.
245 8th Street, San Francisco, CA 94103
phone: +1.415.863.9900
www.nostarch.com; info@nostarch.com

Publisher: William Pollock
Managing Editor: Jill Franklin
Production Manager: Sabrina Plomitallo-González
Production Editor: Sydney Cromwell
Developmental Editors: Alex Freed and Eva Morrow
Cover Illustrator: Gina Redman
Interior Design: Octopod Studios
Technical Reviewer: Doug Couwenhoven
Copyeditor: George Hale
Proofreader: Audrey Doyle

Library of Congress Cataloging-in-Publication Data

```
Name: Kneusel, Ronald T., author.
Title: The art of randomness : using randomized algorithms in the real world / Ron Kneusel.
Includes bibliographical references and index.
Identifiers: LCCN 2023029979 (print) | LCCN 2023029980 (ebook) | ISBN 9781718503243 (paperback) |
 ISBN 9781718503250 (ebook)
Subjects: LCSH: Algorithms. | Numbers, Random. | Python (Computer program language)
Classification: LCC QA9.58 .K635 2024  (print) | LCC QA9.58  (ebook) | DDC 519.2/3--dc23/eng/20231018
LC record available at https://lccn.loc.gov/2023029979
LC ebook record available at https://lccn.loc.gov/2023029980
```

For customer service inquiries, please contact info@nostarch.com. For information on distribution, bulk sales, corporate sales, or translations: sales@nostarch.com. For permission to translate this work: rights@nostarch.com. To report counterfeit copies or piracy: counterfeit@nostarch.com.

[S]

In memory of George Marsaglia (1924–2011), PRNG designer extraordinaire

```
u32 x32(){static u32 _=9;_^=_<<13;_^=_>>17;_^=_<<5;return _;}
```

About the Author

Ronald T. Kneusel has been working with machine learning in industry since 2003 and completed a PhD in machine learning at the University of Colorado, Boulder, in 2016. Ron has six other books: *How AI Works: From Sorcery to Science* (No Starch Press, 2023), *Strange Code: Esoteric Languages That Make Programming Fun Again* (No Starch Press, 2022), *Practical Deep Learning: A Python-Based Introduction* (No Starch Press, 2021), *Math for Deep Learning: What You Need to Know to Understand Neural Networks* (No Starch Press, 2021), *Numbers and Computers* (Springer, 2017), and *Random Numbers and Computers* (Springer, 2018).

About the Technical Reviewer

Doug Couwenhoven is a research scientist with more than 30 years of experience developing algorithms for digital imaging applications. He has a BS in physics and an MS in electrical engineering and signal processing. He spent the first 24 years of his career working at a large imaging company developing software algorithms for digital photography, printing, and imaging systems, and holds 50 US patents in the field. In 2013, he joined an aerospace technology company, and is currently part of a research group there that focuses on developing deep learning and machine learning algorithms for remotely sensed data.

BRIEF CONTENTS

CONTENTS IN DETAIL

5
SWARM OPTIMIZATION   137

6
MACHINE LEARNING   173

10
EXPERIMENTAL DESIGN

11
COMPUTER SCIENCE ALGORITHMS

12
SAMPLING

FOREWORD

For a good part of the past 5,000 years, humans have been—so to speak—at odds with randomness. Whether it was feared, worshipped, or used for divination or gambling, randomness had the upper hand for much of that time.

Even though the ancient civilizations understood randomness (and philosophized about its role in the universe, causality, and free will), they did not possess the understanding of probabilistic behavior and long-term frequencies. In fact, it wasn't until the Renaissance that scientists started treating chance with rigor. The great minds of the 17th and 18th centuries—including Galileo Galilei, Jacob Bernoulli, Pierre de Fermat, Blaise Pascal, Abraham de Moivre, and Pierre-Simon Laplace—realized that the frequencies of random outcomes can be assessed and measured experimentally, allowing for the start of theoretical foundations of probability. Science was finally making progress with chance, much to the chagrin of fortune tellers and soothsayers everywhere.

Later centuries brought even more excitement, as randomness started to permeate almost every scientific field, partly energized by the developments in Brownian motion and stochastic processes, statistical physics, and the ideas based on ensemble averages of statistical samples that allowed scientists to tackle up-to-then impossibly hard questions. However, even though statistical physics held immense promise, the sampling methods of the late 19th and early 20th centuries—often done by countless women "calculators"—could not scale up nor keep up with the demands of science, which were made even more pressing by the world wars.

It was finally due to the development of the 30-ton 18,000-cathode-ray-tube ENIAC (Electronic Numerical Integrator and Computer) that the sampling could be done at 1,000-fold speeds and scales large enough to simulate the physics of a nuclear fission experiment. The whole endeavor was a large effort led by the University of Pennsylvania and Los Alamos National Lab, and included leading physicists, mathematicians, and computer scientists of the time such as Nicholas Metropolis, Stanisław Ulam, John von Neumann, and Enrico Fermi. These efforts finally enabled randomness to find its undisputed place in the sciences. (For a detailed account of the science in Los Alamos at the time, check out the *Los Alamos Science* Special Issue dedicated to Stanisław Ulam from 1987, available online at *https://la-science.lanl.gov/lascience15.shtml.*)

Randomness has since become commonplace, and is now a part of our everyday thinking—from deciding whether to bring an umbrella to work to analyzing whether to raise interest rates. Randomized algorithms have allowed us to develop completely novel paradigms in mathematics and computer science, and to extend science fields such as decision theory, sequential optimization and explore-exploit (bandit) schemes, and artificial intelligence. We can now solve many practical but previously intractable problems, from inferring spatial directions of impending solar storms to optimizing bidding streams in online advertising systems.

Dr. Kneusel's book is about applied randomness, bringing us into that world of scientific thinking and computer algorithms that have been improved by adding randomness, from true/pseudo/quasirandom number generators to the very first Markov Chain Monte Carlo code implementation by Arianna Rosenbluth. The book gives a gentle introduction to probability and entropy, with scientific anecdotes and Python examples sprinkled throughout. It leaves the readers with a set of randomness tools that could be used in everyday life, using open source and free resources, all while not requiring more than high school–level algebra.

This book couldn't be more timely. As our society is embracing artificial intelligence and machine learning at unprecedented scales, the need to understand randomness in the data and algorithms, and many different sources of uncertainty, can only grow. This book is a beginning, opening the door to the exploration of those vast fields of scientific randomness.

Vanja Dukic, PhD
Professor, Department of Applied Mathematics
University of Colorado, Boulder
May 2023

ACKNOWLEDGMENTS

Gratitude is not only the greatest of virtues, but the parent of all others.
—Marcus Tullius Cicero

Thanks, first and foremost, to Alex Freed and Eva Morrow, whose gentle approach to editing makes the entire process a joy. The book is better now than when I wrote it.

Thanks to Doug Couwenhoven, MSEE, for his time and thorough review of the book's technical aspects. Any remaining technical errors are entirely on me for ignoring Doug's wise advice and suggestions.

Special thanks to Monica Kneusel, MPH, for her review of Chapter 10 on experimental design.

Finally, I want to thank Vanja Dukic, PhD, for taking the time to write the foreword and all the good folks at No Starch Press for believing in the book and helping to make it a reality.

INTRODUCTION

In this book, randomness is produced by a *random process*, which returns an output when requested. The output is usually a number in the range 0 to 1, but may be any member of a predefined set of things (integers less than 100, letters, colors, dog breeds, and so on). In a random process, knowledge of previous outputs is of no utility in predicting future outputs. The output of a random process must be *unpredictable*; we can't know ahead of time what value or element of the set will be returned next.

This definition sidesteps any philosophical concerns. In fact, we aren't concerned with whether true randomness is possible; we care only that there are processes approximating this definition with sufficient fidelity to allow us to accomplish our goals.

In the end, this is a book about solving problems with randomness by using random processes.

Who Is This Book For?

This book is for people curious about, or fascinated by, using random processes to accomplish a nonrandom goal. As we'll learn, randomness offers a powerful way to approach problems that might otherwise be beyond our ken.

You'll find something of value in this book if you work in computer science, engineering, or mathematics; if you're a visual artist or a musician; or if you're a scientist of any stripe. This book is for everyone curious enough to pick it up, page through it, and read these words. In short, this book is for you.

What Can You Expect to Learn?

This book is a topical survey of what we might call applied randomness. By the end, you'll know when randomness is a viable tool and how to use it.

The following provides descriptions of each chapter.

Chapter 1: The Nature of Randomness   Explores ways to generate true randomness and approximate it with deterministic algorithms.

Chapter 2: Hiding Information   Discusses ways to use randomness as an essential part of steganography, the art of hiding information so that it isn't even noticed.

Chapter 3: Simulate the Real World   Explores randomness-driven simulation, which is critical to nearly every area of modern science, engineering, and economics.

Chapter 4: Optimize the World   Discusses a surprisingly powerful approach to optimization that involves swarms of agents or the evolution of a solution, both of which use randomness.

Chapter 5: Swarm Optimization   Continues to present examples of optimization involving randomness.

Chapter 6: Machine Learning   Investigates the ways that neural networks, extreme learning machines, random forests, and other forms of machine learning make heavy use of randomness.

Chapter 7: Art   Considers ways to use randomness to create art.

Chapter 8: Music   Attempts to generate music and evolve pleasant melodies from scratch.

Chapter 9: Audio Signals   Explores how compressed sensing can allow reconstruction of a signal from a sparse collection of random samples.

Chapter 10: Experimental Design   Demonstrates how randomness is used in the design of experiments, as well as its importance in producing meaningful results from a study.

Chapter 11: Computer Science Algorithms   Explores the randomized algorithms often used in the world of computer science.

Chapter 12: Sampling   Discusses the ways that drawing samples from complex probability distributions is a fundamental requirement when applying probabilistic models and Bayesian analysis. Randomness is key to this process.

The book ends with a list of resources to help you find additional information about each chapter's topics.

What I Expect You to Know

This is an intermediate-level book. You should be familiar with programming, especially in Python, and somewhat comfortable with Python's standard extensions like NumPy, Matplotlib, and to a lesser degree Pillow, SciPy, and scikit-learn. If these names mean nothing to you, you can find a brief introduction on this book's GitHub site (*https://github.com/rkneusel9/TheArt OfRandomness*).

You should be comfortable with high school–level math and perhaps a bit beyond. When things get math-y, I'll explain; in all cases, you'll profit from reading the code directly, even if some of the math is fuzzy.

How to Use This Book

This is a hands-on book filled with experiments. The code is in Python and uses the previously mentioned standard libraries. I describe additional libraries when needed. I'm assuming a Linux system running Ubuntu 20.04 or later. Everything will run under Windows or macOS, but installing the Python libraries might be a bit more complicated.

To install the core libraries under Ubuntu, or Windows and macOS if `pip3` is installed, try the following commands:

```
> pip3 install numpy
> pip3 install scipy
> pip3 install matplotlib
> pip3 install pillow
> pip3 install scikit-learn
```

Whatever version these commands install is likely to work fine.

If using Windows, install Python from *https://www.python.org*. Any current version of Python 3 will do. When you install, select any option for installing `pip` or `pip3`, and add Python to your path so you can run it from a command prompt. If you do this, the install commands should succeed.

Before reading any further, I recommend downloading all files from the book's GitHub repository at *https://github.com/rkneusel9/TheArtOfRandomness* by either cloning the repository from the command line

```
> git clone https://github.com/rkneusel9/TheArtOfRandomness
```

or using your browser to download everything as a ZIP file (click **Code**).

Chapter 1 is the starting point, in which we learn about both truly random processes and those that are deterministic approximations, along with some hybrid approaches. This chapter also develops the randomness engine, a Python class we'll use for every experiment. The randomness engine supplies random values for the experiments and lets us select between different sources of randomness.

Feel free to read the remaining chapters in any order, though I recommend reading Chapters 4 and 5 sequentially.

Each chapter introduces the topic and then explores it with experiments. Some experiments include low-resolution test images to demonstrate the concepts at work. First, we run the code (after you review the code ahead of time) and interpret the results. Most experiments admit multiple parameter settings to allow you to build intuition. Second, we walk through the essential parts of the code to understand how and where randomness comes into play. Many chapters conclude with an "Exercises" section proposing questions designed to encourage you to continue exploring.

Questions will likely arise. Feel free to contact me. Also, if you find any bugs—or if you create something amazing with the materials in this book—please let me know: *rkneuselbooks@gmail.com.*

1

THE NATURE OF RANDOMNESS

Random processes power the systems we'll develop later in this book. This chapter introduces specific random processes, from those that are truly random to those that are deterministic but still random enough to use—that is, pseudorandom and quasirandom processes.

We'll begin with a brief discussion of the relationship between probability and randomness. After learning how to determine whether a process is random, we'll explore truly random processes, meaning processes strongly influenced by true randomness. We'll also learn the difference between pseudorandom and quasirandom processes. Finally, we'll use Python to create the RE class, the randomness engine we'll use in all of our experiments.

Probability and Randomness

Probability distributions represent the possible values a *random variable* can take and how likely each value is to occur. For us, a random variable is the output of a random process.

Probability distributions come in two varieties. *Continuous* probability distributions return values from an infinite set, meaning any real number in the allowed range. Here, the word *real* means elements of the set of real numbers, denoted $\mathbb{R}$, which are all the numbers on the number line. *Discrete* probability distributions are restricted to returning values from a finite set of values, like the heads or tails of a coin or the numbers on a die.

Random processes generate values, known as *samples*, that come from some kind of probability distribution, be it continuous or discrete. For example, a coin flip delivers samples that are either heads or tails, while a die delivers samples from the set $\{1, 2, 3, 4, 5, 6\}$ (assuming a standard six-sided die).

If a random process returns a single number, how do we know what distribution it's sampling from? In some cases, we have theoretical knowledge, but in other cases, all we can do is generate many samples. Over time, the relative frequency with which each possible outcome appears will become evident and serve as a stand-in for the true probability distribution.

Discrete Distributions

As an example of a discrete probability distribution, suppose that, thanks to the generosity of a local wizard, I have in my possession a one-of-a-kind die with only three sides. The numbers 0, 1, or 2 appear on the faces of my three-sided die. Therefore, each roll of the die results in one of the three sides face up. When I roll my die 50,000 times, keeping a tally of the number of times each side appears face up, I get the results in Table 1-1.

Table 1-1: Rolling a Three-Sided Die 50,000 Times

Face	Count
0	33,492
1	8,242
2	8,266

The results indicate that the probability of getting a 0, 1, or 2 is not equal: 0 appeared 33,492/50,000 = 66.984 percent of the time, 1 appeared 16.484 percent of the time, and 2 appeared 16.532 percent of the time. If I were to repeat the experiment, the exact count for each outcome would be slightly different, but it's clear that many such experiments would lead to outcomes in which 0 appears approximately 67 percent of the time and 1 and 2 each appears 16.5 percent of the time. Notice that 67 percent + 16.5 percent + 16.5 percent = 100 percent, as it should if the only possible outcomes from rolling my magic die are 0, 1, and 2 in the proportion 67 : 16.5 : 16.5.

Rolling the three-sided die is a random process that samples from a probability distribution, one where the likelihood of a 0 is 67 percent and the likelihood of a 1 or a 2 is 16.5 percent each. There are a finite number of possible outputs, so the probability distribution is discrete.

Let's try two more experiments that sample from discrete probability distributions. The first flips a fair coin 50,000 times. The second rolls a standard six-sided die 50,000 times. As before, let's tally the different outcomes, using 0 for tails and 1 for heads. Table 1-2 shows the results for the coin flips.

Table 1-2: Flipping a Fair Coin 50,000 Times

Side	Count
0	25,040
1	24,960

Table 1-3 shows the results for the die rolls.

Table 1-3: Rolling a Standard Die 50,000 Times

Face	Count
1	8,438
2	8,252
3	8,292
4	8,367
5	8,336
6	8,315

Just as my magic three-sided die's outcomes are not equally likely, the coin flips and standard die rolls deliver each outcome with equal probability. The counts are nearly uniform. Such *uniform distributions*, those that produce each possible outcome with equal probability, are far and away the most common type of distribution we'll harness in this book.

NOTE *To run this experiment, I didn't actually flip a coin and roll a die 50,000 times. Instead, I used a pseudorandom generator, discussed later in this chapter.*

Continuous Distributions

Think of continuous probability distributions as the limit of discrete distributions. For example, a six-sided die selects from six possible outcomes. A 20-sided die, sometimes called a D20, selects from 20 possible outcomes. If we could somehow let the number of sides extend to infinity, then such a die would choose, on each roll, from an infinite number of outcomes. This is what a continuous probability distribution does.

Suppose a random process generates real numbers in the range 0 to 1 such that all real numbers are equally likely. In that case, the random process is sampling from a continuous uniform distribution, just as a die samples from a discrete uniform distribution. We'll make heavy use of the continuous uniform distribution as we proceed through the book. Likewise, we'll occasionally use the *normal* (sometimes called *Gaussian*) distribution.

Introducing some math notation at this point will make it easier to understand what comes later. The uniform distribution we'll use most often draws samples from [0, 1), meaning the sample is greater than or equal to 0 and strictly less than 1, that is, $0 \leq x < 1$. When writing a range like [0, 1), note whether a square bracket or a parenthesis is used. The square bracket means the limit is included in the range, but a parenthesis excludes the limit. Therefore, (0, 1) specifies a range where all possible real numbers except 0 and 1 are allowed. Similarly, [0, 1] includes both 0 and 1, while [0, 1) includes 0 but excludes 1. The pseudorandom and quasirandom processes described later in the chapter usually produce outputs in the range [0, 1).

A uniform distribution is straightforward, whether continuous or discrete: each possible outcome is equally likely to appear. However, in a normal distribution, some values are more likely to be produced than others.

The best way to appreciate a normal distribution is to examine its histogram. For example, Figure 1-1 shows the distribution of 60 million samples from a normal distribution.

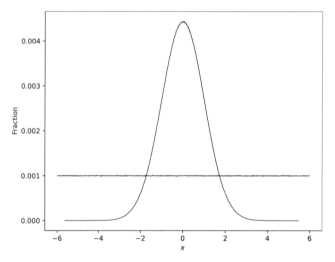

Figure 1-1: A histogram showing normal (curve) and uniform (line) distributions

The values in Figure 1-1 range from approximately −6 to 6. Values around 0 are most likely, with those at the extremes least likely. The normal distribution is widespread; many physical phenomena follow this distribution. Crucially, when we take a large set of samples from *any* distribution, the mean values will follow a normal distribution. This is the *central limit theorem*, and it's fundamental to statistics.

In the previous section, Table 1-2 and Table 1-3 tally the number of heads and tails and the number of times each side of a die appeared. A histogram is a graphical representation of such a table. The possible outputs are placed into bins of some specified width. For a die, the natural bin width is 1 so that each side falls into the bin with the same label.

For a continuous distribution, the bins cover a range. For example, if we have a process that generates numbers in the range [0, 1), and we want

10 bins, then we might make each bin 0.1 wide. A sample of $x = 0.3052$ will then fall into bin 3, counting from 0, because:

$$0.0 \leq x < 0.1 \;\rightarrow\; \text{bin } 0$$
$$0.1 \leq x < 0.2 \;\rightarrow\; \text{bin } 1$$
$$0.2 \leq x < 0.3 \;\rightarrow\; \text{bin } 2$$
$$0.3 \leq x < 0.4 \;\rightarrow\; \text{bin } 3$$

Likewise, a sample of $x = 0.0451$ will fall into bin 0, and so on. When all samples have been placed in a bin, the histogram plots either the count in each bin or the fraction of samples that fell into that bin. The fraction is found by dividing each bin's count by the sum of all the bins. A histogram using fractions per bin approximates the true probability distribution.

Let's return to Figure 1-1. The figure uses 1,000 bins, which explains why the curve looks more like a curve than a bar plot. The figure plots the fraction in each bin, not the count, which lets us compare different distributions without worrying about the number of samples used to generate the histogram. As the number of samples increases, such a histogram becomes a better approximation of the true probability distribution.

The horizontal line in Figure 1-1 represents a continuous uniform distribution selecting values in the range $[-6, 6)$. If each value is equally likely, then on average, each bin will be equally populated and $1/1{,}000 = 0.001$, thereby explaining the y-axis value for the uniform distribution.

We need only remember that the random processes powering our experiments generate values according to some distribution, primarily the uniform or normal distributions. There are many other standard distributions, but we won't explore them in this book.

NOTE *To learn more about probability and statistics, I recommend* Statistics Done Wrong *by Alex Reinhart (2015) or my book* Math for Deep Learning *(2021), both available from No Starch Press. You'll find discussions of probability and statistics, along with differential calculus, all of which we'll touch on at various points throughout this book.*

We need to know how close our random processes come to being, well, random. Before we dive into randomness engines proper, let's consider how to approach testing the output of a random process.

Testing for Randomness

How can we know that the output of a random process is truly random? The short answer is: we can't. However, that shouldn't deter us. We're not attempting to solve deep philosophical issues, as interesting as they might be. Instead, we seek what's good enough to accomplish our immediate goals and no more.

Is this string of binary digits random?

0101010011100000110000011101101111111011100000

Well, it looks kind of random, but how can we tell? Earlier, we used the frequency of each possible output to tell us whether the sample matches expectations. A random process generating 0 or 1 with expected equal likelihood should match as well. In this case, there are 23 zeros and 23 ones. Does that indicate that the sequence is random?

You begin to see what I mean when I say we can't tell whether a process is truly random. All we can do is apply tests to increase our confidence in our belief that the sequence israndom. We can check the expected frequency of the various outputs, but that's not sufficient. For example, the following sequences also have equal numbers of zeros and ones:

0000000000000000000000011111111111111111111111

1010101010101010101010101010101010101010101010

Most of us wouldn't consider either sequence to be particularly random.

In truth, none of the previous three sequences are the output of a random process. The first one is the binary representation of the operation codes for a 6502 microprocessor program to display the letter A on the screen of an old Apple II computer. The bit patterns for the opcodes are not random, but depend critically on the internal architecture of the microprocessor. I made the other two sequences by hand to have equal numbers of zeros and ones.

Over the years, researchers have invented many statistical tests designed, collectively, to detect whether a sequence of values is worthy of being called random. We have no space here to dive into these tests, but they go far beyond frequencies and consider short- and long-term correlations of all kinds. A few such test suites are DieHarder, TestU01, and PractRand. These suites generally require vast collections of values, far more than we can work with here.

So what's a person to do? We cannot prove that a randomness engine is generating random outputs, but we can gain enough confidence to believe more or less strongly. To accomplish this, we'll use a command line program called ent, from the word *entropy*. It applies a small set of statistical tests that might influence our beliefs.

Many Linux distributions include ent, but if yours doesn't, install it using the following command:

```
> sudo apt-get install ent
```

Check the following website for a compiled Windows version (along with a link to its GitHub repository, should you wish to examine ent's source code): *https://www.fourmilab.ch/random*. To install ent on macOS in the previous command, replace sudo apt-get with brew.

The ent program requires a file of bytes, which it assumes are uniformly distributed in the range $[0, 255]$. This means that the random process under test must have its output converted into a set of uniformly distributed bytes. We'll learn how to do this later in the chapter.

For now, the book's GitHub page includes a file we can use to understand ent's output, *ent_test.bin*. Pass it to ent like so:

```
> ent ent_test.bin
Entropy = 7.999996 bits per byte.

Optimum compression would reduce the size
of this 40000000 byte file by 0 percent.

Chi square distribution for 40000000 samples is 241.36, and randomly
would exceed this value 72.09 percent of the times.

Arithmetic mean value of data bytes is 127.5064 (127.5 = random).
Monte Carlo value for Pi is 3.141776714 (error 0.01 percent).
Serial correlation coefficient is -0.000234 (totally uncorrelated = 0.0).
```

The file *ent_test.bin* contains bytes generated by a good, but rarely used, pseudorandom number generator called MWC (multiply-with-carry). We might expect ent to report that the file is random. However, ent doesn't make that determination for us. Instead, ent runs a set of six statistical tests and reports the results, leaving it up to us to decide whether those results warrant a belief toward randomness.

The first test measures the entropy of the bytes. *Entropy* is a measure of disorder in a system. To a physicist, entropy is related to the number of microstates of a system—for example, the position and momentum of the molecules in a gas leading to the same macroscopic values of large-scale properties like temperature and pressure and the ways in which those molecules can be arranged. The entropy reported by ent, however, is deeper than that. It is the *Shannon entropy*, a measure of information content. In this case, it's expressed in bits. There are 8 bits in a byte, so a maximally random sequence would have an entropy of 8.0, meaning information content is maximized. Our test file has an entropy of 7.999996 bits per byte, which is extremely close to 8, a good sign.

We use the entropy reported by ent to estimate how much it's possible to compress the file. It's an alternative presentation of the entropy. Compression algorithms work by taking advantage of the information contained in a file, as measured by its entropy. The lower the entropy, the more redundant the data, and the lower the information content. If the information content is low, there is another way to express the information that takes less space. However, if the file is random and entropy is maximized, there is no alternative way to express the file's contents, so it cannot be compressed.

Next, ent applies a χ^2 test. The important part here is the percentage reported. If this percentage is below 5 percent or above 95 percent, then the expected frequencies—that is, the number of times each byte value appears—is suspect. Here, we have 72 percent, so we're again on solid ground.

If the sequence of bytes is random, we might, correctly, expect the average value of the bytes to be $255/2 = 127.5$. Here, we have an average value of 127.5064, which is quite close.

It's possible to estimate π with random numbers; ent uses this as another test of randomness. In this case, ent arrives at an estimate 0.01 percent off from the number of digits shown. If there is something in the sequence of bytes that biases the simulation, then it should manifest itself in the calculated π value. We'll use random numbers to estimate π in Chapter 3.

The final output line applies a statistical test to measure how related byte n is to byte $n + 1$; that is, it pays attention to the order of the bytes. If the bytes are not serially correlated, at least to the level of one to the next, the resulting coefficient will be zero. Here, it's ever so slightly negative but quite close to zero.

All in all, then, ent's report gives us strong confidence that the contents of the file *ent_test.bin* are worthy of being called random. Would you secure your bank account with this level of confidence? I sincerely hope not, but we're not interested in cryptography; we're interested in random processes, real or synthetic, that are sufficiently random to power our experiments. For that, ent is the only tool we need.

However, ent's output is needlessly verbose, especially since we'll use it often in this chapter. Let's define an abbreviated version. Instead of the output shown earlier, I'll report ent results like so:

```
entropy: 7.999996
chi2   : 72.09
mean   : 127.5064
pi     : 3.141776714 (0.01)
corr   : -0.000234
```

Let's put our nifty randomness detector to work. We'll begin with truly random processes.

Truly Random Processes

In this section, we'll review several sources generally accepted as true random processes: experimenting with coins, dice, electrical noise in different forms, and the decay of radioactive elements. We'll use the datasets we create here for experiments later in the book. The random processes covered in this section also provide a comparison against the next section's pseudorandom processes, which merely give the appearance of randomness.

Humanity has, over the centuries, developed multiple ways of generating randomness, including coin flips and dice rolls. Let's consider these to see if we might trust them as randomness engines.

Flipping Coins

Most people consider a coin flip to be a reasonable source of randomness, but is that really the case? In 2009, two undergraduate students at the University of California, Berkeley, flipped a coin a total of 40,000 times while keeping track of the coin's starting orientation, either heads up or tails up. Table 1-4 shows what they found (data used with permission).

Table 1-4: Flipping 40,000 Coins by Hand

Orientation	Heads	Tails	p-value
Heads up	10,231	9,770	0.0011
Tails up	9,985	10,016	0.8265

A glance at Table 1-4 shows that when the coin was facing heads up, there were more heads at the end of the flip. The same is true for tails up; more tails were measured. We can use the χ^2 test to see if these proportions are in concert with the expected 50-50 split of a fair coin. The resulting p-values are in the rightmost column.

A p-value is the probability of measuring the observed number of heads and tails, given that the null hypothesis is true. In statistical testing, the *null hypothesis* is the hypothesis being tested. In this case, for the χ^2 test, the null hypothesis is that the observed number of heads and tails is consistent with equal likelihoods. The p-value provides evidence for or against this hypothesis. If the p-value is below the standard, somewhat arbitrary rule-of-thumb threshold of 0.05 (5 percent), then we say that the p-value is *statistically significant*, and we claim evidence against the null hypotheses.

The smaller the p-value, the stronger our evidence becomes. For a p-value threshold of 0.05, we might expect to falsely reject the null hypothesis about 1 time in 20, while for a p-value of 0.01, the false rejection rate becomes 1 in 100, and so on as the p-values get smaller and smaller. However, only death and taxes are certain. A small p-value is not *proof* of anything; it's only an indicator, a reason to believe or not to believe, though with perhaps strong evidence.

Look again at the Heads up row in Table 1-4. The p-value is 0.0011, or 0.11 percent. According to the χ^2 test, there is a 0.11 percent probability of the observed counts (or a more extreme difference), given the null hypothesis is true. Therefore, we have evidence in favor of rejecting the null hypothesis. In other words, we have evidence for believing that Subject 1, who did the heads-up portion of the experiment, was not random, but was instead biased toward heads.

Subject 2, however, produced results consistent with the null hypothesis. For her, the p-value was 0.8265, or 83 percent. Again, the p-value, in this case, means the χ^2 test reports a probability of 83 percent for observing the counts, given the null hypothesis is true. This makes perfect sense, so we have evidence supporting the null hypothesis for the tails-up case.

The χ^2 test compared the tallies with the expected 50-50 split. We can do one more test: a *t-test*. The t-test compares two datasets and returns a p-value we interpret as the likelihood the datasets were generated by the same process. In this case, the t-test between the heads-up and tails-up datasets returned a p-value of 0.0139, or 1.39 percent, again below the standard 0.05 threshold. This serves as evidence that the two datasets are likely drawn from different processes.

What does that mean in this case? We have a single set of flips from two subjects, and each subject only flipped coins with the same side up each time. It's conceivable, but not proven, that Subject 1 was highly consistent in her flips and, as such, biased the coin tosses so that heads were favored when heads were the starting condition. For us, this fun example serves as an indication that humans are not to be trusted to act randomly.

We have evidence that Subject 1 biased the coin flips. Are we stuck with the bias? Actually, no. American mathematician and computer scientist John von Neumann came up with a clever algorithm to make a biased coin fair. The algorithm is straightforward:

1. Flip the biased coin twice.

2. If both flips came up the same—that is, both heads or both tails—start again from step 1.

3. Otherwise, keep the result of the first flip and disregard the second.

Applying this algorithm to the sequence of heads and tails generated by Subject 1 leaves us with 2,475 heads and 2,538 tails. The χ^2 test delivers a p-value of 0.37, which is well above 0.05 and strong evidence that the resulting dataset is now acting as expected.

Why does von Neumann's algorithm work? Consider a biased coin where the probability of getting a heads is not 0.5 but 0.8, implying that the probability of a tails is 0.2, since probabilities add to 1. In that case, flipping the coin twice will lead to the four possible combinations of heads and tails with the following probabilities:

$$(0.8)(0.8) = 0.64 \ (HH)$$
$$(0.8)(0.2) = 0.16 \ (HT)$$
$$(0.2)(0.8) = 0.16 \ (TH)$$
$$(0.2)(0.2) = 0.04 \ (TT)$$

Recall that if events are independent, as even the flips of a biased coin are, then the probabilities multiply. Also, the probability of getting heads followed by tails equals that of getting tails followed by heads. Therefore, consistently selecting the first (or the second) in either of these cases must lead to selecting heads or tails with equal likelihood.

The file *40000cointosses.csv* contains the dataset used in this experiment with the associated code in *40000cointosses.py*.

NOTE   *Please review the original web page for other comments on how the experiment was conducted, including suitable warnings about taking the results as anything more than a hint that further experimentation might uncover something interesting: https://www.stat.berkeley.edu/~aldous/Real-World/coin_tosses.html.*

Rolling Dice

Flipping a fair coin is a random process, as is rolling a fair die. But is there really such a thing as a "fair" die? Imperfections in the manufacturing process, a slight deviation in shape, or unequal density throughout the die's body might lead to a bias. Still, overall, and especially for our purposes, we might believe that dice rolls are random enough to be useful.

I gathered 14 six-sided dice and rolled them, en masse, using a dice cup from a game. I then took a picture of the results so I could count how many of each face appeared. I repeated this process 50 times for a total of 700 dice rolls. Table 1-5 shows the results.

Table 1-5: Tallying Dice Rolls

Outcome	Count
1	122
2	98
3	106
4	126
5	119
6	129

As before, if the dice are fair, we expect each count to be the same. For 700 rolls, we expect each possible output to appear $700/6 \approx 117$ times—a spartan number because of the small scale of the experiment, but sufficient for us to acknowledge the chief concern of this section: to master our appreciation of truly random processes.

The counts in Table 1-5 are not 117 but deviate, often by quite a bit. Does this mean the dice are loaded? Perhaps, but we'll never know for sure; we can only gain evidence favoring one answer over another. The χ^2 test is our tool of choice here, as it was for the preceding coin flips. Applying it returns a p-value of 0.28, well above the threshold of 0.05 that is generally accepted as statistically significant. Therefore, we do not reject the null hypothesis and believe that the dice are reasonably fair and, consequently, a potential source of true randomness.

Macroscopic physical systems are likely insufficient if the goal is to create truly random numbers; there are too many biases involved, though we can use an algorithm like von Neumann's to improve the situation. Also, a physical randomness engine, perhaps based on automatic dice rolls, will degrade over time, further introducing bias. Therefore, we must search in a different direction. Let's now consider processes more suitable for a randomness engine.

ROULETTE WHEELS

Roulette is another physical process that comes to mind when contemplating potential sources of randomness. In roulette, people bet on the position in which a marble will ultimately land when rolled against a spinning wheel. It is a favorite target for people trying to game the system because players can make

(continued)

bets while the wheel is spinning. On the face of it, a roulette wheel should be as random as rolling dice, but mechanical defects, especially if the wheel is tilted even the tiniest bit, bias the final ball position to the advantage of the clever.

The first occurrence of gaming a roulette wheel that I could find happened around 1880 when Englishman Joseph Jagger, a textile worker, realized that imperfections in the construction of the roulette wheels in Monte Carlo enabled him to predict the final ball position reliably enough to win more often than he lost. His success forced improvements to the design of roulette wheels.

Around 1960, Edward Thorp, working with Claude Shannon, constructed what is likely the first wearable computer for the sole purpose of gaming roulette. The full account is in Thorp's paper, titled "The Invention of the First Wearable Computer," published in 1998. The computer was small, with only 12 transistors, and was operated by a footswitch concealed in a shoe. As the roulette wheel turned, the footswitch started a timer that delivered one of eight tones to a tiny earpiece, each tone indicting a predicted octant in which the ball was more likely to fall. The system, though fragile, worked and was tried in Las Vegas in 1961 with some success.

In the 1970s, J. Doyne Farmer and Norman Packard essentially repeated that experiment using a microcomputer. Like Thorp and Shannon, they were similarly successful in the casinos; see Farmer's brief page at *http://www.doynefarmer .com/roulette* or the more detailed account in Thomas Bass's book *The Eudaemonic Pie*, which is available online via the Internet Archive (*https://archive.org*).

Using Voltage

Most desktop computers have a microphone input jack. A program like Audacity can record samples from this input device. We might expect recording when no microphone is connected to give us an empty file, but it doesn't. The microphone input is an analog input and is susceptible to electronic noise: minor, random variations in voltage due to the nature of the components involved and other environmental factors. We'll use the variation of this voltage as a randomness engine.

This experiment requires you to record a WAV file. It doesn't matter which tool you use to do so. I used Audacity, an open source sound editor available for most systems. Install it under Ubuntu with the following command:

```
> sudo apt-get install audacity
```

Visit *https://www.audacityteam.org* to install Audacity for Windows and macOS.

We want to record from the microphone input using a single channel (mono) and at a high sampling rate, like 176,400 Hz (samples per second). For Audacity, this means changing the project rate to 176,400 (see the lower left of the screen) and changing the recording channel drop-down menu to **Mono**. We also need to select the microphone as the input source. The name of this device on your system is beyond my powers of clairvoyance, but experimentation with the drop-down menu next to the microphone icon will likely turn up the proper device. To test the input, select **Click to Start Monitoring**. You should see a slowly varying sound bar for only one channel (for example, the left channel). If not, play around a bit more until you do.

To record a sample, click the red record button. I suggest recording for several minutes. To stop, click the square stop button. Use the appropriate option on the File menu to export the recording as a WAV file, setting the output format in the File Save dialog to **32-bit float PCM**. If you can't use this method to record a WAV file, there are several small ones available on the book's website that you can use with the upcoming code.

Audio signals are represented as a continuous voltage that changes with time. Sampling an audio signal means measuring the instantaneous voltage at a set time interval, the sampling rate, and turning that voltage into a number over some range. For example, a compact disc uses 16-bit samples, so each voltage is assigned a number in the range −32,768 to 32,767. The sampling rate decides how frequently the voltage is measured.

When a digitized signal is played back, meaning converted back into a voltage to drive a speaker, for instance, the quality of the sound depends on how many numbers were used to represent the signal and how often it was sampled. For our experiment, the samples are represented as 32-bit floating-point numbers, and the sampling rate is 176,400 Hz. By comparison, a compact disc samples at 44,100 Hz.

Figure 1-2 shows a subset of audio samples and their corresponding histogram.

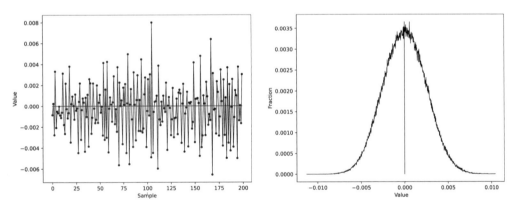

Figure 1-2: The audio samples (left) and corresponding histogram (right)

The *x*-axis in Figure 1-2 is time, that is, the sample number, and the *y*-axis is the floating-point sample value, or the digitized value representing the voltage at that time. The horizontal line is the mean value of all the samples in the file; Figure 1-2 shows only the first 200. As we might expect, the mean value is virtually 0.

The right side of Figure 1-2 shows us the histogram of all the samples in a two-second clip. We've seen this shape before; it's nearly identical to Figure 1-1, and it tells us that the noise samples are normally distributed with a mean value of 0. We'll use this observation to convert the audio stream into a file of random bytes.

We cannot use the recording as is; we must process the samples to make them more random, using the code in *silence.py*. Let's walk through it part by part.

First, we import some library routines:

```
import sys
import numpy as np
from scipy.io.wavfile import read as wavread
```

The sys module provides an interface to the command line; numpy we know. To read WAV files, we need the read function from SciPy. Here, I'm renaming it wavread.

At the bottom of the script, we load the desired WAV file and process it:

```
s, d = wavread(sys.argv[1])
print("sampling rate: %d" % s)
n = len(d)//2
a = MakeBytes(d[:n])
b = MakeBytes(d[n:])
if (len(a) < len(b)):
    c = a[::-1] ^ b[:len(a)]
else:
    c = a[:len(b)] ^ b[::-1]
c.tofile(sys.argv[2])
```

The wavread function returns the sampling rate (s) and the samples themselves as a NumPy vector (d). We display the sampling rate, then split the samples into two halves and pass each half to MakeBytes before assigning the return values a and b, respectively. MakeBytes turns a vector of audio samples into a vector of bytes.

The final set of bytes is in c. This is the exclusive-OR (XOR) of the bytes in a and b. XOR is a logical operation on the bits of an integer. If one of the inputs to XOR is 1 and the other 0, then the output is 1. If the inputs are the same, the output is 0. I remember the phrase "one or the other, but not both." XOR is different from the standard OR operation, which outputs 1 if any input is 1, including if both are 1, as follows:

$$1 \text{ XOR } 1 = 0, \text{ but } 1 \text{ OR } 1 = 1$$

Using one part of the generated stream of bytes to modify the other is a powerful way to alter the bit patterns, which adds to the random nature of the output. In *silence.py*, one of the byte streams is reversed ([::-1]) before XOR to add that much more randomness to the process.

The number of bytes returned by MakeBytes depends on the actual samples passed to it, not the number of samples. Therefore, it is likely that the vector a will be of different length than b, hence the if statement and indexing based on len. When c is ready, it's written to disk in binary via tofile.

All the action is in the MakeBytes function. Converting the stream of audio samples into bytes requires four steps:

1. Subtract the overall mean from each sample.

2. Convert each sample to a bit based on its sign: 1 if positive, 0 if negative.

3. De-bias the bits using the von Neumann algorithm from the previous section.

4. Group each set of 8 bits into an output byte.

The function MakeBytes performs each of these steps, using the following code:

```
def MakeBytes(A):
❶ t = A - A.mean()
❷ thresh = (t.max()-t.min())/100.0
   w = []
   for i in range(len(t)):
       if (np.abs(t[i]) < thresh):
           continue
       w.append(1 if t[i] > 0 else 0)
❸ b = []
   k = 0
   while (k < len(w)-1):
       if (w[k] != w[k+1]):
           b.append(w[k])
       k += 2
❹ n = len(b)//8
   c = np.array(b[:8*n]).reshape((n,8))
   z = []
   for i in range(n):
       t = (c[i] * np.array([128,64,32,16,8,4,2,1])).sum()
       z.append(t)
   return np.array(z).astype("uint8")
```

The code passes in the samples as A, a NumPy vector. Now come the four steps. First, we subtract any mean value (t) ❶. Next, we define thresh to be 1 percent of the maximum range of the samples. I'll explain why momentarily.

Step 2 says to use the sign of each sample as a bit ❷. We observed that the samples are normally distributed, and the mean is now zero because we subtracted it, so roughly half the samples will be negative and half positive. This sounds a lot like a coin flip: either one value or another. Therefore, we make the positive samples ones and the negative samples zeros. But what's the point of thresh?

The bits are collected in the list w, initially empty. The loop after initializing w examines each sample and asks if the sample's absolute value is less than 1 percent of the maximum range. If it is, we ignore that sample (continue); otherwise, we use the sample's sign to add a new bit to w. There may be a tiny, nonrandom signal with a low amplitude in the samples. Ignoring samples close to zero helps remove any such signal. In effect, we are saying we're interested in only "large" deviations from the mean of zero.

We now have a long list of individual bits (w). These bits might be biased, representing a collection of unfair coin tosses. To handle this, we de-bias by using the von Neumann algorithm, which makes a new list of bits in b ❸.

Finally, we convert each set of eight bits in b into a new list of bytes in z ❹ by reshaping b into an $N{\times}8$ array before multiplying each row of that array, column by column, by the value of each bit position in a byte. When all is said and done, the code converts the list of bytes into a NumPy array and returns it.

You may want to review the code again to ensure you follow each step. The central concept is that we took advantage of the fact that the samples are normally distributed to generate a stream of bits that, when de-biased, form the stream of bytes used to make the final output of the code.

To test out this code, I made a 30-minute recording using Audacity and saved the samples in the file *silence.wav*. I then used *silence.py* to convert this large WAV file into *silence.bin*.

NOTE   *The file* silence.wav *is too large to include in the book's repository. However, if your heart is set on having it, contact me, and I'll see what I can do.*

Here's the command line I used to convert the WAV file:

```
> python3 -W ignore silence.py silence.wav silence.bin
```

The wavread function tends to complain about WAV file elements it doesn't understand. Adding -W ignore to the command line suppresses these warnings.

The resulting output file, *silence.bin*, is not quite 6MB. It's a collection of random bytes, so we can pass it to ent to get a report:

```
entropy: 7.999867
chi2   : 0.01
mean   : 127.4948
pi     : 3.136872354 (0.15)
corr   : 0.000610
```

These are reasonably good values. The only value that is not where we might want it to be is χ^2, but we can live with that. Keep *silence.bin* on hand to use as a randomness engine later in the book.

Random Physical Processes

Many physical processes are random, though possibly biased. In this section, we'll explore three physical processes leading to randomness: atmospheric radio frequency noise, plasma and charged particle rates as detected by the

Voyager 1 and *Voyager 2* spacecraft during their 40-plus-year mission, and the decay of radioactive isotopes.

Atmospheric Radio Frequency Noise

In 1997, Mads Haahr, a computer science professor at Trinity College, Dublin, started *https://www.random.org*, a website dedicated to generating truly random numbers from atmospheric radio frequency noise. The site has been running ever since and offers free random numbers to internet users, along with a host of paid services involving random numbers. You can take a look at the services offered at *https://www.random.org*, but for the purposes of this book, we'll stick to the free collection of random bytes.

There are two primary ways to get random data from the site. First, random data in 16KB chunks are available at *https://www.random.org/bytes*. Simply select the desired format and download. However, since we need larger collections of bytes to use as a randomness engine for our experiments, we are better served by using the archive available at *https://archive.random.org*. Downloading files directly from the site requires a small fee. However, if you are comfortable using torrents, the files can be legally acquired for free. Our reference Linux distribution includes Transmission, a client for accessing torrents. The application is also available for other operating systems; visit *https://transmissionbt.com* for download instructions.

As a test, I used Transmission to download several months' worth of random bytes (the Binary Files option). I then combined all the binary files into one using the cat command. For example, a command like this will merge all binary files ending in *.bin* into the file *random_bytes*:

```
> cat *.bin >random_bytes
```

I then passed *random_bytes* (about 126MB) to ent to get the following results:

```
entropy: 7.999999
chi2   : 98.32
mean   : 127.4880
pi     : 3.142011833 (0.01)
corr   : -0.000015
```

We see that ent is quite happy with *random_bytes*. I have a much larger collection of bits from *random.org*, including bits from two previous years. It's over 500MB in size, and ent likes this one as well:

```
entropy: 8.000000
chi2   : 40.68
mean   : 127.5023
pi     : 3.141375348 (0.01)
corr   : 0.000021
```

The code shows that the entropy is maximized, with 8 bits per byte.

Voyager Plasma and Charged Particle Data

In 1977, NASA launched the twin *Voyager* spacecraft to explore the outer solar system. *Voyager 1* encountered Jupiter and Saturn before heading out of the solar system. *Voyager 2* passed by Jupiter, Saturn, Uranus, and Neptune on its grand tour. To date, *Voyager 2* is the only spacecraft to explore Uranus and Neptune. In August 2012, *Voyager 1* became the first human-made object to leave the solar system and enter interstellar space. As of this writing, October 2023, both spacecraft are still performing well and have enough power for perhaps a decade.

The *Voyager*s are best known for the fantastic images they returned. However, both spacecraft carry multiple scientific instruments for measuring the environment through which they travel. This includes devices for measuring plasma protons and the flux of other charged particles and nuclei. These measurements are not entirely random but vary around certain values, much like the microphone input we used previously. Therefore, it's possible to use the *Voyager* data to construct a binary file suitable for use as a source of randomness.

The file *voyager_plasma_lecp.bin* contains bytes formed from a set of *Voyager* datafiles, including plasma proton counts, density, and temperature for 1977 through 1980, and low-energy charged particle fluxes (particles passing through an area over time) from 1977 through 2021. I used the code in *process_vgr_data.py* to merge several smaller files to process the individual plasma and charged particle datasets.

For each type of data—plasma protons, low-energy protons, low-energy ions, and cosmic ray protons—the process was the same: find the median value over the time interval and mark each observation as a 1 bit if above the median and a 0 bit if below. To expand the collection of bits, I repeated this process for the median along with 80 percent of the median and 120 percent of the median. I then de-biased the final collection of bits using the von Neumann algorithm.

The resulting file contains 77,265 bytes, and ent reports the following:

```
entropy: 7.995518
chi2    : 0.01
mean    : 127.5484
pi      : 3.150733867 (0.29)
corr    : -0.001136
```

These are reasonable values.

Figure 1-3 shows sample *Voyager* data from the plasma and low-energy charged particle experiments.

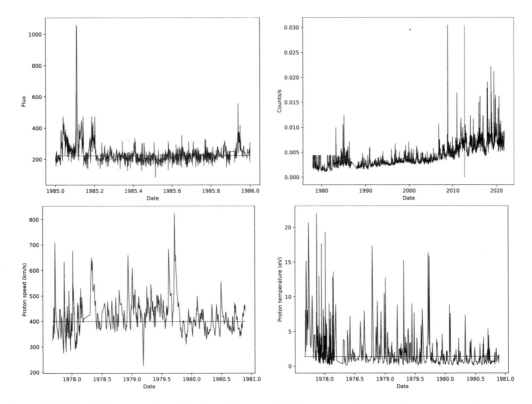

Figure 1-3: Sample Voyager data. Clockwise from top left: low-energy ions, cosmic ray protons, plasma proton temperature, and plasma proton speed.

The dashed line in each plot marks the median value over the dataset. Observations above the median became 1 bits, and those below the median 0 bits. The plot on the upper right shows the flux of cosmic ray protons over the entire mission. The vertical dashed line marks August 2012, the date when *Voyager 1* officially left the solar system. Notice the increase in cosmic ray protons after this time.

NOTE   *The* Voyager *datasets were gathered from multiple websites. They are presented, typically, in text format and required a fair bit of processing on my part to whip them into shape. The files used inconsistent formatting, had typos in places, and, on occasion, used data from the wrong year. If you want the files as I used them, please contact me directly.*

Radioactive Decay

From 1996 through December 2022, the HotBits website (*https://www.four milab.ch/hotbits*) delivered truly random data to the public using the most random of random processes: radioactive decay.

Radioactive elements, like the cesium-137 used by HotBits, are unstable. Eventually, all such atoms decay by some process to another element, in this case, barium-137. The number of protons in an atom determines which element it is. Cesium has 55 protons in its nucleus; that's its *atomic number*. *Isotopes* are versions of an element where the number of neutrons, also in the nucleus, varies. The sum of the two, ignoring the very light electrons, gives the *atomic mass*. If cesium-137 has 55 protons, it must have 82 neutrons. When an atom of cesium-137 decays, one of the neutrons converts into a proton, changing the atomic number to 56 and thereby converting the atom to barium-137, an atom with 56 protons and 81 neutrons. Barium-137 is stable, so the decay process stops. This process varies for different elements. For example, the decay of uranium passes through many stages to reach a stable isotope of lead.

When a neutron becomes a proton, it releases a beta particle (an electron) and an antineutrino. Neutrinos have virtually no mass and are almost undetectable. However, a Geiger counter easily detects beta particles. This is how the HotBits site generates random bits. To be complete, the decay from cesium to barium passes through two stages: the barium nucleus begins in a metastable state, then, about two minutes later, returns to the ground state by emitting a gamma ray. A gamma ray is a high-energy photon, that is, light.

The time when a particular decay will happen is governed by quantum physics and all the "weirdness" we associate with it. For radioactive decay, the weirdness at play is *quantum tunneling*, the fact that even though the cesium atom lacks the energy to, in effect, push itself up and out of the bowl it's in, it nevertheless has a nonzero probability of doing so. There is no classical physics analog for quantum tunneling.

HotBits takes advantage of this unpredictability by using the timing between two pairs of detections to output a 1 or a 0. First, a beta particle is detected, then another. The time interval between the two detections is denoted as T_1. Next, another pair is detected with that time interval labeled T_2. If $T_1 \neq T_2$, a bit is generated. If $T_1 < T_2$, the bit is a 0; otherwise, $T_1 > T_2$ and the output bit is a 1. The sense of the comparison is reversed after each bit to frustrate any systematic bias introduced by the physical setup.

HotBits required an API key to request up to 2,048 bytes at a time. There was a strict limit to the number of bytes granted in a 24-hour period. I downloaded bytes daily as I worked on this book and now have one 3,033,216-byte file. It's pretty good data according to ent:

```
entropy: 7.999935
chi2   : 18.90
mean   : 127.5246
```

```
pi    : 3.144891758 (0.11)
corr  : 0.000100
```

In this section, we've discussed several different processes that generate random data. While all of them are, as we'll see later, useful, none can generate massive quantities of random numbers—and we'll sometimes need millions of random numbers later in the book. We have no choice but to turn to what we might call synthetic random processes and generate random numbers using deterministic means. The use of the word *deterministic* in the previous sentence should bother you, but I suspect you'll be more comfortable with the idea of simulating a random process via a deterministic process by the end of the next section.

Deterministic Processes

As von Neumann once said, "Anyone who attempts to generate random numbers by deterministic means is, of course, living in a state of sin." The very idea of a random process is its unpredictability, such that knowledge of what came before is of no utility in predicting what will come after. By definition, a deterministic process follows a predictable algorithm; therefore, it cannot possibly be a true random process.

Why use deterministic processes, then? Even if the process is deterministic, it can approximate a random process to the level where the outputs are helpful. A *pseudorandom process* approximates a random process. We'll make heavy use of pseudorandom processes throughout the book.

This section also covers how pseudorandom processes are related to *quasirandom processes*. Finally, we'll discuss two hybrid processes likely already available on your computer.

Pseudorandom Numbers

There are many ways to approximate a random process that delivers pseudorandom numbers on demand. Here, I'll introduce one such approach for use in our future experiments.

A *linear congruential generator (LCG)* is a simple approach to creating a sequence of pseudorandom numbers. The numbers generated by an LCG are good enough for a video game and for many of our experiments, but they are statistically weak and not recommended for serious use, like in cryptography.

The best way to understand an LCG is to dive right in:

$$x_{i+1} = (ax_i + c) \bmod m$$

The entire generator is that single equation, where x is the value produced by the generator. The subscript means that the next value in the sequence, x_{i+1}, is derived from the previous, x_i. The initial value, x_0, is the *seed*, the value that primes the generator. Virtually all pseudorandom generators use a seed of some kind.

The equation consists of two parts. The first is $ax_i + c$, where a and c are carefully chosen positive integers. The second part takes the result of the first, y, and calculates $y \bmod m$, where m is another carefully chosen positive integer. The mod operator refers to the modulo, which is nothing more than the remainder. To calculate it, the generator first finds y/m using integer division, then the remainder, which must be a number in the range $[0, m)$. The final result is then used as both the output of the generator and the value used to calculate the following output.

As an example of how this works, let me show you an LCG in action. I'll use small numbers, which helps in understanding, but would be terrible choices in practice, as we'll see:

$$x_{i+1} = (3x_i + 0) \bmod 7$$

The seed must be less than 7, so let's use $x_0 = 4$. Here's the sequence produced by this LCG:

$x_0 = 4$	$x_4 = 2$
$x_1 = 5$	$x_5 = 6$
$x_2 = 1$	$x_6 = 4$
$x_3 = 3$	$x_7 = 5$

The output might seem a bit random, but after x_5, the sequence begins to repeat. All pseudorandom generators repeat eventually. The number of outputs before repeating determines the generator's *period*. Here the period is 6, with 4, 5, 1, 3, 2, 6 repeating forever.

With properly chosen constants, the period can be much larger. For our experiments, we'll use a set of constants popular in the 1980s, which came to be called the *MINSTD (minimum standard generator)*. It produces a sequence of unsigned integers with a period of $2^{31} \approx 10^9$. That's a reasonable period for many applications, but there's more to a good generator than its period. The closer the pseudorandom generator is to a truly random sequence, the better. Statistical test suites like ent, or more professional ones, seek to uncover all manner of correlations in the sequence spit out by the generator. With those test suites, it quickly becomes evident that MINSTD is a poor generator for serious work, but sufficient for our purposes.

I created a file of 100 million bytes using MINSTD and handed it to ent. I got the following results:

```
entropy: 7.999998
chi2   : 75.99
mean   : 127.5058
pi     : 3.141177006 (0.01)
corr   : -0.000144
```

The output is very reasonable, so MINSTD will likely be of use to us.

The generator uses $a = 48{,}271$, $c = 0$, and $m = 2{,}147{,}483{,}647$ to make the generating equation:

$$x_{i+1} = 48271x_i \bmod 2147483647$$

If you're familiar with how computers store numbers internally, you'll notice that m is not using all 32 bits of a 32-bit integer. Instead, $m = 2^{31} - 1$ is the largest positive integer that can be stored in a *signed* 32-bit value. Because of the modulo operation, the output of the generator must be in the range $[0, 2^{31} - 1)$, implying the generator can create at most only some 2 billion unique values.

If we need to generate a sequence of bytes, $[0, 255]$, then we're good. But what if we want to create a sequence of 64-bit floating-point values, what the C language calls a double? In that case, we have to do more work. The simplest approach is to divide x by m, since that must produce a number in the range $[0, 1)$. For many applications, that's sufficient, but it still only delivers a value selected from a set of 2 billion or so numbers. You don't need to know the details of this, but a 64-bit floating-point value can store more than that because it uses a 52-bit base-2 mantissa. If we want to make full use of what a 64-bit floating-point value can give us, we need to generate two outputs from MINSTD and assign 52 bits extracted from both of them to the mantissa of a floating-point value with an exponent of 0. Thankfully, we don't need that much precision, but it's worth noting that simply dividing by m isn't giving you all that you might think it is.

I previously mentioned that pseudorandom generators use seed values, a starting value. This requirement is a double-edged sword. Setting the seed to a specific number causes the generator to repeatedly output the same sequence of values, but the downside is that sometimes we might need to jump through hoops to make sure we aren't using the same seed repeatedly.

For example, MINSTD with a seed of x = 8,675,309 will produce this sequence each time:

0.00304057, 0.77134655, 0.66908364, 0.33651287, 0.8128977, ...

Meanwhile, using a seed of x = 1,234 will consistently produce:

0.02773777, 0.93004224, 0.06911496, 0.24831591, 0.45733623, ...

The floating-point numbers are produced by dividing x by m = 2,147,483,647.

Often, we'll use a pseudorandom generator, like those built into NumPy, without specifying a seed value. In that case, we want an unrepeatable sequence, at least unrepeatable for us, because we won't know the seed selected. If no seed value is supplied, NumPy selects a value from /dev/urandom, a special system device discussed later in the chapter.

The pseudorandom generator we'll use most often is PCG64, the generator NumPy uses by default. It's a good generator, with a period of 2^{128}. NumPy's old default generator was the Mersenne Twister. Statistically, it's still quite good, much better than MINSTD. The Mersenne Twister's period is $2^{19,937} - 1$, which explains its more common name: MT19937. To be precise, its period is a huge number, 6,002 digits, and is utterly without meaning in human terms. PCG64's period isn't anything to scoff at either:

$$2^{128} = 340,282,366,920,938,463,463,374,607,431,768,211,456$$

That's a number similarly without meaning in human terms. None of our experiments will come anywhere close to exhausting the sequence generated by either MT19937 or PCG64, both of which we'll use frequently throughout the book.

The output of these generators follows a uniform distribution. Any other distribution, like a normal distribution, can be formed from a collection of uniformly distributed numbers. For example, it is straightforward to generate normally distributed numbers via the Box-Muller transformation, which maps two uniformly distributed numbers, u_1 and u_2, to two normally distributed numbers:

$$z_1 = \sqrt{-2 \log u_1} \cos(2\pi u_2)$$
$$z_2 = \sqrt{-2 \log u_1} \sin(2\pi u_2)$$

There's much more to say about pseudorandom generators; many computer scientists have spent their careers working with them. However, we now understand all we need for our purposes.

Quasirandom Sequences

For some experiments, we'll want random data that is *space-filling*. By that I mean a generator producing a sequence that appears random but that will, over time, fill space more or less evenly. A purely random process offers us no such guarantee, but a quasirandom sequence does.

The difference between a random or pseudorandom sequence and a quasirandom sequence is best understood via example. Figure 1-4 compares two sequences of 40 points, one generated pseudorandomly and one generated quasirandomly.

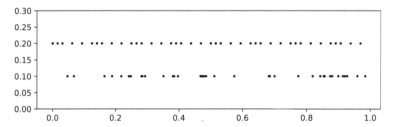

Figure 1-4: Two sets of 40 points, random (at 0.1) and quasirandom (at 0.2)

The lower sequence, at $y = 0.1$, was generated pseudorandomly. It covers the range from 0 to 1 (the *x*-axis), but does so with gaps. The second sequence, at $y = 0.2$, is a quasirandom sequence. It covers the same range, but is more consistent, with no crowding or gaps. Given enough samples, both sequences will eventually fill in the entire range, but the quasirandom sequence does so by scattering values more or less evenly throughout.

Our quasirandom process uses what is known as a *Halton sequence*. A Halton sequence is based on a prime number that divides the interval $[0, 1)$ first by the base (the prime), then the base squared, then cubed, and so on. For a base of 2, the interval is first divided in half, then quarters, then eighths, and so on. This process will eventually cover the interval in the infinite limit and fills the interval approximately evenly. For example, Table 1-6 shows the first few values of the Halton sequence for the given base (rounded to three digits).

Table 1-6: Halton Sequences

Base	Sequence
2	0, 0.5 , 0.25 , 0.75 , 0.125, 0.625, 0.375, 0.875
3	0, 0.333, 0.667, 0.111, 0.444, 0.778, 0.222, 0.556
5	0, 0.2 , 0.4 , 0.6 , 0.8 , 0.04 , 0.24 , 0.44

Each sequence begins with 0 and is entirely deterministic, like a pseudorandom generator with a fixed seed.

If the quasirandom sequence is so predictable, why is it useful? There are times when it is more important to fill space evenly than purely randomly. For example, some of our experiments will involve searching a multi-dimensional space to locate a point in that space that we consider best, for some definition of best. In that case, it often makes more sense to initialize the search so that the locations evaluated at first represent all of the space more or less equally.

Quasirandom sequences are most useful in combination. Figure 1-4 used a single quasirandom sequence with a base of 2 to fill in a one-dimensional space, the interval from 0 to 1. What if, instead, we wanted to fill in a two-dimensional space, like the *xy*-plane? For that, we need pairs of numbers. The seemingly obvious thing to do is sample twice to use the first number as the *x*-coordinate and the second as the *y*-coordinate. Let's see what happens when we apply this approach to pseudorandom and quasirandom numbers.

Figure 1-5 attempts to fill 2D space with points.

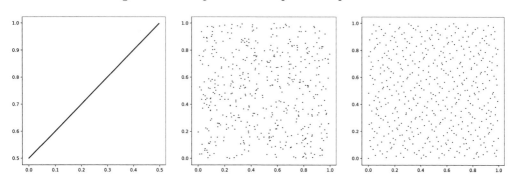

Figure 1-5: From left to right: examples of bad quasirandom, pseudorandom, and good quasirandom sequences

In the middle of Figure 1-5, I plotted 500 pairs sampled from a pseudorandom generator. The points are randomly distributed in space, but there are regions with fewer points and regions with more points, as expected. Next, I repeated this exercise using a quasirandom sequence with base 2.

That is, I asked the sequence for two samples, one after the other, and plotted them as a point. The result is on the left in Figure 1-5. Clearly, something strange is happening.

The plot isn't a mistake; the quasirandom sequence is doing precisely what it's supposed to do. We've defined a random process as one where knowledge of previous values is of no utility in predicting subsequent values. That's what the pseudorandom generator supplied; therefore, we can use sequentially generated values as coordinates for a point in 2D space. However, the quasirandom sequence offers no assurance that previous values do not indicate what comes next. Instead, the sequence is entirely predictable. Pairs of samples will be in a simple relationship to each other, which is what we see on the left in Figure 1-5.

This does not mean we cannot use quasirandom sequences beyond one-dimensional cases. The trick is to use a different base for each dimension. Here, we have two dimensions, so we need *two* quasirandom sequences, each with a different base. The plot on the right in Figure 1-5 was generated this way. The *x*-coordinates are the first 500 samples from a quasirandom sequence with base 2, and the *y*-coordinates are the first 500 from a quasirandom sequence using base 3. Recall that the base must be a prime number. Since the two quasirandom sequences are not using the same base, the values are not (simply) correlated, so the points fill in the 2D space. Likewise, as each sequence is filling in the interval from [0, 1), and we are plotting them in pairs, it makes intuitive sense that they will fill in the 2D space as well, which is what the plot shows.

The moral of the story is that we can use quasirandom sequences, one base per dimension of the problem, to fill in some space in a way that seems random.

Combining Deterministic and Truly Random Processes

Truly random processes are the gold standard, but they tend to be relatively slow, at least given how quickly a computer works. A pseudorandom generator is a good substitute, but it's only a truly random process wannabe. Why not merge the two? This section will examine two approaches to combining truly random processes with pseudorandom generators. I call these *hybrid processes*. We'll discuss /dev/urandom, the Linux operating system's approach, and RDRAND, a CPU-based hybrid processor instruction supported by many newer Intel and AMD processors.

Both /dev/urandom, henceforth urandom, and RDRAND operate in the same way: they use a slow source of true randomness to reseed a cryptographically secure pseudorandom number generator. We learned earlier in the chapter that pseudorandom generators have seeds, something that sets the initial state of the generator. Set the seed the same way, and the sequence of values generated repeats. The hybrid approaches frequently reseed the generator with the intention of altering the sequence to the point where, even if an adversary were to figure out what the pseudorandom generator was doing, it wouldn't matter in practice because the seed would be random, making any knowledge of the state of the generator useless after a short time interval.

I slipped a new phrase into the previous paragraph: *cryptographically secure*. Our experiments are not worried about adversaries and cryptography (well, mostly); we're only concerned that the pseudorandom generator is "pretty good" in the sense that our experiments perform well using the generator.

A cryptographically secure pseudorandom generator is a high-quality pseudorandom generator with the following properties:

- An attacker's knowledge of the generator's state at time t_c offers no ability for the attacker to know anything about the state of the generator at any previous time $t < t_c$.

- For any output bit i, there is no polynomial-time algorithm operating on all previously generated bits, $0, \ldots, i - 1$, that can predict i with better than 50 percent accuracy.

The second property is known as the *next-bit* test. It's what marks the cryptographically secure generator as a high-quality generator.

The phrase *polynomial-time algorithm* comes from the study of algorithms and how they perform regarding time and space. Polynomial-time algorithms are the nice ones; they are the algorithms that might finish sometime before the heat death of the universe. The classic bubble sort algorithm is a polynomial-time algorithm because its runtime scales as the square of the number of elements to sort; that is, sorting n elements takes time on the order of n^2, a polynomial. Any algorithm whose time or space resources are bounded by a polynomial is also a polynomial-time algorithm. For example, the Quicksort algorithm scales as $n \log n$ for n items. That's a function easily bounded by a polynomial, like n^2, so Quicksort is also a polynomial-time algorithm. An algorithm that scales as 2^n is not a polynomial-time algorithm because 2^n is an exponential, not a polynomial. Exponential algorithms are bad because they quickly become intractable, which is precisely what people concerned about attacks on their pseudorandom number generators want.

Algorithmically, a hybrid process uses a strong pseudorandom generator to deliver random values on demand while periodically reseeding said generator from a truly random source. Let's explore the urandom and RDRAND approaches to this problem.

Reading Random Bytes from urandom

Unix systems treat many things that are not files as if they were files, and urandom is no exception. To acquire random bytes from urandom, we need to treat it like a file. For example, let's run Python, open urandom, and dump 60 million bytes from it to a disk file so ent can evaluate them:

```
> python3
>>> b = open("/dev/urandom", "rb").read(60000000)
>>> open("ttt.bin", "wb").write(b)
60000000
```

The output file is *ttt.bin*, a throwaway name. When I ran the code, ent reported the following:

```
entropy: 7.999997
chi2   : 23.16
mean   : 127.4981
pi     : 3.141671600 (0.00)
corr   : 0.000024
```

As expected, urandom performs well and will work for our purposes as a randomness engine.

The Linux kernel maintains an *entropy pool*, a collection of bytes derived from system operations that it uses to update the seed of the pseudorandom generator every 300 seconds, according to kernel file *random.c*, which is available at *https://github.com/torvalds/linux/blob/master/drivers/char/random.c*. The same file also reveals the kernel's entropy sources, if curiosity strikes.

We can monitor the size of the entropy pool by reading the contents of the file *entropy_avail*:

```
> cat /proc/sys/kernel/random/entropy_avail
3693
```

In this case, we're told that at the particular moment I executed the command, there were 3,693 bytes in the entropy pool, meaning that 3,693 bytes were available to reseed the generator when the reseed interval expired.

Linux uses the ChaCha20 pseudorandom number generator. It's a reasonably new generator that performs exceptionally well overall, even when subjected to more intensive test suites. The details don't concern us here, only that urandom delivers the goods.

NOTE *If you spend much time reviewing resources related to /dev/urandom, you'll run across its cousin, /dev/random. The latter device was intended for small amounts of high-quality random data, and it will block until enough entropy is available. This is contrary to our requirements, so we're ignoring /dev/random entirely. Indeed, it was announced in March 2022 that future versions of the Linux kernel will make /dev/random nothing more than another name for /dev/urandom.*

Using the RDRAND Instruction

The RDRAND instruction, if available for your CPU, provides access to high-quality random numbers using much the same approach as the Linux kernel and urandom. The key differences are that the entropy source is part and parcel of the CPU itself and the pseudorandom generator is updated at an interval no longer than after returning 1,022 random values. That is, RDRAND is reseeded based on the number of samples returned, not on a fixed time interval like urandom.

The pseudorandom number generator used is not ChaCha20 but CTR_DRBG, a generator developed by the National Institute of Standards and Technology (NIST), part of the United States Department of Commerce. The close association between NIST and the US government has caused some to distrust RDRAND's output, but as we are not concerned with crypto-graphic security, RDRAND is fair game.

Unlike urandom, which is accessed as if it were a binary file and therefore available immediately to all programming languages, RDRAND is a CPU instruction, so we need to access it via lower-level code or by installing a Python library that uses the instruction for us. As it happens, the rdrand library will do nicely:

```
> pip install rdrand
> python3
>>> from rdrand import RdRandom
>>> RdRandom().random()
0.2047133384122450
>>> b = rdrand.rdrand_get_bytes(60000000)
>>> open("ttt.bin", "wb").write(b)
60000000
```

The class RdRandom returns an interface to RDRAND and supports the random method to return a random float. Alternatively, the rdrand_get_bytes function returns the requested number of random bytes, here 60 million, so we can compare ent's report with the output of urandom.

Passing the bytes in *ttt.bin* to ent gives us the following results:

```
entropy: 7.999997
chi2   : 54.51
mean   : 127.5036
pi     : 3.141309600 (0.01)
corr   : -0.000159
```

For our purposes, these results are indistinguishable from those of urandom; therefore, we'll rely on RDRAND from time to time as well.

If you're looking to access RDRAND from C, the code in *drng.c* will serve as a guide. Note the gcc compile instructions at the top of the file.

In the next section, I'll detail the RE class, a Python wrapper for different randomness sources, and the randomness engine we'll use consistently for all the book's experiments.

NOTE *For other ways to generate pseudorandom numbers, see my book* Random Numbers and Computers *(Springer, 2018). It includes a thorough treatment of all things related to pseudorandom number generation, including code for specific algorithms.*

The Book's Randomness Engine

In this section, we build the RE class, the randomness engine that will power most of our experiments throughout the remainder of the book. If you'd prefer to jump right in, read through *RE.py*, and then come back here to fill in any details you missed.

We want to design a class that will accomplish the following:

- Provide a vector of a specified number of uniform random samples
- Generate pseudorandom output using PCG64, MT19937, or MINSTD
- Generate hybrid random output using urandom or RDRAND
- Generate quasirandom output for any specified prime base
- Produce floats or integers in any range, $[a, b)$
- Produce bytes ($[0, 255]$) or bits ($[0, 1]$)
- Allow a seed value to generate the same output repeatedly
- Substitute a file of bytes as the randomness source

Let's review RE's source code before taking it for a brief test drive.

The RE Class

The RE class is configured via its constructor and provides only one method meant for public use: random. This method takes a single argument, an integer specifying the number of samples to return as a NumPy vector.

The RE class's private methods implement the possible randomness sources, shown in Table 1-7.

Table 1-7: RE's Methods

Method	Description
Fetch	Sample from a disk file
MINSTD	Sample from MINSTD
Urandom	Read from /dev/urandom
RDRAND	Read from RDRAND
Quasirandom	Sample from the Halton sequence
NumPyGen	Sample from PCG64 or MT19937

Each private method accepts a single argument, the number of samples to return. Also, aside from Fetch, each private method returns a floating-point vector of samples in the range $[0, 1)$.

NumPyGen and MINSTD

Let's look first at NumPyGen and MINSTD, as they are straightforward. To save space, I've removed comments and doc strings. They are present in *RE.py*:

```
def NumPyGen(self, N):
    return self.g.random(N)

def MINSTD(self, N):
    v = np.zeros(N)
    for i in range(N):
```

```
    self.seed = (48271 * self.seed) % 2147483647
    v[i] = self.seed * 4.656612875245797e-10
return v
```

Let's begin with NumPyGen, as it's about as simple as you can get. The RE class constructor, which we'll discuss later, creates a NumPy generator, if that's the source desired, and stores it in the g member variable. Don't give me that look; we're experimenting, and g is a perfectly good variable name.

NumPy already knows how to return vectors of floating-point numbers, so all that remains is to tell NumPy we want N of them, which we immediately return to the caller, random, which we'll also discuss later on.

Similarly, MINSTD's job is also to return a vector of floating-point samples. We first create a vector to hold the samples and then loop, applying the LCG equation

$$x_{i+1} = (ax_i) \bmod m$$

with a = 48,271 and $m = 2^{31} - 1 = 2,147,483,647$. To convert x_{i+1} to a float in $[0, 1)$, we divide by m or, as here, multiply by $1/m$.

There are a few things to notice. First, the seed member variable *is x* if the MINSTD generator is selected. That's why it's updated. Second, the MINSTD equation is iterative; to get x_{i+1} we need x_i as pseudorandom values must be generated sequentially (usually). In practice, this means that our implementation of MINSTD is a bit slow, but that's fine for our experiments.

Remember that MINSTD deals with 32-bit integers, meaning floating-point numbers returned by MINSTD are only as precise as 32-bit floats. If you know C, this means it returns float, not double. A Python float is a C double, and likewise uses 64 bits (52 for the mantissa or significand). For most of what we do in this book, the loss of precision isn't important, but you should be aware of it if you wish to use RE in your projects.

RDRAND

The RDRAND method needs to use the CPU's RDRAND instruction, which it does via the rdrand module. When the RE class is imported, Python tries to load rdrand (see the top of *RE.py*). If present, the RDRAND method uses the library to access samples. If the library isn't present, RDRAND still works but falls back to NumPy's PCG64 generator and issues a suitable warning message.

Let's look at RDRAND, ignoring the part where rdrand isn't available:

```
def RDRAND(self, N):
    v = np.zeros(N)
    rng = rdrand.RdRandom()
    for i in range(N):
        v[i] = rng.random()
    return v
```

The rdrand module supports a handful of methods, but we're restricting ourselves to random, which returns a single float in $[0, 1)$, using all 52 bits of the mantissa. However, as random is returning a single number, we need a loop, so RDRAND isn't particularly fast. If you want bytes from rdrand, you'll be better served by using the module's rdrand_get_bytes function directly.

Urandom

The RE class uses /dev/urandom as follows:

```
def Urandom(self, N):
    with open("/dev/urandom", "rb") as f:
        b = bytearray(f.read(4*N))
    return np.frombuffer(b, dtype="uint32") / (1<<32)
```

The code first reads four times as many bytes as requested into a Python byte array. Then, it tells NumPy to treat the byte array as a buffer and read 32-bit unsigned integers from it. The resulting array is divided by 2^{32} to change it into a vector in $[0, 1)$. Like MINSTD, we're treating urandom as a source of 32-bit floats.

Quasirandom

Quasirandom numbers are a bit different, as we saw earlier in the chapter, but the implementation is similar to the other sources:

```
def Quasirandom(self, N):
    v = []
    while (len(v) < N):
        v.append(Halton(self.qnum, self.base))
        self.qnum += 1
    return np.array(v)
```

The output vector (v) is constructed sample by sample. The while loop calls Halton, a function contained within Quasirandom. This function returns a specific number in the Halton sequence for the given prime base. The next number to use is in member variable qnum. We'll return to qnum later when discussing the RE constructor.

Random

The RE class's only public method is random:

```
def random(self, N=1):
    if (not self.disk):
        v = self.generators[self.kind](N)
        if (self.mode == "float"):
            v = (self.high - self.low)*v + self.low
        elif (self.mode == "int"):
            v = ((self.high - self.low)*v).astype("int64") + self.low
        elif (self.mode == "byte"):
            v = np.floor(256*v + 0.5).astype("uint8")
        else:
            v = np.floor(v + 0.5).astype("uint8")
```

```
    else:
        v = self.Fetch(N)
    return v[0] if (N == 1) else v
```

If we're not using a disk file, then the process is the same, regardless of the randomness source: first, generate the requested number of floating-point samples in [0, 1), and then modify them to be in the desired range and of the desired type.

As all randomness sources accept the same argument and return the same [0, 1) vector, we store references to the particular methods in the dictionary generators. Then, to get v, we need only call the appropriate method indexing by kind and passing the number of samples (N), which defaults to 1.

With v in hand, we modify it according to the selected configuration. If we want floating-point, we multiply by high and add low, which default to 1 and 0, respectively. This returns a float in the range [low, high).

For integers, we first map v to [0, high) before adding low to give an integer in the range [low, high). The upper limit is not present in both cases, thereby following Python convention.

To get bytes, multiply by 256 and round. Finally, to get bits, round v.

Fetch

What about Fetch, you ask? It's a mix of Urandom and random:

```
def Fetch(self, N=1):
    if (self.mode == "byte"):
        nbytes = N
    else:
        nbytes = 4*N

    b = []
    n = nbytes
    while (len(b) < nbytes):
        t = self.file.read(n)
        if (len(t) < n):
            n = n - len(t)
            self.file.close()
            self.file = open(self.kind, "rb")
        b += t

    if (self.mode == "byte"):
        v = np.array(b, dtype="uint8")
    else:
        v = np.frombuffer(bytearray(b), dtype="uint32")
        v = v / (1 << 32)
        if (self.mode == "float"):
            v = (self.high - self.low)*v + self.low
        elif (self.mode == "int"):
            v = ((self.high - self.low)*v).astype("int64") + self.low
```

```
        elif (self.mode == "byte"):
            v = np.floor(256*v + 0.5).astype("uint8")
        else:
            v = np.floor(v + 0.5).astype("uint8")
    return v
```

The code is split into three paragraphs. The first calculates the number of bytes to read from the disk file (nbytes). As with Urandom, we're restricting ourselves to 32-bit floats.

The second paragraph reads the bytes from disk and stores them in b. If we run out of file, we start again from the beginning. Remember this to ensure you don't ask for far more samples than the file can supply. Think of the size of the file divided by four as the period of the generator.

The third paragraph massages the data accordingly. If we want bytes, we're done; we simply convert the list b into a NumPy vector and return it. Otherwise, we first treat the bytes as a buffer and read them as unsigned 32-bit integers, which we divide by 2^{32} to make v in the range $[0, 1)$. We then convert the data to the final output format as in the random method. There is some code duplicated between random and Fetch, but pedagogically, the clarity this provides is worth it.

The Constructor

The RE class constructor configures the randomness engine. The first part defines defaults and builds the dictionary of private methods:

```
def __init__(self, mode="float", kind="pcg64", seed=None, low=0, high=1, base=2):
    self.generators = {
        "pcg64"  : self.NumPyGen,
        "mt19937": self.NumPyGen,
        "minstd" : self.MINSTD,
        "quasi"  : self.Quasirandom,
        "urandom": self.Urandom,
        "rdrand" : self.RDRAND,
    }

    self.mode = mode
    self.kind = kind
    self.seed = seed
    self.low  = low
    self.high = high
    self.base = base
    self.disk = False
```

The mode defines what RE returns. It's a string: float, int, byte, or bit. Use kind to define the source. Possible values are the keys of generators or a pathname to a disk file. Case matters here.

Use low and high to set the output range, ignored for bytes and bits. Remember, the output does not include high. To get deterministic sequences, set the seed. Finally, use base if working with a quasirandom sequence.

The second part handles source-specific things:

```
if (self.kind == "pcg64"):
    self.g = np.random.Generator(np.random.PCG64(seed))
elif (self.kind == "mt19937"):
    self.g = np.random.Generator(np.random.MT19937(seed))
elif (self.kind == "minstd"):
    if (seed == None):
        self.seed = np.random.randint(1,93123544)
elif (self.kind == "quasi"):
    if (seed == None):
        self.qnum = 0
    elif (seed < 0):
        self.qnum = np.random.randint(0,10000)
    else:
        self.qnum = seed
elif (self.kind == "urandom") or (self.kind == "rdrand"):
    pass
else:
    self.disk = True
    self.file = open(self.kind, "rb")
```

I leave parsing the second part to you. I'll just point out that for quasirandom sequences, the seed value is used to set the initial Halton sequence value, which is random if a negative seed is given. Use a negative seed to spice up the deterministic Halton sequence with a bit of randomness.

With that, we're finished with the implementation of RE. Now let's learn how to use it.

RE Class Examples

The following examples illustrate how we'll use the RE class.

The first example imports RE and defines a generator using all the defaults—PCG64 with floating-point output in [0, 1):

```
>>> from RE import *
>>> g = RE()
>>> g.random(5)
array([0.44018704, 0.98320526, 0.61820454, 0.3124574 , 0.32110503])
>>> g.random(5)
array([0.47792361, 0.67769858, 0.50001674, 0.35449271, 0.92454641])
```

The second example creates an instance of the MT19937 generator and uses it to return five floating-point values in $[-3, 5)$:

```
>>> RE(kind='mt19937', low=-3, high=5).random(5)
array([ 4.51484908,  2.31892577,  0.98488816, -1.36846592,  1.70944267])
```

Next, we use urandom to return integers in $[-3, 5)$ and use RDRAND to sample bytes:

```
>>> RE(kind='urandom', low=-3, high=5, mode='int').random(5)
array([2, 2, 4, 3, 2])
>>> RE(kind='rdrand', mode='byte').random(9)
array([ 67, 173, 207, 230,  10, 127, 241,  21, 213], dtype=uint8)
```

Here's an example that specifies a seed value to return the same sequence each time:

```
>>> RE(kind='minstd', seed=5, mode='bit').random(9)
array([0, 0, 0, 0, 1, 1, 1, 1, 0], dtype=uint8)
>>> RE(kind='minstd', seed=5, mode='bit').random(9)
array([0, 0, 0, 0, 1, 1, 1, 1, 0], dtype=uint8)
```

For a quasirandom sequence, if no seed is given, the sequence begins at zero each time. If the seed is less than zero, the starting position is set randomly:

```
>>> RE(kind='quasi', base=2).random(6)
array([0.   , 0.5  , 0.25 , 0.75 , 0.125, 0.625])
>>> RE(kind='quasi', base=2).random(6)
array([0.   , 0.5  , 0.25 , 0.75 , 0.125, 0.625])
>>> RE(kind='quasi', base=2, seed=-1).random(6)
array([0.3458252, 0.8458252, 0.2208252, 0.7208252, 0.4708252, 0.9708252])
>>> RE(kind='quasi', base=2, seed=-1).random(6)
array([0.74029541, 0.49029541, 0.99029541, 0.01373291, 0.51373291,
       0.26373291])
```

This example sets kind to a filename to sample from a disk file, here called *hotbits.bin*:

```
>>> RE(kind='hotbits.bin').random(5)
array([0.58051941, 0.79079893, 0.91321132, 0.26857162, 0.49829243])
```

Now that we know how to use RE, let's start experimenting.

Summary

This chapter focused on generating randomness, that is, on random processes. First, we explored the relationship between probability and randomness and learned that random processes sample from probability distributions, either continuous or discrete. We learned that there is generally no concrete answer to the question, How do we know if the output of a process is random? However, there are ways to test sequences to enhance our belief one way or the other. In practice, we declared the output of ent as our standard, since our experiments do not need state-of-the-art random processes.

Next, we discussed truly random processes. We began with classical approaches like coin flips and dice rolls, and then shifted our attention to physical processes such as random fluctuations in an analog signal, radio frequency noise due to atmospheric effects, and the decay of radioactive elements.

Pseudorandom and quasirandom sequences are random process mimics. Though they pretend to be the output of truly random processes, they are not. However, they do shadow truly random processes with sufficient fidelity to make them the primary drivers for our experiments. After learning the basics of pseudorandom and quasirandom generators, we covered hybrid generators, the marriage of pseudorandomness and truly random processes. Hybrid generators provide cryptographically secure sequences.

The chapter concluded with a walkthrough of the design and code for the RE class, the randomness engine that will power all of our experiments.

2

HIDING INFORMATION

Steganography, from the Greek term for "covered writing," is the art of concealing a message so that adversaries aren't aware that the message is present. Unlike cryptography, which relies on the difficulty of undoing the algorithm that generated the encrypted message, steganography relies on secrecy. Anyone can read the covered message if they know how the message is hidden. If the art of hiding a message is called steganography, then *steganalysis* is the process of detecting hidden messages. It's a classic arms race: steganographers improve their techniques while steganalysis attempts to thwart them.

While steganography exists anywhere a 0 and 1 can be hidden, it is most commonly found in text, binary files, audio, video, and images. Excepting video, these are also the targets of our experiments.

In Strings

This section presents two experiments with text. The first hides a message in a string of words using a fixed offset, and the second uses random offsets.

A BRIEF HISTORY OF DECEPTION

Steganography has been around for a long time, in close association with cryptography.

In the 5th century BCE, the ancient Greek author, Herodotus, related a story about Histiaeus, exiled tyrant of Miletus, in what is now modern-day Turkey. He claims that Histiaeus shaved the head of a slave and tattooed a message on his scalp, then waited for the man's hair to grow back before sending him to Miletus. Friends of Histiaeus shaved the man's head and read the message— instructions to revolt against the Persians. Ultimately, the revolt failed and Histiaeus's head was sent to the Persian king, Darius.

In 1605, Sir Francis Bacon developed a secretive communication method that combines steganography and cryptography. First, Bacon devised a five-digit binary code for each letter of the alphabet. Then, he used the cipher to hide messages in ordinary text with slight modifications to the font to indicate a 0 or 1 (Bacon used "a" and "b" in place of 0 and 1).

For example, we might hide the message "EAT AT JOES" using this cipher as:

```
alIce openeD thE Door anD foUNd ThaT iT LEd inTo a SmaLl passage
aabaa aaaaab aab baaa aab aabba baab ab bba aaba a baaba
```

The first line shows the text, a line from *Alice's Adventures in Wonderland*, where lowercase letters represent "a" and uppercase represent "b." In practice, the font used for "a" letters would be almost, but not quite, the same as the font used for "b" letters. The actual Bacon cipher for each letter of the message is on the second line. To figure out the message, we organize the letters in groups of five like so:

E	aabaa	A	aaaaa	J	abaab
A	aaaaa	T	baabb	O	abbba
T	baabb			E	aabaa
				S	baaba

Fixed Offset

During WWI, a German spy sent the following message from New York:

> Apparently neutral's protest is thoroughly discounted and ignored. Isman hard hit. Blockade issue affects pretext for embargo on by-products, ejecting suets and vegetable oils.

The missive contains a hidden message. Extracting the second letter of each word gives us:

> Pershing sails from NY June 1.

Note that we're interpreting the final "i" as "1." The message was useless in the end as Pershing sailed from New York on May 28.

While using the second letter is better than the first as it's more challenging to accidentally spot the message, security is still low. However, this idea provides a starting point for hiding messages in a string of words. Let's begin by developing a script to embed a text message using words from a *pool text*, a large text document from which we select words, and a selected letter offset from the beginning of each word. There is no randomness to this example, but it sets us up for the following experiment.

Here's our approach:

1. Select a letter offset from the beginning of the word. The German spy used an offset of one for the second letter of each word. Recall that computer scientists use zero-based numbering, starting at 0 rather than 1.

2. Hide a source message, text only, by selecting words from a pool text—for example, a book—such that the current letter of the word is the offset letter of the selected word.

3. Write the resulting list of words to disk as the hidden message.

The source code we need is in *steg_simple.py*. I suggest reviewing it before continuing. Notice that the file does not import RE; there is nothing random here.

To learn how it works, here's an example that hides the contents of *message.txt* using words selected from *alice.txt* with an offset of 2:

```
Three may keep a secret, if two of them are dead.
```

Use the command line

```
> python3 steg_simple.py encode 2 message.txt alice.txt output.txt
```
---

to produce the following in *output.txt*:

```
GET SCHOOLROOM VERY THERE ONE COME THAT SAYING LIKE THE
SHE HAPPENED THAT WAS THE NECK WORDS THE LITTLE ALICE
TOFFEE HOT NOW THOUGHT DOOR OFF INTO ASHAMED GREAT
MOMENT TEARS LARGE DEEP AND THE HEARD AND
```

The offset is zero-based, meaning an offset of 0 uses the first letter, so an offset of 2 uses the third letter. With that in mind, we see the embedded message as:

```
GET SCHOOLROOM VERY THERE ONE COME THAT SAYING LIKE THE SHE HAPPENED
THAT WAS THE NECK WORDS THE LITTLE ALICE TOFFEE HOT NOW THOUGHT DOOR
OFF INTO ASHAMED GREAT MOMENT TEARS LARGE DEEP AND THE HEARD AND
```

Every underlined letter of which is the message:

```
THREE MAY KEEP A SECRET IF TWO OF THEM ARE DEAD
```

First, *steg_simple.py* reads the message file, removes any characters that aren't letters, and then uppercases the remaining characters. The result is a Python list of the words in the message. The pool text receives similar treatment.

Next, *steg_simple.py* processes each word of the message, letter by letter, scanning the pool text for words where the offset letter matches the current message letter. Each match is appended to the output list of words. When all message letters have been similarly processed, the output list is written to disk.

Changing the offset modifies the output list; for example, making it 3 in the previous listing results in the following:

```
SITTING NOTHING THERE ITSELF LATE SEEMED STRAIGHT VERY LOOK TRIED
MAKE DROP MANAGED HERSELF AFTER WHICH NEAR MILES SORT LATITUDE ROOF
LITTLE FLOWERS THROUGH THROUGH HALF ANOTHER WITH WISE THEM USUALLY
CHERRYTART PINEAPPLE SAID LIKE CHEATED FOND
```

Now that we've embedded the message, let's get it back:

```
> python3 steg_simple.py decode 2 output.txt tmp.txt
```

The *tmp.txt* file contains the following:

```
THREEMAYKEEPASECRETIFTWOOFTHEMAREDEAD
```

We've lost punctuation and spaces, and the message is screaming at us in all caps, but it's discernible.

Now let's walk through the source code for this example. Both the input message and, if encoding, the pool file are run through ProcessText to remove non-letters and return a list of words, as shown in Listing 2-1.

```
def ProcessText(s):
    s = s.upper().split()
    text = []
    for t in s:
        z = ""
        for c in t:
            if (c in "ABCDEFGHIJKLMNOPQRSTUVWXYZ"):
                z += c
        text.append(z)
    return text
```

Listing 2-1: Converting a string of text to a list of words

The outer loop processes the list of words generated by the first line (s). This list still contains non-space characters. The inner loop over the characters of each word in s builds the letters-only version and appends it to text. When all is said and done, the function returns text.

The bottom of *steg_simple.py* interprets the command line according to the mode, encode or decode. The code is shown in Listing 2-2.

```
offset = int(sys.argv[2])
if (sys.argv[1] == "encode"):
    sfile = sys.argv[3]
    pfile = sys.argv[4]
    dfile = sys.argv[5]
    Encode(offset, sfile, pfile, dfile)
elif (sys.argv[1] == "decode"):
    dfile = sys.argv[3]
    sfile = sys.argv[4]
    Decode(offset, dfile, sfile)
else:
    print("Unknown option")
```

Listing 2-2: Parsing the command line

All the action is in Encode and Decode. Let's begin with Encode, as in Listing 2-3.

```
def Encode(offset, sfile, pfile, dfile):
❶   msg = ProcessText(open(sfile).read())
    pool= ProcessText(open(pfile).read())
    enc = []
    idx = 0
❷   for word in msg:
        for c in word:
            done = False
            while (not done) and (idx < len(pool)):
                if (len(pool[idx]) <= offset):
                    pass
                elif (pool[idx][offset] != c):
                    pass
                else:
                    enc.append(pool[idx])
                    done = True
                idx += 1
❸   with open(dfile, "w") as f:
        f.write(" ".join(enc)+"\n")
```

Listing 2-3: Encoding the message

First, Encode reads and processes both the message and pool text ❶. Next comes a triply nested loop ❷, first to process each word of the message (msg), then character by character (c), and lastly to search the pool text (pool) for a word where the offset character matches. The index into the pool text (idx) is set to zero initially and only ever incremented afterward, so the per-character search does not begin from the start each time but continues moving through the pool text file. In theory, this means the set of selected words is more diverse, as the same word is not selected repeatedly to match the same character of the message.

After processing the entire message, enc contains the selected words in order. All that remains is to dump them to disk ❸. Notice the Python idiom to convert a list of words into a single string by calling the join method on a string constant (a space) before appending a final newline character.

Decoding a message is less involved, as shown in Listing 2-4.

```python
def Decode(offset, dfile, sfile):
    enc = ProcessText(open(dfile).read())
    plain = ""
    for w in enc:
        plain += w[offset]
    with open(sfile, "w") as f:
        f.write(plain+"\n")
```

Listing 2-4: Decoding the message

Decode processes the encoded message file to make a list of words (enc). Then the function examines each word, extracts the offset letter, and adds it to plain, the output string. Finally, Decode dumps the output string to the decoded message file.

Random Offset

The previous experiment works, but it's not too difficult to defeat. A bit of trial and error will locate the offset and reveal the hidden message. We can do slightly better by changing the offset at random so that the third letter of the first word is important, the fifth letter of the second, the first followed by the third again, and so on. The result is still not particularly secure but is likely to confound accidental discovery of the message.

The file *steg_text.py* implements this approach and introduces us to a technique we'll use throughout the remainder of the chapter, namely using a randomness engine with a fixed seed to generate a deterministic sequence of randomly selected offsets. Let's take a look at what this approach buys us.

First, we run *steg_text.py* on the same input message as the quotation about not sharing secrets. In this case, the command line is:

```
> python3 steg_text.py message.txt alice.txt output.txt
Your secret key is 499377
```

Take note of the "secret key," which is, in reality, a pseudorandom number generator seed. We won't be able to recover the hidden message without it.

Each run of *steg_text.py* produces a new *output.txt* and secret key. For the previous run, *output.txt* became:

```
IT WHAT THERE THE HEAR SEEMED NATURAL VERY SKURRIED HE ALICE UP
DEAR THIS DIFFERENT VOICE CROCODILE WATERS IT THEIR BEFORE THAT
RAILWAY WOODEN SOON HALF SITS CATCHING THE TREMBLING TAIL YOUARE
WENT HUNDRED THE ANIMALS BIRDS
```

This corresponds to per-word offsets of:

```
11321313114123431313234223242314143243
THREEMAYKEEPASECRETIFTWOOFTHEMAREDEAD
```

The offsets are selected pseudorandomly using the reported seed, which here is 499,377.

To recover the original message, use:

```
> python3 steg_text.py 499377 output.txt tmp.txt
> cat tmp.txt
THREEMAYKEEPASECRETIFTWOOFTHEMAREDEAD
```

With *steg_text.py*, the output is harder to parse because the offsets are random. If we knew there was a hidden message, it would require effort to recover it because of the many paths through the letters of each word. For example, Figure 2-1 shows every path through the first three words, for a total of 40 combinations.

Figure 2-1: All the paths through the first three words

Each combination is as follows:

IWT	IWG	IWE	IWR	IWE	IHT	IHH	IHE
IHR	IHE	IAT	IAH	IAE	IAR	IAE	ITT
ITH	ITE	ITR	ITE	TWT	TWH	TWE	TWR
TWE	THT	THH	THE	THR	THE	TAT	TAH
TAE	TAR	TAE	TTT	TTH	TTE	TTR	TTE

The combinations ITH, ITE, TWE, THE, THR, TAR, TAH are the only possible beginnings of common English words, for example: Ithaca, item, tweet, theme, three, tarnish, and tahini. The fourth word, THE, adds one more letter to each of the existing prefixes, of which only THRE and TART seem plausible. The following word, HEAR, makes it clear that the first word of the message is likely THREE or the beginning of the next word after TART, assuming the message isn't itself encrypted.

The source code for this example is similar to *steg_simple.py* and uses the same ProcessText function shown in Listing 2-1. The important parts for us are the encoder and decoder functions. Let's start with the encoder, shown in Listing 2-5.

```
def Encode(mfile, pfile, ofile):
    msg = ProcessText(open(mfile).read())
    pool= ProcessText(open(pfile).read())
❶ key = RE(mode='int', low=10000, high=1000000).random()
    rng = RE(mode='int', low=1, high=5, seed=key)
    enc = []
    idx = 0
    for word in msg:
        for c in word:
          ❷ offset = rng.random()
            done = False
            while (not done) and (idx < len(pool)):
                if (len(pool[idx]) <= offset):
                    pass
                elif (pool[idx][offset] != c):
                    pass
                else:
                    enc.append(pool[idx])
                    done = True
                idx += 1
    with open(ofile, "w") as f:
        f.write(" ".join(enc)+"\n")
❸ print("Your secret key is %d" % key)
```

Listing 2-5: Encoding with random offsets

As with *steg_simple.py*, the message and pool files are loaded and processed. The secret key is selected using the default randomness engine, PCG64 ❶. With key in hand, a new generator is initialized with key as the seed (rng).

Each letter of the message is encoded by searching the pool for a matching word, but this time, the generator ❷ returns the offset to use. After all message letters have been processed, the function writes the final output file and reports the secret key ❸. Compare Listing 2-5 with Listing 2-3; the fixed offset is now replaced by a randomly generated one.

The decoder is shown in Listing 2-6.

```
def Decode(key, ofile, mfile):
    enc = ProcessText(open(ofile).read())
    rng = RE(mode='int', low=1, high=5, seed=key)
    plain = ""
    for w in enc:
        plain += w[rng.random()]
    with open(mfile, "w") as f:
        f.write(plain+"\n")
```

Listing 2-6: Decoding with random offsets

The secret key forms the seed for the generator; then, the message is decoded word by word using the next value from the generator before writing the message to disk.

Some may object at this point as the experiments select words that, while actual words, do not form a meaningful sentence, which might tip off an adversary. I agree. My defense is that a simple program cannot generate a meaningful sentence, but we are still embedding a message and therefore following the spirit of steganography, if not the absolute law of it. Fear not; experiments later in the chapter produce unambiguous results.

Steganography with text is merely a warm-up exercise, as we can make better use of what the digital age offers. So let's leave text behind and progress to hiding arbitrary files in other files.

In Random Data

In Chapter 1, we used ent as a tool to help us decide whether a file contains random data. Here we'll embed an arbitrary file inside a file of random data so that tools like ent still lead us to believe the file contains random data, even though it no longer does. This is true steganography in that the file appears random both before and after embedding the data we wish to hide.

To accomplish our goal, we need to think in terms of bits, not characters and words. The file we want to hide, as well as the pool file, is merely a stream of bits; we don't care what those bits represent. The idea, then, is to randomly scatter the bits of the source file throughout the bits of the pool file.

For example, if we want to hide source bits 11011011 in

11010101100010101001011010101100101001011100010111

then we need to scatter the bits randomly

11010101100110101001001010101100111100010111

in such a way that we can recover the random bit positions later on to reconstruct the original file.

This approach works if the pool file is substantially larger than the file we wish to hide. For any source bit, the probability that the pool file bit we select is already the same value is 50 percent because we assume the pool file to be random. Therefore, with enough pool file bits and random placement of the source file bits, we don't expect most tools like ent to pick up that the file has been modified. However, the proper ratio between hidden bits and pool file bits is difficult to ascertain. Fewer hidden bits are better, but how many can we insert until they become noticeable? We'll perform a simple experiment shortly to try to figure out the answer.

We're almost ready to think about coding. Our plan is to randomly scatter the source file bits throughout the bits of the pool file. We can get the random bit positions by using RE with a fixed seed value to generate a sequence of random offset values, one for each source bit.

But while we can undoubtedly encode the source file with what we have, can we get it back? How many bits do we read out of the encoded file? We don't know how long the source file was, so we need to encode not only the source file but also its length.

For example, if the source file is 10,356 bytes long, we need to encode not only all 8 × 10,356 = 82,848 bits of the file but also the number of bytes we encoded. We'll use 32 bits to encode the file length, so we will encode 82,848 + 32 = 82,880 bits, the first 32 representing the encoded file length in bytes. Then, to recover the file, we read 32 bits and form the length to know how many additional bits to extract.

Let's lay out the steps. To encode a file:

1. Read the source file and convert it to a list of bits.

2. Prepend the list of bits by the 32 bits from the length of the source file.

3. Read the pool file and convert it to a list of bits.

4. Generate a random list of offset positions, one for each source bit, using the supplied key as the seed.

5. Set the pool bit at each offset position to the corresponding source bit value.

6. Convert the pool bits back to a set of bytes and write them to disk.

To decode a file:

1. Read the encoded file and convert it to a list of bits.

2. Generate 32 offset positions using the supplied key as the seed.

3. Calculate the file length and number of bits to read.

4. Generate offset positions for that many bits.

5. Collect the bits at those positions and convert them to bytes.

6. Write the bytes to disk as the extracted file.

It seems complicated, and there is some accounting to keep track of, but in the end it's straightforward: we'll randomly scatter the source file bits throughout the pool file and then collect them to extract the source file.

Before we get to the code, we'll first learn how to run it so you can experiment. To use the code, we need a file to embed and a pool file. The book's GitHub repository contains *RandomDotOrg_sm.bin*, a 5MB file of random data from *random.org* that is used with permission. It will be the pool file for this example. As we can see, ent likes this file:

```
entropy: 7.999969
chi2   : 84.34
mean   : 127.4992
pi     : 3.140270649 (0.04)
corr   : 0.000322
```

As for the source file, look in the directory *test_images*, which contains a collection of standard image-processing test images of various types and sizes. We'll use these again later in the chapter when we experiment with hiding files in images. Presently, we want *boat.png*, shown in Figure 2-2.

Figure 2-2: The boat image

The code we need is in *steg_random.py*. To encode a file, use a command line like this:

```
> python3 steg_random.py 12345 test_images/boat.png output.bin data/RandomDotOrg_sm.bin
```

The code takes about 30 seconds to run. When it finishes, the *output.bin* file contains the hidden image. Does ent still like this file? It does:

```
entropy: 7.999969
chi2    : 86.38
mean    : 127.4903
pi      : 3.140005417 (0.05)
corr    : 0.000403
```

To recover the image, run *steg_random.py* a second time using the same key (12345):

```
> python3 steg_random.py 12345 output.bin tmp.png
```

The file *tmp.png* now contains the boat image.

The code in *make_random.py* generates files of random bytes using the random number generators available in the RE class. I recommend using this code to create pool files for your experiments. For example, use the following to generate a file of 5 million bytes using the Mersenne Twister:

```
> python3 make_random.py 5000000 none mt19937.bin mt19937
```

As with most programs, running *make_random.py* without arguments tells us how to use the code.

How Much Can You Hide?

Let's try to figure out how many bits we can hide before tools like ent give suspicious results. To do so, we need the code in *steg_random_test.py*. I'll let you read the code; it's a simple script that does the following:

1. Uses *make_random.py* to make a random file of 10 million bytes

2. Uses *steg_random.py* to embed ever-larger files that are nothing but the letter *A* repeated

3. Runs ent on the result and extracts the estimated value of π

4. Plots the estimate of π as a function of the embedded file size

Repeating the letter *A* creates a file that is the antithesis of random; it's as nonrandom as possible, with an entropy of zero. Therefore, it's the worst possible file to embed, which will have the most negative effect on the output. There's nothing special about the letter *A*; any single-byte value will do.

Figure 2-3 shows the result.

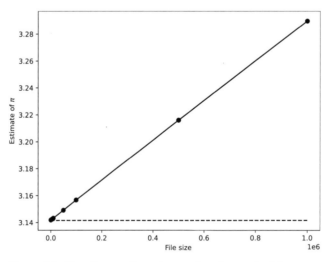

Figure 2-3: The effect on the estimate of π when embedding ever-larger files of A repeated

There's a linear relationship between the estimate of π and the embedded number of *A*s. The pool file on its own gives an estimate that is 0.01 percent off, which is good. Hiding 50,000 *A*s bumps the error to

0.05 percent, which is also not likely to raise eyebrows. However, 100,000 As changes the error to 0.5 percent, which is starting to look suspicious.

Since the pool file had 10 million bytes, 100,000 bytes is equal to $10^5/10^7 = 0.01 = 1$ percent, implying that hiding files that are less than 1 percent the size of the pool file should, in general, result in an output file that won't be noticed.

For example, Table 2-1 shows the estimated π values from ent for a pool file (using urandom) of 10 million bytes along with the output of *steg_random.py* when hiding 100,000 bytes that are all As, the first 100,000 characters of *alice.txt*, or another random file of bytes using RDRAND.

Table 2-1: Estimated π Values by Hidden File Type

Hidden file type	Estimated π values
urandom pool	3.142167657 (0.02%)
Hide RDRAND	3.142429257 (0.03%)
Hide *alice.txt*	3.146091658 (0.14%)
Hide As	3.157525263 (0.51%)

These results hint that, for a hidden file that's 1 percent the size of the pool file, we should expect a tool like ent to show a slight increase in metrics related to nonrandomness, like a larger error in estimating π. Note that *alice.txt* is a text file, so we can expect a binary file to do even better. The moral of the story is: hide your data in a random file that is at least 100 times larger than the data itself.

Now let's take a look at the code.

The steg_random.py Code

The code in *steg_random.py* is slightly more complex than the code we've used so far. I suggest reviewing *steg_random.py* before continuing. You'll see two utility functions, MakeBit and MakeByte. We'll use these functions here and in the following sections, so we'll define them before moving on to the encode and decode functions.

Converting Between Bits and Bytes

We intend to work with bits and bytes, so it helps to have functions that convert between the two. This is the point of MakeBit and MakeByte. Listing 2-7 shows MakeBit.

```
def MakeBit(byt):
    b = np.zeros(8*len(byt), dtype="uint8")
    k = 0
    for v in byt:
        s = format(v, "08b")
```

```
        for c in s:
            b[k] = int(c)
            k += 1
    return b
```

Listing 2-7: Converting bytes to bits

The argument (byt) is an array of bytes that we want to turn into an output array of bits (b). There are eight bits per byte, so we define b up front and then fill it in by processing the bytes in byte one by one using the for loop.

For each byte, we ask Python to convert it to a binary string (s), being careful to insist on leading zeros. Then, we place each bit into the output array using an ever-increasing index (k) before returning to the caller.

We also need to convert an array of bits back to an array of bytes; MakeByte is shown in Listing 2-8.

```
def MakeByte(b):
    n = len(b)//8
    t = b.reshape((n,8))
    byt = np.zeros(n, dtype="uint8")
    for i in range(n):
        v = (t[i] * np.array([128,64,32,16,8,4,2,1])).sum()
        byt[i] = v
    return byt
```

Listing 2-8: Converting bits to bytes

The argument is an array of bits (b), so its length will be a multiple of eight. Therefore, n is the number of bytes we need in the output array. To get the bytes, we first reshape b to group every eight bits. In other words, if the array is a vector of 32 bits

```
11010101001011101101011001010100
```

it becomes

```
11010101
00101110
11010110
01010100
```

in t, a 4×8 array of bits representing 4 bytes.

Next, we multiply each row of t by the vector representing the place value of each bit in a byte. Finally, we sum the resulting vector to give the actual byte value (v), which we place into the output array, byt.

Now, let's learn how to hide bits in a random file.

Encoding a File

Encoding a file is accomplished by the cleverly named Encode function:

```
def Encode(key, sfile, dfile, pfile):
    src = np.fromfile(sfile, dtype="uint8")
    s = format(len(src), "08x")
    b3 = int(s[0:2],16);  b2 = int(s[2:4],16)
    b1 = int(s[4:6],16);  b0 = int(s[6:8],16)
    src = MakeBit(np.hstack(([b3,b2,b1,b0],src)))

    step = RE(mode="int", low=1, high=16, seed=key).random(len(src))
    idx = [step[0]]
    for i in range(1, len(step)):
        idx.append(idx[-1]+step[i])

    pool = MakeBit(np.fromfile(pfile, dtype="uint8"))
    if (len(pool) <= idx[-1]):
        print("Pool file is too small")
        exit(1)

    for i in range(len(src)):
        pool[idx[i]] = src[i]

    dest = MakeByte(pool)
    dest.tofile(dfile)
```

Our goal is to randomly scatter the bits of the source file (sfile) among the bits of the pool file (pfile) and dump the result to the destination file (dfile). The seed for the generator is in key.

In the first code paragraph, we read the source file into an array of bytes (src). As discussed previously, we also need to encode the length of the file, which we do by representing the number of bytes as a hexadecimal string (s). We're using an unsigned 32-bit integer as the file length, so we can only hide files less than 4,294,967,296 bytes. That shouldn't be a problem for us. The individual bytes representing the length are extracted from s and interpreted as hexadecimal numbers from the most significant (b3) down to the least (b0). As src is still a vector of bytes, we add the 4 bytes of the length to the beginning using NumPy's hstack function. Finally, we use MakeBit to convert src to a vector of bits.

In the second code paragraph, we construct idx, a vector of ever-increasing offsets, one per bit of src. These are the positions in the pool that will be updated with src's bit values. First, step is a vector of offsets from a low of 1 to a high of 15 bits. Note the use of key to set the seed. The idx vector is constructed from the offsets by adding each offset to the last element of idx before appending the new offset to idx.

The third paragraph reads the pool file as a vector of bits (pool). If the pool file is too short to accommodate all the bit offset positions in idx, the program delivers a terse message and abruptly ends.

At this point, we have all we need: src as bits, pool as bits, and idx telling us where the bits of src go. Therefore, the fourth paragraph is a simple loop setting the proper bit of pool to the corresponding bit of src. We march through src sequentially, and idx likewise, but the values in idx are the positions in pool that get updated.

The final paragraph converts the now updated pool into a vector of bytes and writes it to the destination file (dfile).

We can officially encode a file; now let's get it back.

Decoding a File

Decoding a file uses the similarly named Decode function:

```python
def Decode(key, sfile, dfile):
    src = MakeBit(np.fromfile(sfile, dtype="uint8"))

    rng = RE(mode="int", low=1, high=16, seed=key)
    step = rng.random(32)
    bits = []
    idx = [step[0]]
    for i in range(1, len(step)):
        idx.append(idx[-1]+step[i])
    for i in range(len(idx)):
        bits.append(src[idx[i]])
    n = MessageLength(bits)

    offset = idx[-1]
    step = rng.random(8*n)
    idx = [offset + step[0]]
    bits = []
    for i in range(1, len(step)):
        idx.append(idx[-1]+step[i])
    for i in range(len(idx)):
        bits.append(src[idx[i]])
    dest = MakeByte(np.array(bits))
    dest.tofile(dfile)
```

Here, sfile is the name of the file generated by Encode and dfile is the name of the desired output file. Note that key must match the value used to encode.

The code first reads the encoded file and immediately converts it to a vector of bits (src).

In the following code paragraph, the properly seeded generator is created (rng). Notice the RE instantiation: integer, default PCG64 generator, low of 1 and high of 15, with the seed. This is the configuration used for encoding, as it must be.

The first 32 encoded bits are extracted by adding the 32 values in step to each final offset in idx, as for Encode. The resulting bits are passed to the helper function, MessageLength, as in Listing 2-9.

```
def MessageLength(bits):
    b = MakeByte(np.array(bits))
    n = 256**3*b[0] + 256**2*b[1] + 256*b[2] + b[3]
    return n
```

Listing 2-9: Converting 32 bits to an unsigned integer

This function makes the 32-bit vector a vector of 4 bytes (b); multiplies each byte, highest-order byte first, by the proper place value; and sums to get the actual encoded file length in bytes (n). Working with the bytes of the 32-bit value is the same as treating it like a base-256 number, hence the exponents on 256.

In the final paragraph, the bit positions for the remainder of the encoded file are calculated from step and idx as before, but we need to add the extra offset to account for the 32 bits already read. Once idx is constructed, it is immediately used to extract the bits. All bits collected are converted to bytes (dest) and finally dumped to the output file (dfile).

Whew! Much is happening in *steg_random.py*. Thankfully, most of it transfers to the remaining two sections of this chapter, where we hide data in audio files and images. Let's learn to whisper carefully so that no one hears us unless we want them to.

In an Audio File

Digital audio is ubiquitous and a natural target for steganography. This section explores how to embed files in digital audio, specifically WAV files. Our approach is similar to the previous section: we'll scatter the bits of the file across the digital samples. However, instead of altering random bits, we'll alter the least-significant bit to minimize the impact and make the message essentially silent.

The file we need is *steg_audio.py*. Be sure to review the file before proceeding. You'll notice it uses MakeBit and MakeByte, which we have worked with before—see Listings 2-7 and 2-8, respectively. Likewise, you'll also see MessageLength (Listing 2-9).

We discussed digital sampling in Chapter 1 when generating random bits from the microphone input. At that time, we wanted 32-bit floating-point samples. Here, we'll use more common signed 16-bit samples in the range [−32,768, 32,767]. Such samples use two bytes each. We'll restrict ourselves to changing only the least-significant bit, thereby altering the sample value by at most 1, which no one will hear.

WAV files are not compressed and, therefore, might be rather large. Most internet audio is in the form of MP3s, which employ *lossy compression*. The human auditory system doesn't need all the information in raw audio samples—about 90 percent or so can be discarded, which is the "loss." However, as we're planning on using the simplest of audio steganography approaches, and our hidden file will be lost if the file is converted to an MP3, we must use the full WAV file. There are more sophisticated types of audio steganography that employ characters, like echoes, that survive the compression process, but this is out of the scope of this book.

The algorithm's needs are quite close to that of the previous section:

1. Read the source file and convert it to bits along with the file length in bytes.

2. Read the original WAV file, ensure the source file will fit, and work with only one channel if it's stereo.

3. Generate a random but ever-increasing list of indices into the WAV samples, one for each bit of the source file.

4. Alter bit 0 of the selected samples to match the corresponding source file bit.

5. Write the altered samples to a new audio file.

We'll first use the code to hide files, then we'll explore the code itself.

A Quiet Live Performance

Let's play around with *steg_audio.py* before walking through the code. The book's GitHub repository includes several sample WAV files courtesy of composer Paul Kneusel (*https://www.paulkneusel.com*). First, let's hide an image in *Fireflies*, a live performance in 2016 by pianist Kristen Kosey. *Fireflies* is a quiet piece, thereby increasing the probability that we'll hear any effect from altering the samples. Here's the command line:

```
> python3 steg_audio.py 2718281828 test_images/tulips_gray.png tmp.wav Fireflies.wav
Using 3712480 samples to store the file
```

First, we supply the secret key, 2718281828, then the file to embed, *tulips_gray.png*, followed by the output filename, *tmp.wav*, and finally the source WAV file, *Fireflies.wav*. The code informs us that it used 3,712,480 samples to store the bits of *tulips_gray.png*. Bonus points for readers who recognize the secret key.

As before, we recover the hidden file with the same code:

```
> python3 steg_audio.py 2718281828 tmp.wav tmp.png
```

This produces *tmp.png*, the original image file shown in Figure 2-4.

Figure 2-4: The recovered image

Play the output file, *tmp.wav*. Listen closely and try to hear the difference between *tmp.wav* and the original *Fireflies.wav*. The image file is well-hidden and utterly imperceptible. This is steganography as it should be: unnoticed.

Okay, *steg_audio.py* works, but what did it really do? We need to change the least-significant bit of each sample to match the corresponding bit of the file we're hiding. Let's look at the first eight bits of *tulips_gray.png* along with the corresponding sample offsets and the samples themselves, as in Table 2-2.

Table 2-2: The First Eight Bits of *tulips_gray.png* with Offsets and Samples

	0	1	2	3	4	5	6	7
Bits	1	0	0	0	1	0	0	1
Offsets	35	36	50	68	85	90	92	99
Samples	65,533	5	7	65,532	65,535	65,532	5	65,534

Sample 35 has value 65,533, and we want the least-significant bit of sample 35 to be a 1. Likewise, sample 36 is 5, and we want its least-significant bit to be a 0, and so on.

We previously discussed that WAV files use signed 16-bit integers in the range [−32,768, 32,767]. This is true, but when we read the audio samples, we interpret them as unsigned integers in the range [0, 65,535] instead to avoid working with negative numbers. We'll change the interpretation back to signed 16-bit integers before writing the output WAV file.

We want sample 35, currently 65,533, to be such that the first bit is a 1. We have a few options to accomplish this. For one, we can use low-level bit operations. But notice that when bit 0 of an integer is a 1 the number is odd, and when it's a 0 the number is even. In this case, 65,533 is odd, meaning bit 0 is already a 1, so we leave the sample as is.

The second sample is currently a 5, but we want bit 0 to be a 0. As 5 is an odd number, we subtract 1 to make the new sample value a 4 with bit 0 a 0, as desired. This process repeats for each bit of the file we're hiding. Either the sample's least-significant bit is already correct, or it's off by one, requiring us to add or subtract 1. To summarize: we add 1 only if we need the sample to be an odd number; likewise, we subtract 1 only if we want the sample to be an even number.

Let's return to my arrogant statement about the image file being "utterly imperceptible." Suppose an adversary were to acquire the original *Fireflies* sound file. A bit-by-bit comparison of the samples might offer a clue that something is up, but we changed samples only when they needed to be changed. Therefore, merely extracting samples that differ between the suspect WAV file and the original won't be enough to discover the hidden file because, on average, only half the samples used to encode the file will need to be altered. In other words, it's doubtful the original file can be recovered without knowledge of the distribution of the samples used.

What if the adversary had only the altered WAV file? In that case, a savvy adversary suspecting steganography might decide that the least-significant bit of the samples is the one that is likely changed because altering other bits would produce noise when the WAV file is played, and none is present. This suspicion might prompt the adversary to examine the distribution of bit 0 across the samples and compare it with other WAV files.

If you run *steg_audio_test.py*, you'll be presented with this output:

```
Fireflies: [5104781 5103286] 0.6398434737379124
Attitude : [1863322 1860129] 0.09798008575083986
Fun-Key  : [1664208 1661616] 0.15522993473071336
Encoded  : [5093639 5114428] 7.681149082012241e-11
```

Each line reports the distribution of bit 0 across the samples in the corresponding WAV file. For instance, *Fireflies.wav* has 5,104,781 even samples and 5,103,286 odd samples (for channel 0 only). The final value in each row is the p-value for a χ^2 test asking whether the even and odd sample counts are as expected if the evenness or oddness of a sample's value is random. Only the first few digits of the p-value are meaningful.

Recall that p-values above 0.05 indicate support for the null hypothesis, the claim that the samples are equally likely to be even or odd. The p-values for Fireflies, Attitude, and Fun-Key are all in good agreement with the null hypothesis.

The final line, Encoded, shows the distribution for *tmp.wav*, the version of Fireflies with *tulips_gray.png* hidden inside it. Here, the p-value is 7×10^{-11}, which is as good an approximation of 0 as any. In other words, the χ^2 test tells us that bit 0 is *not* in accordance with the null hypothesis—massively so. Upon seeing these results, our adversary would be convinced that something fishy is going on.

So, is my arrogance justified? No, not at all. A simple statistical test, combined with a bit of thought about how one might hide information in a WAV file, has given strong support to the notion that this particular WAV file has been manipulated. Also, even though our adversary strongly suspects we've manipulated the WAV file, they're still unable to extract the information because they don't know *which* bits were altered.

I encourage you to experiment with *steg_audio.py* and WAV files you have on hand. When you're ready, read on to walk through the code.

The steg_audio.py Code

The file *steg_audio.py* consists of five functions and a small driver at the bottom. We're already familiar with the MakeBit, MakeByte, and MessageLength functions (see Listings 2-7, 2-8, and 2-9).

The remaining two are the Encode and Decode functions. You might notice a trend here. The overall structure of *steg_audio.py* matches the structure of *steg_random.py*. Let's peruse Encode to follow the process in this case. It's best to absorb the function piecemeal, so Listing 2-10 shows the first part.

```
def Encode(key, sfile, dfile, pfile):
    src = np.fromfile(sfile, dtype="uint8")
    s = format(len(src), "08x")
    b3 = int(s[0:2],16);  b2 = int(s[2:4],16)
    b1 = int(s[4:6],16);  b0 = int(s[6:8],16)
    src = MakeBit(np.hstack((([b3,b2,b1,b0],src))))

    sample_rate, wav = wavread(pfile)
    if (wav.ndim == 2):
        samples = wav[:,0].astype("uint16")
    else:
        samples = wav.astype("uint16")

    if (len(src) > len(samples)):
        print("The input WAV file is too short")
        exit(1)
    else:
        print("Using %d samples to store the file" % len(src))
```

Listing 2-10: Hiding data in an audio file (part 1)

The first paragraph of Listing 2-10 is identical to the corresponding code in *steg_random.py*. The file to embed is loaded and converted to bits after prepending the file size in bytes (src).

The second paragraph reads the audio file, keeping the sampling rate (sample_rate) and the raw samples as signed 16-bit integers (raw). Next, the if interprets the samples as unsigned 16-bit integers while also selecting channel 0 if more than one channel exists. The net result is samples, the actual values we'll update. The third paragraph is a sanity check to see if there are enough samples to hide the source file.

Listing 2-11 shows the remainder of Encode.

```
step = RE(seed=key, mode="int", low=1, high=5).random(len(src))
idx = [step[0]]
for i in range(1, len(src)):
    idx.append(idx[-1]+step[i])

if (len(samples) <= idx[-1]):
    print("Audio file too short")
    exit(1)

for i in range(len(src)):
    if (src[i] == 0):
        if ((samples[idx[i]] % 2) == 1):
            samples[idx[i]] -= 1
    else:
        if ((samples[idx[i]] % 2) == 0):
            samples[idx[i]] += 1

out = samples.astype("int16")
if (wav.ndim == 2):
    wav[:,0] = out
    wavwrite(dfile, sample_rate, wav)
else:
    wavwrite(dfile, sample_rate, out)
```

Listing 2-11: Hiding data in an audio file (part 2)

With the samples prepared, we're ready to hide some data. First, we create a vector of offsets using the supplied key (step). We did the same with *steg_random.py*. The offsets are used to build idx, a vector of samples to modify. A final sanity check confirms there are enough samples.

The second for loop sets the least-significant bit of the selected samples, altering it by 1 as necessary to make it odd or even depending on the current bit (src[i]).

The loop body needs some explanation. If the current source bit is a 0, there are two possibilities for the current sample (samples[idx[i]]). If the sample is odd, meaning the remainder after dividing by 2 is 1, we subtract 1 from the sample to make it even. Otherwise, the bit we want to store is a 1, meaning we check to see if the sample is currently even, and if it is, we add 1 to make it odd.

The final paragraph reinterprets the samples as signed integers (out) and writes them to disk using the original sample rate. If the audio file is stereo, channel 0 is updated, leaving the other channels as they were.

To extract a hidden file, we reverse the process with Decode, as shown in Listing 2-12.

```
def Decode(key, sfile, dfile):
    sample_rate, wav = wavread(sfile)
    if (wav.ndim == 2):
        samples = wav[:,0].astype("uint16")
    else:
        samples = wav.astype("uint16")

    rng = RE(mode="int", low=1, high=5, seed=key)
    step = rng.random(32)
    bits = []
    idx = [step[0]]
    for i in range(1, len(step)):
        idx.append(idx[-1]+step[i])
    for i in range(len(idx)):
        bits.append(samples[idx[i]] % 2)
    n = MessageLength(bits)

    offset = idx[-1]
    step = rng.random(8*n)
    idx = [offset + step[0]]
    bits = []
    for i in range(1, len(step)):
        idx.append(idx[-1]+step[i])
    for i in range(len(idx)):
        bits.append(samples[idx[i]] % 2)

    msg = MakeByte(np.array(bits))
    msg.tofile(dfile)
```

Listing 2-12: Decoding data hidden in an audio file

There are four steps: read the encoded WAV file as unsigned integer samples, generate the same set of indices as used to encode the file, extract the bits, and write them to disk.

In the second paragraph, the pseudorandom generator is configured as in Encode before generating the first 32 step values to extract the encoded file length. We need the length to know how many samples to query to reconstruct the hidden file. Notice the call to MessageLength, Listing 2-9.

In the third paragraph, the actual bits of the file are extracted now that we know how many there are. To finish, the bits are converted to bytes and written to the output file (dfile).

Let's try hiding data in images next. The process is similar to hiding data in a file of random bytes or in an audio file, but we'll dispense with the notion of a secret key.

In an Image File

For the final experiment of this chapter, we'll hide files in images. The method should be familiar: we'll scatter the bits of the file across the pixels of the image by making small changes to the existing pixel colors. This is the most compact experiment in terms of code, but we need to discuss a few things about digital images before diving in. Specifically, we need to understand how computers store and interpret images, after which I can describe the algorithm. Experiments follow before a code walkthrough. You know the drill.

Defining Image Formats

The phrase "image format" has multiple meanings, especially when working in Python.

The first references how the image is stored on disk. There are many options that fall primarily into two camps: lossy and lossless.

Lossy image formats compress by throwing information away, much like an MP3 file. The information retained is sufficient to reconstruct the image to some level of fidelity with the original. The most common lossy image format is JPEG, which usually has a *.jpg* file extension. This format uses discrete cosine transforms across small patches to retain essential levels of detail while discarding those that don't contribute substantially to the reconstruction of the image.

Lossless image formats either retain all pixel information or compress the image losslessly, meaning the recovered data is a bit-for-bit match with the original, uncompressed image. The most common lossless image format is PNG (*.png*). We'll restrict ourselves to using only PNG files, as lossy formats like JPEG aren't acceptable for the same reason MP3 files aren't: the subtle changes made to the image will be lost. Advanced steganographic techniques are robust to lossy formats, but they are far beyond what we can explore here.

"Image format" also refers to the pixels themselves, with the main divisions being grayscale and color.

A grayscale image is the simplest: each pixel is a single number, a byte, representing the intensity of the pixels from black (0) to white (255) and shades of gray in between. Grayscale images with 16-bit pixels are rare, so we'll ignore them here.

Computers have many options when representing color images. It's most common to separate the color of each pixel into a combination of red, green, and blue. That is, a pixel showing as a whitish red can be represented as a mix of red, green, and blue that arrives at that color. The red, green, and blue *channels* are typically bytes, with 0 meaning none of that color and 255 being maximum intensity for that color. For example, the whitish red color is A94141, which is a triplet of hexadecimal numbers for the amount of red, green, and blue. Converting the hexadecimal to decimal, we have 169 out of 255 red, 65 green, and 65 blue.

Using NumPy and PIL

To NumPy, a grayscale image is a 2D array of bytes. Likewise, a color image, or RGB image, is a 3D array of bytes—one 2D array for each color channel. We'll use RGB images for the upcoming experiments. If a grayscale image is loaded, it will be converted to RGB by duplicating the grayscale values for each channel.

NumPy will use the images as arrays, but how do we get them into Python in the first place? For that, we need the *Python Imaging Library (PIL)*, now called *Pillow*. It's included with most Linux distributions and is available for Windows and macOS. For Linux and macOS, enter the following:

```
> pip install pillow
```

Any version 8.4 or later should work. We need it only to read and write image files.

The book's GitHub page includes a collection of standard test images. We've used several already without bringing them into Python. Let's remedy that now:

```
>>> import numpy as np
>>> from PIL import Image
>>> im = Image.open("apples.png")
>>> d = np.array(im)
>>> d.shape
(352, 375, 3)
>>> im = Image.open("barbara.png")
>>> d = np.array(im)
>>> d.shape
(256, 256)
>>> im = Image.open("barbara.png").convert("RGB")
>>> d = np.array(im)
>>> d.shape
(256, 256, 3)
```

Here, I'm showing the Python prompt and the commands entered manually. The first two lines load NumPy and PIL's Image class. We only need Image.

The next two lines load the file *apples.png* into im, an instance of the Image class. To make the image a NumPy array, pass it to np.array to create d. Notice d's shape: (352, 375, 3). The image has 352 rows, its height, and 375 columns, its width. The final dimension is 3 for the red, green, and blue channels.

The following two lines load *barbara.png* and convert it to a NumPy array. In this case, the array has only two dimensions: height and width, both 256. This is a grayscale file with one channel.

The final three commands load *barbara.png* a second time but immediately pass the image to the Image's convert method to turn it into an RGB image. The NumPy array now has three dimensions, though as a grayscale image, each color channel is the same.

While many books have been written on image processing in Python, fortunately all we need to know is how to open an image file, ensure that it's RGB, turn it into a NumPy array, and write a NumPy array to disk as an image. Continuing with the previous example:

```
>>> d = np.array(Image.open("apples.png"))
>>> t = d[:,:,0]
>>> d[:,:,0] = d[:,:,1]
>>> d[:,:,1] = t
>>> im = Image.fromarray(d)
>>> im.save("bad_apples.png")
```

The first line rereads *apples.png* and then immediately changes it into a NumPy array. We'll use this idiom consistently.

The next three lines swap the red and green color channels. Read through them carefully. Since NumPy makes the third index the channel, we need to explicitly mention the height and width.

The next line uses Image's fromarray method to change d into an Image object: im. The final line writes the altered image to disk as *bad_apples.png* using the save method. PIL uses the file extension to specify the format, in this case a PNG file. Open *bad_apples.png* to learn why I picked that name.

Hiding Bits in Pixels

In order to scatter the bits of the hidden file throughout the pixels of the image, we need to load the source file and convert it to bits using *MakeBit*. Then we load the image file, confirm it's an RGB image, and convert it to a NumPy array.

Our previous experiments altered bits of the pool file to make them match the bits of the source file. The sequence of random offsets into the pool file was reproducible because we set the randomness engine's seed to the secret key. While we could do the same here, we'll take a different approach. Variety is the spice of life, as they say.

We'll use the unaltered image as our key. When hiding a file, we'll initialize the randomness engine without setting the seed, meaning each run of the code hides the same file in different locations in the image. Because we're altering specific bits, we'll know which bits we altered by using the unaltered image. To simplify things, we'll do much as we did for the audio samples: one bit of the hidden file per byte of the image file.

While we could set the least-significant bit of the pixel, as we did for the audio file, we'd never be able to detect where the image pixel already had the proper bit value. And since we're abandoning the secret key to get a unique set of altered pixels per run, we'd never know which pixels happened to have the proper bit value by chance.

Instead of looking for bit 0 of the pixel to be 0 or 1, or equivalently asking if the pixel is even or odd, we'll encode 0 bits by adding 1 to the pixel and 1 bits by adding 2. In this scenario, any pixel of the unaltered image that isn't the same as the corresponding pixel of the altered image holds a bit of the hidden file.

In RGB images, each pixel is actually three bytes; we can alter any of them. When we implement the algorithm, we'll *unravel* the image by asking NumPy to turn the three-dimensional array into a single vector, which is how the image data is stored in the computer's memory. Where specific red, green, and blue bytes end up in the unraveled array is irrelevant as long as NumPy makes the process repeatable, which it does.

With the image bytes as a vector, we select a subset of locations—as many as there are bits in the hidden file—and alter them by adding 1 if the corresponding bit is a 0, or 2 if the bit is a 1.

Bytes are unsigned numbers in the range $[0, 255]$, meaning if the pixel value is 255 and we add 1 to encode a 0, we'll wrap around to 0 (or 1 if adding 2 for a 1 bit). That won't do at all, so we make an executive decision: before encoding, we'll alter all image pixel bytes greater than 253 and set them to 253 so that adding 1 gives us 254 and adding 2 gives 255; hence no overflow.

We're able to alter the image pixel bytes for two reasons: the code will do this for the unaltered image each time, even when decoding, so it's reproducible; and because the bytes represent color values, the difference between 255 and 253 is too subtle to see, especially when there's a global change in overall color intensity or shade.

The resulting code is *steg_image.py*, which I encourage you to study before reading further. I'll walk through the code after we experiment with it.

This approach is just as fragile as the audio approach. We'll throw away the encoding if we store our encoded image using a lossy image format like JPEG. We can use a JPEG as the original image because NumPy will decode it the same way each time, but even that might be a risk because a future version of NumPy might alter the algorithm used to decode the image (this is unlikely, but possible). Again, we'll restrict ourselves to using only losslessly compressed PNG files.

Now, let's have some fun!

Hiding One Image in Another

For our first experiment, we'll hide an image (*cameraman.png*) within a second image (*apples.png*). The command line we need is:

```
> python3 steg_image.py encode test_images/cameraman.png test_images/apples.png tmp.png
```

We tell *steg_image.py* we want to encode, then provide the file to encode (*cameraman.png*) followed by the reference image (*apples.png*) and the output image file, *tmp.png*.

Open both *apples.png* and *tmp.png* and look carefully, flipping rapidly between the two if your software allows. Do you see any differences? A meticulous examination might reveal some, but my eye doesn't notice anything.

To recover *cameraman.png*, use:

```
> python3 steg_image.py decode tmp.png test_images/apples.png output.png
```

The *cameraman.png* file is 16 percent the size of *apples.png*, and was easily hidden among the apples. How large can the file be before we notice it?

If you run *steg_image_test.py*, then you'll get a new output directory, *steg_image_test*, containing multiple images with names like *apple_A_0.60.png* and *violet_rand_0.40.png*. The files are images, each hiding either a specific number of *A*s or random data read from RDRAND. The first part of the name is the test image source, *apples.png* or *violet.png*. The latter is a 512×512-pixel image of a light violet color.

The fraction in each name is the fraction of the reference pixel images used to hold the encoded data. For example, *apple_A_0.60.png* is hiding the letter *A* so that 60 percent of the pixels in *apples.png* have been modified. The fractions start at 5 percent and go up to 99 percent. Take a look at these output files, especially the violet ones. Even when 99 percent of the pixels have been modified with either the same byte repeated or highly random bytes, we're unable to see a difference between the output image file and the original. This is an example of imperceptible steganography. For brevity, I won't discuss *steg_image_test.py* in detail, but do take a look at it.

We've hidden an image in another image. What's stopping us from hiding a message in an image, then hiding the image with the hidden message in another image, and so on, to produce a final output image with a set of Russian doll images inside it? Why, nothing at all. Let's do it.

The script we want to run is in *russian_dolls_example*, which is shown in Listing 2-13.

```
echo "Encoding..."
python3 steg_image.py encode kilroy.txt         test_images/apples_32.png    /tmp/encode0.png
python3 steg_image.py encode /tmp/encode0.png test_images/peppers_128.png /tmp/encode1.png
python3 steg_image.py encode /tmp/encode1.png test_images/fruit2.png       /tmp/encode2.png
python3 steg_image.py encode /tmp/encode2.png test_images/tulips.png        russian_dolls.png
echo "Decoding..."
python3 steg_image.py decode russian_dolls.png test_images/tulips.png        /tmp/decode0.png
python3 steg_image.py decode /tmp/decode0.png  test_images/fruit2.png        /tmp/decode1.png
python3 steg_image.py decode /tmp/decode1.png  test_images/peppers_128.png /tmp/decode2.png
python3 steg_image.py decode /tmp/decode2.png  test_images/apples_32.png     output.txt
```

Listing 2-13: Encoding and decoding Russian dolls

This is a shell script you can run with:

```
> sh russian_dolls_example
```

When finished, take a look at the image with all the nested images and the original message, *russian_dolls.png*. Then compare the source file, *kilroy.txt*, with *output.txt* to see that they are the same. The intermediate images with each stage of the encoding process are in the system *tmp* directory.

Figure 2-5 shows the process of building a stack of Russian dolls.

"Kilroy was here"

Figure 2-5: Building a stack of Russian dolls

First, the text message "Kilroy was here" is hidden in the small version of the apples image. Then, that image is hidden in the peppers image, which is hidden in the fruit image, all of which are hidden in the tulips image as the final output. Reversing the process to decode step by step delivers the original message: Kilroy was here.

NOTE *The phrase "Kilroy was here" was used by American and British troops in WWII, partly as a joke and partly to mark places where they had already been. It caught on after the war, becoming an early meme of sorts, often with a silly character face to accompany the message.*

Let's walk through the code to see where Kilroy has been.

The steg_image.py Code

Of all our experiments, *steg_image.py* is perhaps the simplest. It uses both MakeBit and MakeByte, and has a simple driver to call Encode and Decode. Let's focus on those two functions, beginning with Encode, shown in Listing 2-14.

```
def Encode(mfile, sfile, dfile):
 ❶ msg = MakeBit(np.fromfile(mfile, dtype="uint8"))
    simg = np.array(Image.open(sfile).convert("RGB"))
    simg[np.where(simg > 253)] = 253
    row, col, channel = simg.shape
```

```
❷ simg = simg.ravel()
  if (len(msg) > len(simg)):
      print("Message file too long")
      exit(1)
❸ rng = RE(kind="mt19937")
  p = np.arange(len(simg))
  np.random.shuffle(p)
  p = p[:len(msg)]
  p.sort()
❹ for i in range(len(p)):
      simg[p[i]] += msg[i]+1
  simg = simg.reshape((row, col, channel))
  Image.fromarray(simg).save(dfile)
```

Listing 2-14: Hiding a file in an image

The file to hide (`mfile`) is read as bytes and immediately converted into a vector of bits ❶. The reference image is read next, as RGB if not already, and then converted to a NumPy array (`simg`).

As mentioned previously, we need to prevent overflow when encoding bits, so we use NumPy's `where` function to locate the places in the image that are greater than 253. The `where` function returns a collection of indices, regardless of the dimensionality of `simg`, that are immediately used to change said locations to 253, the new maximum. This covers all red, green, and blue parts of the image.

We intend to use the image as a vector, so we next grab the actual row, column, and channel (= 3) numbers to use later when re-forming the image.

Next, we tell NumPy to make the three-dimensional image array a vector by calling ravel ❷. In NumPy, the word *ravel* is a contronym, a word that is its own antonym. In this case, ravel means to *unravel*, to change the array into a vector, a one-dimensional beast. After ravel is called, `simg` is a vector of the bytes of the image. We don't care about where each specific red, green, and blue byte landed; NumPy applies ravel consistently.

We have the image as a vector of bytes and the source file as a vector of bits. To scatter the bits across the bytes, we need a vector of locations; that's the purpose of rng ❸. First, we initialize rng using the Mersenne Twister. The following four lines build the vector of indices, the bytes we'll modify (`p`).

We want `p` to have as many elements as bits in src. First, we set `p` to $0, 1, 2, \ldots, n - 1$ for the n number of bytes in the image, `simg`. The resulting vector is shuffled to create a unique permutation of the values.

The following two lines keep as many elements of `p` as we need to place each bit of the source file (`msg`) and sort `p` so the list of indices is in order numerically. This last step is necessary because we're not setting rng's seed, so we must apply some order to the image bytes updated to preserve the order of the bits in `msg`. You'll see what I mean when we review Decode.

To scatter the bits of `msg` we need only iterate over them and use the corresponding index in `p` ❹. We add the bit value plus 1 so that 0 bits become 1 and 1 bit becomes 2. This change lets us detect all altered bits by using the reference image as the key.

Finally, we reshape the updated image bytes to create the 3D output array. Notice that reshape applies the proper reinterpretation of the bytes in memory, an interpretation that matches the order used by ravel. In other words, reshape ravels the array in the opposite sense of the word. The array is passed to the PIL Image class to be output as a PNG file.

As shown in Listing 2-15, Decode recovers hidden files using the reference image as the key.

```
def Decode(dfile, sfile, mfile):
    dimg = np.array(Image.open(dfile)).ravel()
    simg = np.array(Image.open(sfile).convert("RGB")).ravel()
    simg[np.where(simg > 253)] = 253
    i = np.where(dimg != simg)
    d = dimg[i] - simg[i] - 1
    b = MakeByte(d.astype("uint8"))
    b.tofile(mfile)
```

Listing 2-15: Decoding an image

Decode loads two image files, the one with the hidden data (dimg) and the reference image (simg), as NumPy arrays. The function then immediately converts both image files to vectors of bytes, and the reference image is passed through convert to make it an RGB image, as Encode did. Also as in Encode, reference image byte values above 253 are set to 253.

NumPy's where returns all the locations where the two images disagree. Because we assume the reference image is *exactly* the same image used when creating dimg, any differences represent bits of the hidden file.

The difference between the encoded image and the reference image gives us the bit values, but as 1 and 2, not 0 and 1. Therefore, we subtract one more to make d the vector of hidden bit values. Because we insisted that the bytes were altered in numerical order when encoding, the order in which the difference appears in the image matches the proper order for the output bits. In other words, we don't need to sort anything; the indices returned by where are already in sorted order.

The remaining two lines convert d to bytes and write them to the output file (mfile).

Exercises

Steganography offers many possibilities. Here are some to explore and contemplate:

- Bacon's cipher uses subtle changes in the type to embed binary codes representing letters. What needs to change to embed arbitrary binary data in a text block? What about using spaces between words with a proportional font?

- Make your own "spaceship cipher" to hide secret messages in drawings. This might be a good activity for the kids.

- Our text experiments neglected to encode the spaces between words. Remedy that oversight.

- Can you think of a simple sentence structure with many word options and use that to encode a text message à la *steg_text.py*? Generate random sentences using a large pool of adjectives, adverbs, nouns, and verbs, so the output reads as grammatically correct sentences.

- We used channel 0 when hiding data in audio files. Use both stereo channels to double the amount of data that can be encoded.

- Replicate hiding binary data in a file of random bytes, but substitute an audio file of white noise (random samples). This removes the restriction on changing only the lowest-order bit of each sample. How much data can you fit using both channels before the sound changes? What does a file of random bytes sound like in the first place? Might sound be a good way to detect randomness in data?

- Lossy image compression destroys our simple approach of changing byte values to encode information. How might we add redundancy so that the message persists even after using JPEG compression?

- What happens to image encoding when the image is resized? Can you think of a way to preserve the message using nearest-neighbor sampling? Nearest-neighbor sampling duplicates existing pixels when expanding the image and throws pixels away when shrinking the image, that is, it does not interpolate.

Let me know what you think and how it goes.

Summary

This chapter explored randomness and steganography, the hiding of information. We first defined the term, and then contemplated a smattering of historical uses. That section alone could easily be a separate book, but it set the stage for the experiments.

First, we hid text in other text using random selections of letters in a list of discrete words. Pseudorandom generators proved helpful with the seed, the secret key necessary for correctly decoding the hidden message.

The following experiment worked with a file of random bytes. The hidden data was scattered, bit by bit, throughout the random data, again using the pseudorandom generator seed as the secret key. We found that a ratio of about 100:1 between the size of the random data and the hidden data generally preserved sufficient randomness in the output to properly conceal our message.

Then we scattered the bits of the hidden file across bit 0 of a set of WAV sound samples; this led to an output audio file that gave no auditory indication it had been altered. However, a statistical test on the distribution of bit 0 for the WAV file revealed evidence of tampering, though extraction of the hidden data remained extremely difficult.

Finally, we used the color information in a reference image to embed a file. In this experiment, no secret key was necessary; the reference file itself was the key. Each encoding generated a different collection of altered image bytes at the expense of requiring the entire reference image as the key. We learned how computers store and manipulate color images and how to work with image data in NumPy. We'll use this knowledge again in Chapter 5.

It isn't difficult to imagine illegitimate uses of steganography. I'm reminded here of the wisdom of Ben Parker: with great power comes great responsibility. If you're inclined to use steganography for something other than simple "I wonder if it would work" experiments, think carefully about it, put this book down, get a cup of coffee, take a walk, and forget about it.

The next chapter introduces us to the vast, randomness-heavy world of simulation. I think it'll be fun. I hope you agree.

Read *poem.txt*. It's an Easter egg.

3

SIMULATE THE REAL WORLD

Computer simulations are programs that use randomness to simulate real-world events and processes. More specifically, computer simulations manipulate *models*, programmatic stand-ins for the real world.

We'll begin this chapter by defining what a model is. Then we'll get our feet wet with two straightforward simulation examples: estimating π by throwing darts and gathering people together in a room to estimate the probability that at least two of them share a birthday. Once we've done that, we'll wade in further to explore Darwinian evolution via simulation, capturing essential characteristics of natural selection and genetic drift.

Introduction to Models

We can define a model in many ways, but I like this definition from Daniel L. Hartl in *A Primer of Population Genetics and Genomics* (Oxford University Press, 2020):

> A model is an intentional simplification of a complex situation designed to eliminate extraneous detail in order to focus on the essentials.

Think of a model as an approximation of something that we're interested in exploring or characterizing. There are no requirements for what that something is or for how we model it.

In this chapter, a model is a piece of code that attempts to capture the essential character of a real-world process, like what happens to the probability of shared birthdays as more and more people gather in a room, or how natural selection and genetic drift affect the genomes of a population. Simulation lets us control the experimental world while allowing random behavior, to understand what has happened or might happen, especially as critical parameters (environmental factors) are varied.

Consider the following statement, which is attributed to British statistician George Box:

> All models are wrong, but some are useful.

Unless particularly trivial, all models are wrong in some way, especially those of the real world. If the model is well conceived and well implemented, it might lead to valuable conclusions about the modeled process. The word *process* implies a sequence of events, that is, time. Many models simulate processes unfolding in time; for example, we'll explore fundamental evolutionary processes acting at a population level over time.

A good model captures enough of the thing being modeled to generate conclusions worthy of confidence tempered with reality. Blind faith in a model's output isn't recommended. At best, a model falls into the "trust, but verify" category—a good rule of thumb for all scientific claims.

Let's slide into simulation by throwing darts to estimate π.

Estimate Pi

We'll generate an estimate of π, the ratio between the circumference of a circle and its diameter, by throwing darts at a board. Doing this in real life would be time consuming, so we'll simulate the process instead; that is, we'll make a model.

Using a Dartboard

First, let's learn how throwing darts at a board tells us about the value of π. For that, we need a diagram (Figure 3-1).

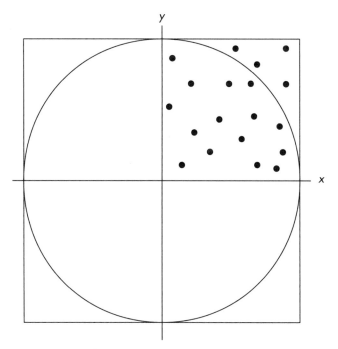

Figure 3-1: Simulating dart throws

Figure 3-1 shows a square with a circle inside it. The circle's diameter isn't marked explicitly, but we'll say it's 2, meaning the radius is 1. The diameter also matches the length of the sides of the square; therefore, the areas of the square and circle are

$$A_{\text{square}} = s^2 = 2^2 = 4 \quad \text{and} \quad A_{\text{circle}} = \pi r^2 = \pi(1^2) = \pi$$

implying:

$$\frac{A_{\text{circle}}}{A_{\text{square}}} = \frac{\pi}{4}$$

We calculate π by dividing the area of the circle by the area of the square and multiplying that by 4.

If we throw many darts, or pick many random points, they'll eventually cover the circle and the square. We can use the number of darts that land inside each shape as a proxy for the areas. We now have an algorithm: throw darts and count the number that land inside the circle (N) and inside the square (M), then divide N by M and multiply by 4 to get an estimate of π.

The previous figure's example points are all in the first quadrant, which works well for our estimate because the ratio of the area of the square to the circle is the same as the ratio of the portion of each shape in the first quadrant. Specifically, the first quadrant is 1/4 the size of the full shapes, so the areas are divided by 4. But both the circle and square areas are divided by 4, meaning their ratio remains the same, $\pi/4$. This means we need only to throw darts that land in the first quadrant.

Now that we understand how to estimate the respective areas and π by throwing darts, how should we actually "throw" them? The answer lies in the previous comment about the first quadrant.

Here's the algorithm:

1. Randomly pick two numbers in $[0, 1)$ and call them x and y. These become the point where the dart lands, (x, y).

2. Increment M, the counter for the number of points in the square.

3. If $x^2 + y^2 \leq 1$, increment N, the number of points inside the circle.

4. Repeat steps 1 through 3 for as many darts as desired.

5. Return $(4N)/M$ as the estimate of π.

We pick points in $[0, 1)$ so that the points are all in the first quadrant and land inside the square. Therefore, if we throw n darts, $M = n$, and we need only ask if the same points are also inside the circle.

In step 2 we ask a question about the circle, with $x^2 + y^2 \leq 1$ coming from the Pythagorean theorem: $a^2 + b^2 = c^2$, where c is the side opposite the right angle. Here the triangle sides are x and y, meaning the radius ($r = 1$) is the hypotenuse, $x^2 + y^2$. Any point forming a hypotenuse less than $r = 1$ is inside the circle.

Before we proceed, we should ask whether this is a fair model of the process of throwing darts, and whether we've made any unfair assumptions. After all, a model attempts to mimic *what's most important* about a process. We're using two uniformly selected random numbers in $[0, 1)$ to represent the location where a dart might land, and we've made only one assumption: that *all* darts land in the first quadrant. Limiting the random values to $[0, 1)$ eliminates out-of-range darts, so we count every dart throw as landing in at least the square.

With these questions answered, we're ready to test.

Simulating Random Darts

The code we want is in *sim_pi.py*. To run it, supply the number of simulated darts and the desired randomness source. For example:

```
> python3 sim_pi.py 10000 pcg64
pi = 3.16120000
```

This throws 10,000 darts using PCG64. The result is $\pi \approx 3.1612$. The correct value rounded to four digits is 3.1416, so we're in the ballpark. The estimate uses only four decimal places because we're approximating π with a fraction that has a denominator of 10,000. Ten more runs gives:

$$3.1496, 3.1188, 3.1468, 3.1700, 3.1292, 3.1372, 3.0916, 3.1608, 3.1236, 3.1140$$

Combining all 11 runs gives $\pi \approx 3.1366$, which is about 0.16 percent off from the four-digit value.

Let's increase the number of darts:

```
> python3 sim_pi.py 1_000_000 pcg64
pi = 3.14157600
```

That's more like it. The correct value to six places is 3.141593.

Let's go for broke—this should nail it:

```
> python3 sim_pi.py 100_000_000 pcg64
pi = 3.14180732
```

Odd. We threw 100 times as many darts as the previous run, but the result wasn't as good. Nothing's wrong with our approach; that's how the random cookie crumbles. A second run using PCG64 and 100 million darts returned 3.14160636, which is better. Still, it raises the question: Why such variation? That's the nature of a random generator, and it serves as a reminder to repeat simulations multiple times to convince ourselves that they're producing reasonable output and to get numerous estimates.

Individual runs using the other randomness sources supported by the RE class gave:

```
> python3 sim_pi.py 100_000_000 mt19937
pi = 3.14151340
> python3 sim_pi.py 100_000_000 urandom
pi = 3.14148696
> python3 sim_pi.py 100_000_000 rdrand
pi = 3.14139680
> python3 sim_pi.py 100_000_000 minstd
pi = 3.14156084
> python3 sim_pi.py 100_000_000 RandomDotOrg.bin
pi = 3.14161204
```

The last run uses *RandomDotOrg.bin*, a 510MB file of random data from *random.org*. All randomness sources produce reasonable estimates of π but are still not satisfying. Why aren't they closer to the actual value of 3.14159265 ... ?

Understanding the RE Class Output

Let's reconsider what we want the random throwing of darts to simulate. We're looking to compare *areas*, so we want the darts to cover the areas as evenly as possible.

Figure 3-2, which is a duplicate of Figure 1-5, shows us what is happening. The middle plot in the figure shows the placement of points when using a random generator. There are gaps and places where the points are concentrated. It's a reasonable coverage of the area, but not a uniformly dense one.

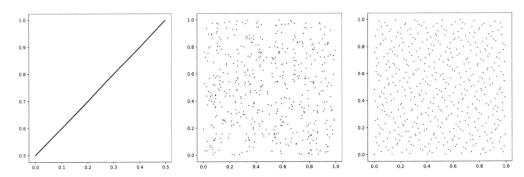

Figure 3-2: Bad quasirandom (left), pseudorandom (middle), and good quasirandom sequences (right)

The right plot in Figure 3-2, from a pair of quasirandom sequences, is much more uniform over the area. Let's try using that instead. The code we want for this case is in *sim_pi_quasi.py*:

```
> python3 sim_pi_quasi.py 10000 2 3
pi = 3.14480000
> python3 sim_pi_quasi.py 100000 2 3
pi = 3.14208000
> python3 sim_pi_quasi.py 1_000_000 2 3
pi = 3.14157200
```

The first argument is the number of darts; the other two are the bases for the quasirandom sequence. To cover a 2D plane, we need two different bases, here 2 and 3. As the number of darts increases, so does the quality of the estimate. With 1 million darts, it has already matched the first run of *sim_pi.py* with PCG64. There's no randomness here; every run with the same number of darts and the same bases results in the same output. Also, because we're generating the quasirandom sequence in pure Python, the runtime increases dramatically with the number of darts. For example, this run

```
> python3 sim_pi_quasi.py 10_000_000 2 3
pi = 3.14159680
```

is correct to five decimals but took 12 minutes to run on my reference Intel i7 system. Asking for 100 million samples produces $\pi \approx 3.14159184$ after a two-and-a-half-hour run.

Let's focus momentarily on the performance of each pseudorandom generator supported by the RE class. The file *sim_pi_test.py* estimates π 50 times each for PCG64, MT19937, MINSTD, urandom, and RDRAND using 2 million simulated dart throws. The result is the box plot in Figure 3-3.

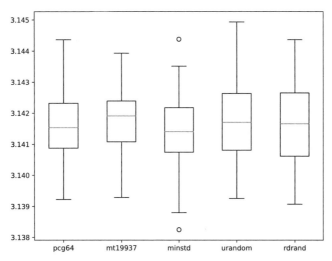

Figure 3-3: A box plot showing the distribution of π estimates by randomness source

A *box plot* is a diagram summarizing a set of data; in this case, the 50 estimates of π, that is, the 50 separate runs of *sim_pi.py* for each pseudorandom generator. Each generator's output produces a box with a horizontal bar across it. The bar represents the median value, or the 50th percentile. Half the estimates were below this value and half above. The box's lower and upper limits are the 25th and 75th percentiles, respectively. So, 75 percent of the estimates were below the upper part of the box and the remaining 25 percent were above it.

The *whiskers*, called fliers by Matplotlib, extend beyond the box. The height of the box, the difference between the 75th percentile and the 25th, is known as the *interquartile range (IQR)*. The whiskers are the box quartiles plus or minus 1.5 times the IQR. Any data values outside the whiskers are candidates for *outliers*, values that are atypical when compared to the rest of the data. Outliers might be errors or the exciting thing we're looking for; context is everything.

The five boxes in Figure 3-3 are statistically identical. There are two potential outliers for MINSTD, but another run of *sim_pi_test.py* generates a new plot with a different set of boxes and potential outliers, even from RDRAND, which is as close to a true source of randomness as we can get for repeated sampling. Random processes sometimes produce strange output; there's no meaning attached to it. This phenomenon is partly why detecting true cancer clusters can be tricky.

Implementing the Darts Model

Our dart-throwing simulation works. Now, let's review the code to understand how:

```
import sys
from RE import *
```

```
def Simulate(N, rng):
    v = rng.random(2*N)
    x = v[::2]
    y = v[1::2]
    d = x*x + y*y
    inside = len(np.where(d <= 1.0)[0])
    return 4.0*inside/N

N = int(sys.argv[1])
kind = sys.argv[2]
rng = RE(kind=kind)
pi = Simulate(N, rng)
print("pi = %0.8f" % pi)
```

The code at the bottom parses the command line to get the number of darts to throw (N) and the type of randomness source to use (kind). A generator is created (rng) and passed, along with the number of darts, to Simulate, which returns an estimate of π that is then printed.

All the action is in Simulate. We need N points, the locations where our darts landed. We can either use rng twice—first to get x-coordinates and then again to get y-coordinates—or generate twice as many points as we need and partition them in pairs. I chose the latter. Therefore, v contains $2N$ values. The first and then every other point become x, while the second and every other after that become y.

By design, all the points are inside the square. We need only to decide which are also inside the circle. For that, we need to know if $x^2 + y^2 \leq r^2$ for radius $r = 1$. To this end, we set d to $x^2 + y^2$ and use NumPy's where to find the indices that are less than or equal to 1. The count of those indices tells us how many points are inside the circle. Finally, the function returns the estimate of π as four times the number inside divided by the number of darts thrown.

Our dart simulation is complete. Now, let's simulate a party to see how many people we need in a room to have a better-than-50 percent chance that at least two of them share a birthday.

Birthday Paradox

How many people need to be in a room for the probability of at least two of them sharing a birthday to be above 50 percent? There's a mathematical way to determine this probability, but if we don't know the math, we can find it by extensive experimentation: we can throw many parties, with differing numbers of people invited, and at each one figure out if at least two of them share a birthday. While this approach will work, it'll be terribly slow and expensive.

Simulating 100,000 Parties

Assuming we aren't willing to write a grant proposal for a million dollars to conduct this experiment with actual people, to say nothing of gaining review board approval and informed consent from thousands of people, is there any other way to approach the problem? You guessed it: simulation.

Every person has a birthday, so we'll simulate the number of people in the room and assign each a randomly selected birthday. Then we'll look at each possible pair and ask if they have the same birthday. There are 365 days in a year, ignoring leap years, so we'll represent birthdays by picking a day of the year as a proxy for an actual birthday. In other words, each simulated person is assigned an integer in $[0, 364]$. If any two have the same number, they share a birthday.

We want the probability of a match for a given number of people, meaning one simulation isn't sufficient. We need many, many simulations for a fixed number of people in the room. The number of times there's a match divided by the number of simulations converges to the probability we seek.

Here's our algorithm:

1. Fix the number of people in the room (K).
2. Assign each of the K people a birthday ($[0, 364]$).
3. Check each possible pair. If they share a birthday, increment M.
4. Repeat from step 2 N times.
5. The estimated probability for K people in a room is M/N.

We'll vary K from 2 to 50. N should be a large number, like $N = 100{,}000$ to simulate 100,000 parties with K people. We can always make N larger and try again, as varying simulation parameters and observing what happens is part of what makes simulations worthwhile. If things blow up when we make a subtle change, we might have a bug in our code or, worse yet, a logic flaw in our design.

Testing the Birthday Model

The code we need is in *birthday.py*. Let's run it to understand its output before walking through it. For example, here's the output for a run asking for the probability of at least one match in a room with 11 people:

```
> python3 birthday.py 11 minstd
11 people in the room, probability of at least 1 match = 0.140430
```

We're told that the probability of at least one birthday match for a room of 11 people is about 14 percent. The second argument is the randomness source to use, here MINSTD. Feel free to experiment with other sources.

If we add a third argument, we can store the output and bring it into Python to understand what it means:

```
> python3 birthday.py 11 minstd 11.npy
11 people in the room, probability of at least 1 match = 0.142610
```

```
> python3
>>> import numpy as np
>>> d = np.load("11.npy")
>>> d
array([85739, 13462,   663,   125,    10,     0,     0,     1])
```

The first line runs the code a second time. Notice that the probability changes slightly; random selection of birthdays will produce varying results that eventually converge to a mean after many repetitions of the simulation. We'll experiment with this fact in a bit.

Next, I ran Python (and ignored the startup message) before importing NumPy and the output file, *11.npy*. The d array contains a histogram of the number of times that many birthday matches were found for 100,000 simulations where the index into d is the number. Table 3-1 shows the number of matches and how often they appeared.

Table 3-1: Number of Matches and Frequency of Appearance

Matches	Count	Percent
0	85,739	85.739
1	13,462	13.462
2	663	0.663
3	125	0.125
4	10	0.010
5	0	0.000
6	0	0.000
7	1	0.001

In 85.7 percent of the cases, when 11 people were in the room no two shared a birthday. Likewise, 13.5 percent of the cases resulted in a single pair sharing a birthday. Finally, in one run out of 100,000, seven pairs of people shared a birthday. This is the nature of randomness: sometimes remarkable things happen.

Next, I ran *birthday.py* five times, once for each of the randomness sources built into RE, and always with 23 people in the room. The average probability returned was 0.507478 or 50.7 percent. This is the first number of people to return a probability greater than 50 percent; therefore, to answer the question at the beginning of this section, we need 23 people in a room, on average, to have a greater than 50 percent chance that at least two of them share a birthday.

Let's try to visualize what's happening here (Figure 3-4).

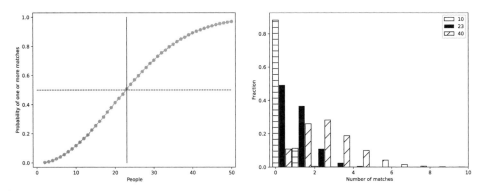

Figure 3-4: The probability of at least one match as a function of number of people (left) and the histogram of matches by people in the room (right)

The left-hand side of Figure 3-4 shows the probability of one or more matches as a function of the number of people in the room. The vertical line is 23 people, and the dashed horizontal line is 50 percent. As claimed, 23 people is the minimum number needed to exceed 50 percent.

The right side of Figure 3-4 presents three histograms showing the fraction of runs returning the indicated number of matches. The bars are offset to prevent overlapping, but the leftmost bar is on the actual number of matches. When there are only 10 people in the room, the probability of no match is high and more than one match is essentially zero. For 23 people, one match is relatively common, two less so, and three pairs happen about 3 percent of the time. With 40 people, we're past the 23-person transition, so it's more likely than not to have multiple matches.

Implementing the Birthday Model

Let's take a walk through *birthday.py*, shown in Listing 3-1.

```
import sys
import numpy as np
from RE import *

def Simulate(rng, M):
    matches = []
    for n in range(100_000):
        match = 0
        bdays = rng.random(M)
        for i in range(M-1):
            for j in range(i+1,M):
                if (bdays[i] == bdays[j]):
                    match += 1
        matches.append(match)
```

```
    matches = np.array(matches)
    return np.bincount(matches)

people = int(sys.argv[1])
rng = RE(kind=sys.argv[2], low=0, high=365, mode="int")
matches = Simulate(rng, people)
prob = matches[1:].sum() / matches.sum()
print("%d people in the room, probability of at least 1 match = %0.6f" % (people, prob))
if (len(sys.argv) == 4):
    np.save(sys.argv[3], matches)
```

Listing 3-1: Simulate checking birthdays for multiple people in a room

As with *sim_pi.py*, all the action is in Simulate. The code at the bottom of Listing 3-1 parses the command line to get the number of people in the room along with the randomness source, one of those supported by RE or a filename, and, if specified, the name of an output file (a NumPy array). Note that rng is configured to return integers in $[0, 365)$.

The randomness source (rng) and the number of people in the room are passed to Simulate. The return value is a histogram of the number of times that many matches occurred in the fixed 100,000 simulations (matches). The first element of matches is the number of times there were no matches, so the sum of all remaining elements divided by the sum of all elements is the probability of one or more matches (prob). The code then displays the probability and writes the histogram to disk if requested.

In Simulate, M is the number of people in the room, fixed for all 100,000 simulations. Matches will hold the outcome of each simulation, or the number of matches found.

The first for loop covers the simulations. For each simulation, a random set of birthdays is selected (bdays), one for each of the M persons in the room. Then, the double loop over i and j compares the ith person's birthday with all others, counting each match. The loop limits for i and j avoid double-counting; if the ith person's birthday matches the jth person's, then the jth person's will match the ith person's, which has already been counted. When all pairs of people have been compared, the count in match is appended to matches.

Finally, after all 100,000 simulations, the list matches is turned into a NumPy array and passed to np.bincount to count the occurrences of each number of matches.

Is *birthday.py* a fair simulation? Does it do what we expect? As Hartl says, does it "eliminate extraneous detail in order to focus on the essentials"? The essential task here is to pick birthdays fairly once the number of people in the room is fixed. We assumed that birthdays are uniformly distributed throughout the year—a reasonable assumption.

The simulations we've discussed so far are warm-ups. Let's kick things up a notch and explore what is, surely, the most important process in the world, at least to the myriad of lifeforms on this planet: evolution.

Simulating Evolution

The evolution of organisms is a complex process affected by genetic and environmental factors. In this section, we'll explore two factors: natural selection and genetic drift.

Natural selection, described by Darwin in the 19th century, is often characterized as "survival of the fittest." It posits that in an environment, organisms whose *genotype* (genetic code) leads to an increased likelihood of survival and reproduction are those more likely to pass their genes on to succeeding generations. In this way, over time, the characteristics of the organism are modified, often eventually leading to organisms that can no longer breed with each other—that is, new species.

While natural selection relates to improved likelihood of survival and reproduction, *genetic drift* is an effect caused by environmental changes that isolate a small population of organisms from the larger population. In genetic drift, the subpopulation of organisms present during the time of isolation will have a different mix of genes that can cause rapid changes in the overall gene pool, often leading to new species.

We want to simulate the essential components of natural selection and genetic drift. Let's begin with the former; once we simulate natural selection, simulating genetic drift becomes that much clearer.

Natural Selection

Here are the requirements for simulating natural selection:

1. We need a population of organisms, each consisting of a collection of genes. An organism's genes determine its fitness to the environment.

2. We need an environment and some way to characterize it in terms of how well organisms are adapted to it. Additionally, we need a measure of fitness for this environment.

3. We need to simulate natural selection's two most important tools: breeding between organisms (*crossover*) and random *mutation*. This simulation must be affected by an organism's level of fitness to the environment.

4. We need to step from generation to generation so we can monitor the population over time.

5. Finally, we need to easily visualize the characteristics of the population as it evolves across generations.

Let's work through each statement.

The Organisms

Our organisms have six genes in their genomes, each with 16 possible variants or *alleles*. Therefore, an organism is a vector of six elements, each [0, 15]. These numbers will become clear as we proceed.

The Environment

We'll define our environment by a set of genes that correspond to the "ideal" organism, the one best suited to the environment. In nature, most organisms are well suited to their environment; if they weren't, they'd quickly go extinct. However, in the spirit of simulation, we'll pick a set of genes to be the "best" and use them as a proxy for the environment.

We'll use the distance between the environment's gene vector and an organism's gene vector as a measure of the organism's fitness. The smaller this distance, the fitter the organism. While there are many possible definitions of "distance" when discussing vectors (points), we'll use the *Euclidean distance*: the straight line distance between two points. We'll imagine each gene vector to be the coordinates of a point in a six-dimensional space.

If the environment's gene vector is $\mathbf{e} = (e_0, e_1, e_2, e_3, e_4, e_5)$ and the organism's is $\mathbf{x} = (x_0, x_1, x_2, x_3, x_4, x_5)$, then the Euclidean distance between them is:

$$d = \sqrt{(x_0 - e_0)^2 + (x_1 - e_1)^2 + (x_2 - e_2)^2 + (x_3 - e_3)^2 + (x_4 - e_4)^2 + (x_5 - e_5)^2} \qquad (3.1)$$

In other words, it's the square root of the sum of the squares of the differences, coordinate by coordinate. Here, each coordinate is an integer in [0, 15] to represent the selected allele for that gene. In addition, we'll define a minimum distance to interpret as "good enough."

Crossover and Mutation

Sexual reproduction is a brilliant method for mixing genes and creating diversity in the gene pool. Our organisms will breed by crossover, which picks a position along the genome and copies all the genes of the first parent up to that point, followed by all the remaining genes from the second parent. The new combination becomes the genetic code of the offspring. Finally, we'll apply random mutation by picking a gene and randomly changing its value. Figure 3-5 illustrates the crossover and mutation process.

Figure 3-5: Crossover and mutation producing a new offspring organism (frog image in the public domain, courtesy of Wikimedia Commons)

Our organisms aren't frogs, but you get the idea. Two organisms create an offspring using the first two genes of one and the final four of the other. Then, mutation changes one of the genes from 10 to 2.

To simulate fitness influencing reproduction, we'll bias the selection of organisms such that those with smaller fitness values are more likely to breed. We'll do this with a *beta distribution*, which is included with NumPy. A beta distribution uses two parameters to affect the shape reflecting the overall histogram of samples. If both parameters, *a* and *b*, are equal to 1, the beta distribution mimics a uniform distribution. If the *b* parameter is increased slightly, the distribution is modified, making it more likely that values closer to zero will be selected.

Therefore, when breeding the next generation, we'll select population members with indices closer to zero. We'll sort the population by fitness, with fitter organisms nearer to the top of the 2D array of organisms, in which each row is an organism and each column a gene.

The net result is that fitter organisms are more likely to breed. Therefore, generation by generation, we expect the entire population to inch closer to the ideal fitness for the environment.

The Population from Generation to Generation

I previously alluded to keeping the population in a 2D array. We'll fix the population size at 384 organisms; why will become apparent in time. Therefore, a population of organisms becomes a 2D array of 384 rows and 6 columns. Each generation will breed another 384 organisms. Put another way, our organisms are seasonal; they live a season (time step) and die after spawning the next generation. Population geneticists often use such a discrete model.

Therefore, a simulation implements each of these steps:

1. Select the initial population at random.

2. Select an environment at random.

3. For *N* generations, calculate the per-organism fitness, sort the population by fitness, and breed each member of the population using crossover and mutation.

4. Create output based on the sequence of populations.

Visualization

Every generation produces a population of 384 organisms, each with 6 genes represented by one of 16 alleles. Now we'll learn why the population is always 384 organisms with 6 genes. We want to produce as output an image where each row of the image shows the population, one organism per pixel, along with the environment. Therefore, the output image will have 384 columns plus additional columns to show the environment. Each pixel gets its color from the genetic code of the corresponding organism with genes mapping to 4 bits of a 24-bit RGB color value, as in Figure 3-6.

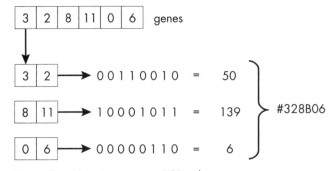

Figure 3-6: Mapping genes to RGB colors

In the figure, the gene vector becomes a forest green pixel. As the generations evolve, we expect the population to move closer to the color of the environment. Of course, random mutation will have some say in the matter, as well as fitness; we'll experiment with both.

Let's begin with a static environment.

Static World

We'll dive into code after our experiments. To run an experiment with a static environment, use *darwin_static.py*:

```
> python3 darwin_static.py 500 60 0.01 4 minstd darwin_static.png 73939133
```

There are several parameters, common to most of our experiments:

500 Number of generations (rows)

60 Fitness bias, [0, 1000]

0.01 Mutation probability

4 "Good enough" threshold

minstd Generator name, or filename

darwin_static.png Output image name

73939133 Seed value (optional)

While the simulation runs, you'll see the average fitness per generation. As the generations evolve, the fitness decreases until it hovers around the "good enough" value. When the simulation ends, then take a look at *darwin_static.png*. Color is required, but the image begins something like Figure 3-7.

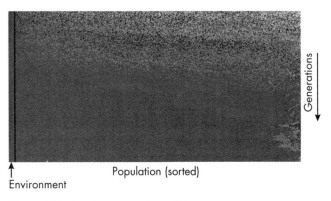

Figure 3-7: Visualizing a static world

Even with the seed specified, there will be variation between runs because we're using the NumPy beta distribution function, and it doesn't pay attention to our seed value.

Read the image from top to bottom. The top row is the initial, randomly generated population of 384 organisms. Each subsequent row is another generation bred from the one above, each time sorted by fitness so that fitter organisms are closer to the left edge. The stripe on the far left is the environment, represented by the ideal genome's color.

As you follow down the rows of the image, the population becomes more like the ideal environment. However, it never collapses to match the environment exactly. Three command line arguments affect how quickly and consistently the population matches the environment: the fitness bias, the mutation probability, and the "good enough" threshold. Let's understand each.

The Fitness Bias

The fitness bias is an integer in the range 0 to 1,000. As we'll learn in the code, this value is divided by 1,000 and added to the second beta distribution parameter. The purpose is to increase the fitness of organisms with genomes that are better suited to the environment. If the fitness bias is 0, there's no reproductive benefit to being better suited to the environment. As the bias increases, the reproductive benefit increases to cause the population to approach the environment's ideal more rapidly.

As an example, run *darwin_static.py* a second time and change only the fitness bias from 60 to 600. The population should approach the environment's ideal in only a few generations. Change the fitness bias to 0 and run again. What do you notice now? The population isn't able to improve because a fitness bias of 0 means no reproductive benefit based on an organism's genome. If you make the fitness bias 15, you might need about 1,500 generations, but you should see the population eventually adapt to the environment. Even a small reproductive advantage matters over the long haul.

The Mutation Probability

Now, set the fitness bias to 60 and adjust the mutation rate, expressed as a probability. For example, a mutation rate of 0.01 gives each newly bred organism a 1 percent chance of undergoing random mutation. A 1 percent mutation rate is exceptionally high compared to living animals, but we need to see the effects we're after without millions of generations.

Change the mutation rate of *darwin_static.py* to 0; this means each new generation will be created by crossover only. Run a few times and look at the output. What do you notice? The population fitness should hit 4 (a distance of 4 from the ideal genome) and remain there indefinitely. It can't do anything else because the genomes are already "ideal," so there's nothing left to change; pick any two for crossover, and, regardless of the crossover point, the offspring's genome will still be identical to the parents'.

Let's see how sensitive the population is to mutation. Alter the mutation rate from 0.01 to 0.1 (10 percent) and run a few more times. Notice that the population adapts to the environment, but never completely. Indeed, as you look down the rows of the output image, you'll likely see regions where many members of the population were adapted, but a new mutation appeared and quickly altered the balance so that the population had to adapt again in the following generations.

My runs that used a mutation rate of 0.1 generally ended with the population mean fitness in the 7.5 to 8.5 range, much higher than the 4 found by no mutation. If you change the mutation rate to 0.2 or even 0.8, the population should have a harder time adapting to the environment because mutations continually push the population away from the ideal. Changing the mutation to a lower rate, say 0.005, leads the population to adapt well, but over time (that is, moving down the rows of the output image), you'll see that small groups of mutants appear, then adapt, then appear again with another mutation. In the output image, these groups appear as splashes of color on the right side of the image—the least fit organisms with the lowest probability of breeding.

The "Good Enough" Threshold

The final command line argument is the mysterious "good enough" value, the minimum distance between the environment's ideal genome and an organism's. When calculating the population's fitness, any distance less than this value is set to this value. Experiment by changing the "good enough" value while holding the fitness bias and mutation rate fixed (for example, at 60 and 0.01, respectively). The higher the "good enough" value, the coarser the population's adaptation to the environment. Set it to 0 and the population will collapse to the ideal, quickly if the fitness bias is larger; if the mutation rate is 0, the population will stay there.

I recommend experimenting with *darwin_static.py* until you develop an intuitive feel for how changes to the fitness bias, mutation rate, and "good enough" value affect the outcome. Try to predict what you expect to see in the output image ahead of time. When using simulations, it's vital to vary parameters, especially by pushing them to their limits. Not only does this help with understanding the processes the simulation is attempting to capture, it serves as a sanity check on the simulation itself, as a way to perhaps uncover weaknesses or errors making the results less valid.

In the real world, environments are not static, at least not for timeframes over which evolution typically acts (though rapid evolution is possible). Let's add a new feature to the simulation, allowing the environment to change slowly with time to understand how such change affects the population.

Gradually Changing World

The code in *darwin_slow.py* is almost identical to the code of the previous section, but it introduces a new feature: the environment will change, slightly, at an interval specified on the command line. For example:

```
> python3 darwin_slow.py 500 60 0.01 4 0.01 mt19937 darwin_slow.png 66
```

The new parameter is the 0.01 before the pseudorandom generator (mt19937). It says to slightly modify the environment on each generation with a probability of 1 percent. The *darwin_slow.py* example leads to an output image where the environment changes four times. The output image is similar to the static case, but each environment transition is marked with a black line on the left. For example, the first two transitions appear as in Figure 3-8.

Figure 3-8: The environment in transition

Detail will be visible only in the full-color image; see *darwin_slow.png* on the book's GitHub repository. The initial, random population is adapting to the environment when the first transition occurs. The population then quickly adapts to the new environment when the environment changes again.

If you look at the full *darwin_slow.png* image—hopefully using an image view allowing full resolution horizontally—you'll notice that after the second change to the environment, the population adapts quite well, but it takes several generations. The visual effect is to smear the colors to the right, where the less fit organisms are listed. I recommend running the simulation several times without the fixed seed of 66 to observe the overall effect with different colors. Then, explore how modifications to the fitness bias and mutation rate play out as the smooth changes to the environment happen. To get you started, observe what happens on multiple runs with the parameters listed in Table 3-2.

Table 3-2: Parameters to Try

Fitness bias	Mutation rate	Environment probability
600	0.01	0.300
60	0.01	0.300
1	0.01	0.005

The first two parameter sets illustrate the effect of a rapidly changing environment with both strong and not so strong advantages to fitter organisms. The final set of parameters uses the weakest of reproductive advantages, but couples it with an almost static environment.

Slowly varying environments give organisms time to adapt. However, throughout the Earth's long history, not all environmental changes were slow; some were quite sudden, even happening overnight. Let's simulate a catastrophe.

Catastrophic World

One fine day, some 66 million years ago, the lifeforms of Earth were minding their own business when a giant asteroid rudely interrupted and, as a consequence, ended the 100-million-year-plus reign of the nonavian dinosaurs. Bad news for them; good news for us. A catastrophe happened, and life responded and appeared quite different after the impact. The same thing happened about 252 million years ago during The Great Dying when life nearly perished. The world after the extinction looked very different from the world before.

We've simulated gradual environmental change; now, let's give the simulation a hard knock and see what happens. We need *darwin_catastrophic.py*. Give it a go:

```
> python3 darwin_catastrophic.py 500 60 0.01 4 0.01 pcg64 darwin_catastrophic.png 12345
```

The parameters are identical to those of *darwin_slow.py*, the difference being that whenever the environment should change, it does so by picking an entirely new ideal environment genome. The transitions are stark. The image generated is similar to that of *darwin_slow.py*, but without the horizontal black line to mark the transitions. In this case, the transitions are, generally, quite obvious. To see what I mean, run the code, or look at *darwin_catastrophic.png* on the book's GitHub page.

Experiment with the simulation parameters to explore the consequences. For example, change the number of generations to 2,000 and look at the output image in its entirety by zooming out. The population's delayed response to each environmental catastrophe is plain to see.

Our final simulation introduces genetic drift. How do suddenly split populations fare in adapting to new environments?

Genetic Drift

A *population bottleneck* happens when a population experiences a sudden reduction in size. One kind of population bottleneck is the *founder effect*, which occurs when a small population splits from a larger population. The random mix of alleles in the new, smaller population might dramatically affect the long-term survival and evolution of the organisms. The code in *darwin_drift.py* simulates genetic drift due to the founding of a new, smaller population.

The code is similar to the previous examples, but with a twist. First, a larger population (all 384 organisms) evolves for generations trying to adapt to its environment. Then, a specified fraction of the whole population "splits off" to become a new population. The two populations now evolve separately. Imagine a colony of organisms stranded on an island and no longer able to breed with their mainland cousins. To make things interesting, after the split, a catastrophe happens, so we can observe how well (or poorly) the two populations cope with the sudden change in environment.

For example:

```
> python3 darwin_drift.py 500 60 0.01 4 0.2 pcg64 darwin_drift 1337
```

The 0.2 before pcg64 now refers to the fraction of the population that will split off to form the new population; that is, 20 percent of the population, randomly selected, breaks off to create the new population, leaving the other 80 percent as the larger population.

Unlike our other simulations, *darwin_drift.py* expects a base filename (darwin_drift) instead of a complete filename. The code outputs an image, *darwin_drift.png*, along with a plot of the mean population fitness by generation, *darwin_drift_plot.png*. As before, there are 500 generations using a fitness bias of 60 and a mutation probability of 1 percent.

So what does our image look like? Yet again, I suggest you review the color image from the book's GitHub repository, but part of the output is in Figure 3-9.

Figure 3-9: Visualizing genetic drift

This snippet of the larger image shows part of the simulation after the split. The smaller population is on the left, with the larger on the right. Also, if you examine the environment, there is a sudden catastrophe about one-sixth of the way down. The two populations respond differently to the disaster. This is particularly visible in the color version of the image.

Before the catastrophe, both populations were reasonably well adapted to their environment. However, the smaller population cannot recover after the catastrophe or successfully adjust to the new environment. The larger population does eventually adapt. The color version of the image clearly shows that the smaller population sometimes produces generations where organisms are adapting to the new environment, but they never last long. This effect mirrors reality; small populations are often very fragile and easily harmed by rapid environmental change because they lack the genetic diversity to adapt in time.

Figure 3-10 tracks mean population fitness as a function of generation.

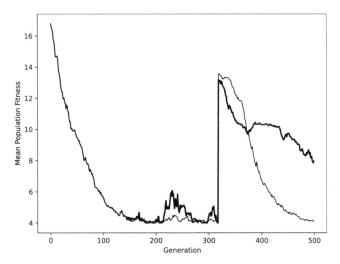

Figure 3-10: Mean population fitness by generation before and after the catastrophe

The founder effect event happens around generation 145, where two lines appear. The thicker line follows the smaller population. At first, both populations are relatively fit to the environment, which hasn't yet changed, though it could be argued that there is more variation in fitness in the smaller population.

The catastrophe occurs around generation 318. Immediately afterward, both populations' fitness scores increase dramatically. Remember, a lower fitness score is better. Subsequent generations begin to adapt to the new environment, but not at the same rate. The larger population, still 80 percent of the original size, adapts to the new environment over time; however, the smaller population fails to do so, at least for the 500 generations simulated. For most runs of *darwin_drift.py*, the smaller population fails to adapt to the new environment as well as the larger population. Sometimes the reverse happens. We'll talk about this effect in the discussion.

What happens if the population splits in half (0.5)? Or, what if the new population is a tiny fraction, say 5 or 10 percent? It's probably easiest to experiment in these cases by fixing the fitness bias and mutation rate. Then, fix the population fraction and vary those parameters.

In Figure 3-9, when the population splits, the members of the new, smaller population are selected at random. Therefore, they typically have an uneven representation of the genotypes in the larger population due to chance. That difference might mean that uncommon genotypes now have an opportunity to become more common.

This effect is illustrated by the code in *drift.py*. First, a "population" of 10,000 digits, [0, 9], are selected. Then 20 subpopulations are constructed by choosing 50 members of the larger population at random. Finally, the mean of the larger population is displayed along with the mean of 20 subpopulations. If the mix of digits is the same in each, then the means will be quite close to each other. They are not:

```
Population mean = 4.562900
Sub-population means:
    3.60, 4.42, 4.52, 4.82, 4.36, 5.40, 4.72, 4.54, 4.24, 4.28,
    4.76, 4.66, 4.98, 4.90, 4.50, 4.50, 5.10, 4.44, 4.30, 4.62
```

The subpopulation means range from a low of 3.60 to a high of 5.40. Uniformly selected digits should give a population mean of 4.5, which is close to the larger population mean. The subpopulations, due to random chance, represent very different collections of digits (genomes). Now we understand why population bottlenecks lead to genetic drift.

Testing the Simulations

We used four different sets of code for the previous simulations. Rather than detail each, thereby committing the literary equivalent of "death by PowerPoint," let's walk through one of them here and present snippets from the others as needed.

Look at Listing 3-2, which contains the critical parts of *darwin_static.py*.

```
❶ ngen = int(sys.argv[1])
  advantage = int(sys.argv[2])
  mutation = float(sys.argv[3])
  good = float(sys.argv[4])
  kind = sys.argv[5]
  oname = sys.argv[6]
  if (len(sys.argv) == 8):
      seed = int(sys.argv[7])
      rng = RE(kind=kind, seed=seed)
  else:
      rng = RE(kind=kind)

❷ npop = 384
  pop = np.zeros((npop, 6))
  for i in range(npop):
    ❸ pop[i,:] = (16*rng.random(6)).astype("uint8")
  environment = (16*rng.random(6)).astype("uint8")
  hpop = np.zeros((ngen,npop,6))
  henv = np.zeros((ngen,6))

❹ for g in range(ngen):
    ❺ fitness = np.zeros(npop)
```

```
    for i in range(npop):
        d = np.sqrt((((pop[i]-environment)**2).sum()))
        if (d < good):
            d = good
        fitness[i] = d

❻  idx = np.argsort(fitness)
    pop = pop[idx]
    fitness = fitness[idx]

❼  hpop[g,:,:] = pop
    henv[g,:] = environment
    print("%6d: fitness = %0.8f" % (g, fitness.mean()))

❽  nxt = []
    for i in range(npop):
        nxt.append(Mate(pop,fitness,advantage))
    pop = np.array(nxt)
```

Listing 3-2: Simulating a static environment

I've excluded comments and code related to generating the output image. Please review the file itself if you're curious about how those parts work. I recommend reading through at least MakeRGB to understand the mapping from genes to RGB color values.

The code falls naturally into three parts: parsing the command line ❶, setting up the simulation ❷, and running the simulation ❹. In the first part, the randomness engine (rng) is configured, with or without a seed, to return floats in [0, 1). The engine is used for different things, so it's better to use only the basic range and adjust the bounds and data type as needed.

Then the initial population (pop) of 384 organisms (npop) is created ❷. Each organism is given a randomly generated genome ❸. The environment is similarly defined. The final two variables, hpop and henv, track the evolution of the population and, for other variants of the code, the environment. Note the use of a 3D array imagined here as a vector of 2D arrays, each holding the population for that generation. The output image is produced using both hpop and henv.

The simulation is now ready; the initial population and environment have been defined. The main loop ❹ evaluates each generation. The loop body has four paragraphs: calculate the per-organism fitness ❺, sort the population by fitness ❻, keep a copy of the population for image generation ❼, and, finally, breed the next generation ❽.

Let's go through each of those steps. To calculate fitness, we subtract each organism's genome from that of the environment and square and sum the result across all genes before applying the square root. This is the NumPy version of Equation 3.1. Fitness in hand, idx sorts both the population and the fitness vector so that fitter organisms are closer to the beginning of the population ❻. Then hpop stores the sorted population and environment for the output image ❼. The mean population fitness is also printed. As the population evolves, this mean value should decrease, depending on the mutation rate and the fitness bias.

The last thing to do for this generation is replace it ❽. Repeated calls to Mate breed a new population of npop organisms from the existing population as shown in Listing 3-3.

```
def Mate(pop, fitness, advantage):
    a = advantage / 1000
    i = int(len(pop)*np.random.beta(1,1+a))
    j = i
    while (j == i):
        j = int(len(pop)*np.random.beta(1,1+a))
    c = int(6*rng.random())
    org = np.hstack((pop[i][:c], pop[j][c:]))

    if (rng.random() < mutation):
        c = int(6*rng.random())
        org[c] = int(16*rng.random())
    return org
```

Listing 3-3: Producing the next generation

Here, Mate is given the current population and associated fitness (both sorted), along with the fitness bias (advantage). As mentioned, the fitness bias is divided by 1,000.

The function needs to select two distinct organisms, indices i and j, and then produce a new organism by crossover. A call to NumPy's beta function returns a value in $[0, 1)$, which, when scaled by the size of the population, will return an integer in $[0, 383]$. The while loop runs until a distinct second organism is selected (j).

Figure 3-11 shows beta distributions for different fitness bias values.

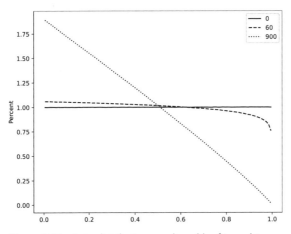

Figure 3-11: Beta distribution as altered by fitness bias value. Fit organisms are more likely to breed if the left portion of the distribution is higher than the right.

If the bias is 0, the beta distribution acts as a uniform distribution. There are 100 bins in the plot, so each will appear about 1 percent of the time (solid line). A relatively weak fitness bias of 60 favors small values, or fitter organisms, while strongly rejecting the least fit. Similarly, a bias of 900 selects most fit to least fit linearly.

Crossover selects a gene position, $[0, 5]$, and constructs the offspring (org) by keeping the first c genes of one parent and adding in the remaining genes from the second. Then, if a random value is less than the global mutation threshold, a randomly selected gene is given a new value, $[0, 15]$. Finally, the function returns the new organism's genome.

To recap: configure, then loop over generations evaluating the population's current fitness before using that information to breed the next generation. Once the dust settles, generate the output image.

The file *darwin_slow.py*, which changes the environment as the population evolves, is nearly identical to *darwin_static.py*. The loop has one additional code paragraph after breeding the next generation:

```
if (rng.random() < eprob):
    offset = 2*rng.random(6)-1
    environment = environment + offset
    environment = np.maximum(0,np.minimum(15,environment))
    environment = (environment + 0.5).astype("uint8")
```

Here, eprob is the probability of the environment changing, read from the command line. If the environment is to change, we add an offset vector to alter the ideal genome by ± 1 for each gene. Compare this with the catastrophic environmental change from *darwin_catastrophic.py*:

```
if (rng.random() < eprob):
    environment = (16*rng.random(6)).astype("uint8")
```

The final program, *darwin_drift.py*, is structurally similar to *darwin_catastrophic.py*, but after a set number of generations, the population splits into two. After this split, the environment is altered catastrophically. Although bookkeeping is involved, conceptually nothing new is happening.

Exercises

Use the following exercises as a springboard to expand your appreciation for the power of simulation. When working through them, entertain any "what if" questions that pop into your head:

- Alter *sim_pi.py* to make two calls to rng, first to get the *x*-coordinates, then to get the *y*-coordinates. Do you notice any difference? Did you expect to?

- Modify *birthday.py* to use $N = 1{,}000{,}000$ or $N = 10{,}000$. Is there a noticeable difference?

- We assumed that birthdays are uniformly distributed throughout the year. This isn't strictly true, at least for Western countries. For those countries, September birthdays are more common. What happens to the true probability of two randomly selected people sharing a birthday in that case? Does it increase or decrease? You may wish to explore the file *birthday_true.py*, which uses data from the United Kingdom.

- How many people need to be at the party to have a 99 percent probability of at least one birthday match?

- Our evolution simulations assumed that all members of a generation bred and then died to produce a new generation that was the same size as the last. What happens if the least fit 10 percent die and do not reproduce while the fittest 2 percent breed twice?

Bonus points:

In his 1889 book, *Calcul des probabilités*, Joseph Bertrand outlined three approaches for calculating the probability that a randomly selected chord of a circle is longer than the side of an equilateral triangle inscribed in the circle. The files *bertrand0.py*, *bertrand1.py*, and *bertrand2.py* implement simulations corresponding to the three approaches:

bertrand0.py   Use the chord defined by two randomly selected points on the circumference of the circle.

bertrand1.py   Use the chord perpendicular to a randomly selected point along a randomly selected radius of the circle.

bertrand2.py   Use a randomly selected point inside the circle as the midpoint of the chord.

Run all three approaches to select random chords of the circle. For example

```
> python3 bertrand0.py 500 b0.png mt19937 359
Probability is approximately 154/500 = 0.3080000
```

produces the estimated probability along with a plot showing the selected chords, as shown in Figure 3-12.

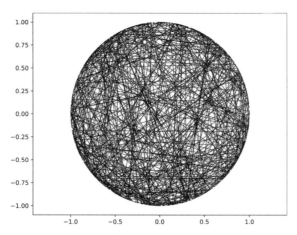

Figure 3-12: Selected chords when choosing points on the circumference of the circle

What is the estimated probability for each approach? Which one is correct? This is known as *Bertrand's paradox*, and it serves as a cautionary tale to be careful when defining what we want to simulate and how we go about it. Review the code to see how the chords are selected.

Summary

In this chapter, we began an introductory exploration of models and simulations; we'll continue to encounter various models throughout the book.

We started slowly, with simulations estimating the value of π by throwing darts and the number of people in a room, on average, necessary to have a better than 50 percent chance of at least two sharing a birthday. We then constructed a model to simulate two essential aspects of biological evolution: natural selection and genetic drift. We learned that even an incomplete model can be a useful tool that offers helpful insights.

Our simulations captured some essence of important evolutionary mechanisms, like natural selection and genetic drift, but a huge part of reality was missing: death and extinction. For example, many small populations kept evolving when they ought to have gone extinct. Extinction is natural; virtually every species that has ever lived is extinct (though there's no reason for us to hurry the process along). Adding death and extinction would create a level of complexity to the simulation that doesn't fit with what we can accomplish in this book. Regardless, the simulations of this section are practical and illustrative as far as they go. All analogies fail at some point—that doesn't make them useless.

The next chapter continues our investigation of useful randomness by diving into the world of optimization. Can randomness be put to work in service of locating the best of something?

NOTE   *There is no formal resolution to Bertrand's paradox. Many people, including me, feel that p = 1/2 is the most reasonable answer.*

4

OPTIMIZE THE WORLD

Optimization is the process of finding the best set of something, usually of parameters defining a function or an algorithm. In mathematics, optimization typically involves functions and uses their derivatives to locate minima or maxima. In this chapter, we'll take a different approach that involves randomness. The algorithms we'll use fall into two broad categories: swarm intelligence and evolutionary algorithms. Collectively, these are known as *metaheuristics*.

Optimizing with metaheuristics is more flexible than calculus-based optimization. What we're optimizing doesn't need to be a mathematical function; it could be an algorithm or another process. In fact, any problem that can be cast as locating the best position in a space, where the space represents the problem in some form, is amenable to swarm intelligence and evolutionary algorithms. I use both kinds of algorithms frequently for everything from curve fitting to evolving neural network architectures. Once you understand the process of formulating tasks as generic optimization problems, you'll begin to see them everywhere.

In this chapter, we'll use swarm intelligence and evolutionary algorithms to fit data to a known function. Then, we'll evolve the best fit function from scratch. We'll begin, however, with a (very) short primer on swarm intelligence and evolutionary algorithms. We've already used an evolutionary algorithm, though it wasn't named so at the time. The algorithm implemented in Chapter 3 to explore natural selection and genetic drift is a genetic algorithm, one of the two kinds of evolutionary algorithms we'll encounter in this chapter.

Optimization with Randomness

Imagine a large haystack where each position within the haystack corresponds to a possible solution to the problem. We wish to locate the part of the haystack that offers the best solution—that is, we want to find a needle. The question is: How do we go about finding it?

We'll use the following generic algorithm to search the haystack:

1. A swarm (population) of "agents" are randomly scattered throughout the haystack.

2. Each agent investigates its immediate location and assigns a number to how well that location solves the problem.

3. The agents report their numbers to headquarters.

4. After each agent has reported in, headquarters evaluates all the numbers and stores the best position currently known, updating it if, on this iteration, any agent finds a better position.

5. Headquarters then orders each agent to a new position in the haystack based on the information received.

6. The process repeats from step 2 until a best position has been found or we've run out of time (iterations).

We have a lot of flexibility when it comes to possible implementations. Indeed, there are literally hundreds of published algorithms based on this approach. Many claim inspiration from nature, but such claims are often quite dubious and generally unnecessary.

Is this an example of a swarm intelligence algorithm or an evolutionary algorithm? It's both. The difference between the two depends on what happens in step 4: the process headquarters uses to decide where the agents should go next. The distinction is vital for researchers but less so for us.

In a swarm intelligence algorithm, the agents, called *particles*, work collectively to locate new positions in the space to explore. They are actively aware of each other and "learn" from each particle's experiences to move the swarm, as a whole, to ever better places in the space, thereby locating increasingly better solutions to the problem.

An evolutionary algorithm, on the other hand, applies techniques like crossover and mutation to breed new agents (organisms). In Chapter 3, we defined an organism's fitness as the distance between its genome and the genome of an ideal organism for the current environment. Here, fitness is a measure of how well the solution represented by the organism's genome (position in the haystack) solves the problem. Generations of breeding fitter solutions, with a dash of random mutation, should move the population closer to the best solution to the problem.

In practical terms, all we need to know is that the two kinds of algorithms search a space to find the best position in it. We'll configure our problems such that the best position is translatable into a best solution.

With the hundreds of swarm and evolutionary algorithms out there, which ones should we use? Each algorithm has strengths and weaknesses, and might work best for certain kinds of problems. You need to try several.

For this chapter, we'll use five algorithms: two swarm intelligence, two evolutionary, and one that is so obvious many don't consider it a swarm algorithm at all. We don't have space to walk through each to understand the code; I'll leave that as an exercise (as always, please contact me with questions). We'll learn about the algorithms and the framework using them as we go.

The two swarm intelligence algorithms are *particle swarm optimization (PSO)* and *Jaya*. PSO is the grandfather of swarm intelligence algorithms, and many nature-inspired algorithms are PSO in disguise. Jaya is a newer algorithm that has no parameters to adjust—either it works well or it doesn't. Although there are many flavors of PSO, we'll use two here: canonical and bare-bones.

The two evolutionary algorithms are the *genetic algorithm (GA)*, a variation on what we used in Chapter 3, and *differential evolution (DE)*, another old-school and widely used technique. DE is one of my go-to algorithms, but it has the sometimes annoying habit of converging too quickly to local minima.

The last algorithm is *random optimization (RO)*. In RO, the particles don't communicate; they conduct a local search and move to a new position whenever they find one, but are completely unaware of what other particles have discovered. Headquarters monitors each particle to track the best position found overall, but never issues orders based on that knowledge.

We learn best by doing, so let's begin fitting a function to data.

Fitting with Swarms

A common task in science and engineering is to fit a function to a set of measurements, where *fit* means finding the best set of parameters for a known type of function—the set that makes the function approximate the data as well as possible. For this task, we know the functional form; we need only to learn the parameter values to tailor the function to the data. In the next section, we'll start with the data and let the swarms tell us what the best-fit function and parameters are (hopefully!). I'm using the word *swarms* in a general sense to mean both the swarm of particles manipulated by a swarm intelligence algorithm and the population bred and evolved by an evolutionary algorithm. Again, the distinction is minor for us.

Let's start with a simple example: some data, a function, and the parameters that best fit the function to the data. The code for the example is in *curfit_example.py*. It generates a set of points from a quadratic function, with random noise added. Then, it uses NumPy's polyfit routine to fit a quadratic: $ax^2 + bx + c$. Figure 4-1 shows the plot and fit function.

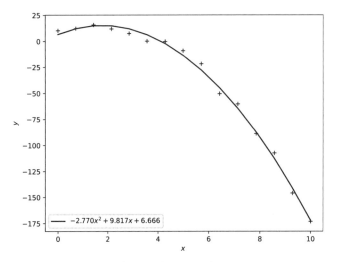

Figure 4-1: Fitting a polynomial to some data

With the fit function, we can approximate y for any x, which is typically why we fit the data in the first place.

You may be wondering why we'd bother with swarms if polyfit can fit the data. Unfortunately, polyfit fits only polynomials, or functions that are sums of powers of x. If your function isn't a polynomial, there are other functions you could use, like SciPy's curve_fit. However, we're not solely interested in curve fitting; we're using it as a warm-up exercise. SciPy won't be of much use for the other optimization problems we'll explore later in this chapter and in the next.

Curves

Now that we have an idea of what curve fitting entails, let's try it using a swarm. The code we want is in *curves.py*. We'll use it first, then look at parts of it. I strongly recommend you read through the code to get a feel for things.

The code expects a datafile that contains the measured points along with the function to fit. We'll use *curves.py* to fit the previous example. The input file we need is *curfit_example.txt*:

```
3
p[0]*x**2+p[1]*x+p[2]
10.2772497 0.0000000
12.2926738 0.7142857
15.7968918 1.4285714
```

```
11.9787533 2.1428571
7.5707351 2.8571429
0.2314503 3.5714286
-0.1762932 4.2857143
-9.0166104 5.0000000
-21.6965056 5.7142857
-50.3670945 6.4285714
-60.2153079 7.1428571
-88.6989830 7.8571429
-107.3679996 8.5714286
-145.8216296 9.2857143
-173.1300077 10.0000000
```

The first line is the number of parameters followed by the function to fit. The function is given as Python code where the fit parameters are elements of a vector, p, and the data points are represented by x.

We want to fit a function like $ax^2 + bx + c$, a three-parameter function, so we use p[0]*x**2+p[1]*x+p[2]. If you want something like $\sin x$, use np.sin(x) (use NumPy). Note that the data points are listed as y then x.

Let's use *curves.py* and this file to fit the data using differential evolution:

```
> python3 curves.py curfit_example.txt -10 20 20 1000 0 DE pcg64
Minimum mean total squared error: 16.430381313  (curfit_example.txt)
Parameters:
  0:   -2.7702810873939598
  1:    9.8170736277919577
  2:    6.6657767196319488
(73 best updates, 20020 function calls, time: 1.618 seconds)
```

The output tells us several things, but look first at the parameters. These are the elements of p, the best set of parameters found. Compare them with Figure 4-1. The fit is quite good.

The generic algorithm says that particles need to evaluate where they are in the haystack to determine how good a solution their current position represents. There are three parameters in the fit function; therefore, our haystack is a three-dimensional space, and the particles are initially scattered randomly throughout this space. Each point in the three-dimensional space corresponds to a p vector, a set of three parameters. The best position found during the search is reported by *curves.py*.

For every particle, at every position in the haystack, we calculate the value of the *objective function*, the fitness function that tells us about the quality of the solution at that position. For curve fitting, our objective function measures the mean squared error between the measured points, (x, y), and the y values the function returns for the same x positions. If $\hat{y} = f(x, p)$ is the output of the function at x for some particle position, p, then the *mean squared error (MSE)* is:

$$\text{MSE} = \frac{1}{N} \sum_{i=0}^{N-1} (\hat{y}_i - y_i)^2 \tag{4.1}$$

The sum is over all the measured points, (x_i, y_i).

The closer the MSE is to zero, the better the function is at fitting the measured data, meaning the particle position giving us the smallest MSE is the best fit found. The swarm algorithm keeps adjusting particle positions until it finds the minimum or we run out of iterations.

The *curves.py* file accepts many command line parameters:

`curfit_example.txt`   Datafile

`-10`   Lower bound

`20`   Upper bound

`20`   Number of particles

`1000`   Number of iterations

`0`   Tolerance

`DE`   Algorithm (DE, Jaya, PSO, GA, RO)

`pcg64`   Randomness source

The first argument is the name of the file that has the measured data points. Again, the first line is the number of parameters in the fit, followed by the code to implement the fit function. The remainder of the file are the actual data points, *y* then *x*, one pair per line.

The next two arguments specify the bounds of the search. These limit the size of the space the swarm can move through. Specifying a scalar applies that value to all dimensions; otherwise, specify each dimension separated by x. In this case, we tell *curves.py* to limit its search space to the cube from $(-10, -10, -10)$ to $(20, 20, 20)$. The bounds are often helpful, but must enclose the actual best values; otherwise, the search will return only the best position within the given bounds.

The following argument, also 20, specifies the size of the swarm, or the number of particles to scatter throughout the haystack. It's generally better to have a smaller swarm and more iterations—the next parameter, here 1,000—but that's only a rule of thumb; exceptions abound.

We are minimizing the MSE. The search stops early if the MSE is less than the given tolerance. By setting the tolerance to 0, we're telling *curves.py* to search for 1,000 iterations of the swarm positions or to stop early if we find a position with no error. The last parameter is the randomness source for RE. A final parameter, the name of an output image file showing the data points and the fit, is also allowed.

Swarm algorithms are stochastic, meaning they change their output from run to run because they randomly assign initial particle positions and use random values during the search. For many problems, the changes are subtle and inconsequential, but sometimes the swarm simply gets lost. Therefore, it's best to repeat searches several times, if possible, to be sure the results are meaningful.

To try the other swarm algorithms—PSO, Jaya, GA, and RO—specify them by name. I suspect you will find that PSO, Jaya, and even RO give results as good as DE. GA, however, is another story. The output, numerically, is poor, though if you plot the result, it often looks at least somewhat reasonable. Does this mean GA is a flawed algorithm? No, it's simply not well suited to this task. In general, GA is best for non-numerical optimization problems and problems with a higher number of dimensions (parameters). In this example, using GA means asking it to evolve a population of organisms with three genes each. That doesn't leave much for evolution to work with.

Let's look at one more curve fitting example. The function to fit is in *sinexp.txt*:

$$y = p_0 \sin(p_1 x) + p_2 e^{\frac{-(x - p_3)^2}{2p_4}} \quad (4.2)$$

This function is the sum of a sine and a normal curve and has five parameters: we're in a five-dimensional search space and each particle is a point in this space. I can't picture a five-dimensional haystack, but we're looking for a needle in one regardless.

Let's try *curves.py* using Jaya:

```
> python3 curves.py sinexp.txt -3 23 20 1000 0 Jaya mt19937 fit.png
Minimum mean total squared error: 0.000000015  (sinexp.txt)
Parameters:
 0:    1.9999892608106149
 1:    3.0000001706464414
 2:   20.0001115681148427
 3:    7.9999997934624147
 4:    0.6000128598331004
(137 best updates, 20020 function calls, time: 1.412 seconds)
```

The first time I tried the code, the fit failed and returned a minimum MSE of 0.1656, which is orders of magnitude larger than the previous fit. A plot of the good result is in Figure 4-2.

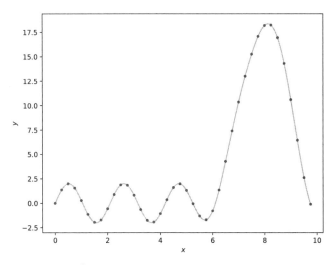

Figure 4-2: A fit to Equation 4.2 using Jaya

The search used 20 particles and 1,000 iterations as before. I limited the search space to the range −3 to 20 in all five dimensions. In this case, the dataset was generated directly from the function with parameter values of 2, 3, 20, 8, and 0.6, respectively. This explains the extremely low MSE: there is no noise in the measurements.

NOTE *You may encounter runtime warnings when using the code. These are the result of the swarm algorithm using parameter values that are too large for the exponential. Adding -W ignore to the command line after python3 will suppress the warnings.*

The curves.py Code

Let's look at some code to get a feel for what the swarm algorithms are doing; it will also help you see how to put the pieces together. Start at the bottom of *curves.py*. The essence of the code, as you'll see when you read through it, is the following:

```
rng = RE(kind=kind)
b = Bounds(lower, upper, enforce="resample", rng=rng)
i = RandomInitializer(npart, ndim, bounds=b, rng=rng)
obj = Objective(X, Y, func)
swarm = DE(obj=obj, npart=npart, ndim=ndim, init=i, tol=tol, max_iter=niter, bounds=b, rng=rng)
swarm.Optimize()
res = swarm.Results()
```

The first line configures an instance of RE using the source given on the command line (kind). The next four lines set up the search for differential evolution. The code is the same for Jaya, PSO, RO, and GA, with the addition of one more argument to the constructor in the case of PSO. Let's go line by line; these are the steps to configure any swarm search using the framework, so we'll see them again as we proceed.

First, swarm searches are bounded, so we need an instance of the Bounds class, or a subclass if we need to override one of its methods, typically Validate. The arguments are the lower and upper limits, the randomness source, and a parameter called enforce that's set to resample. Referring back to the generic search algorithm, step 5 states that headquarters orders the agents to move to new positions based on the objective function values for their current positions. At times, these new positions might be outside the specified boundaries. The enforce parameter decides what to do in these cases. By setting it to resample, any particle dimension that is out of bounds is replaced by a randomly selected value along that dimension. The other option is clip, which clips the offending dimension to the minimum or maximum allowed. Most of the time, this isn't what we want.

The RandomInitializer parameter provides an initializer to configure the swarm. It's given the number of particles in the swarm (npart), the dimensionality of the search space (ndim), and the bounds configured in the previous line (b).

The search also needs to know how to evaluate the objective function, an instance of Objective. For the curve fit example, X and Y are the measured points and `func` is the function to fit, all read from the datafile given on the command line. I'll show you the objective function in a moment.

We're now ready to create the swarm algorithm object (`swarm`), here an instance of `DE`. We give the objective function, number of particles, dimensions, initializer, and bounds, along with the randomness source. We also specify the tolerance (`tol`) and the number of iterations (`max_iter`).

Given all the configuration, using the swarm object is straightforward: call the `Optimize` method. When the call returns, the search is over. Call `Results` to return a dictionary with information about the search.

The most important elements of `res` are `gpos` and `gbest`. Both return lists tracing the collection of best positions the swarm found during its search. Therefore, the final element of these lists returns the best position (`gpos`) and the corresponding objective function value (`gbest`). The position is a vector with one value for each dimension of the search space; here each dimension is a parameter value for the function we're fitting to the data. The gbest value is a scalar, the MSE for this set of parameters.

Let's look at the `Objective` class. The object passed as the objective function must have, at a minimum, a method called `Evaluate`. The details aren't very important, but because Python uses duck typing, any object with an `Evaluate` method that accepts a single argument is permitted. Here's the code *curves.py* uses:

```
class Objective:
    def __init__(self, x, y, func):
        self.x = x
        self.y = y
        self.func = func
        self.fcount = 0

    def Evaluate(self, p):
        self.fcount += 1
        x = self.x
        y = eval(self.func)
        return ((y - self.y)**2).mean()
```

The constructor keeps references to the measured points, x and y, along with the string representing the function to fit (`func`). Many applications do not have ancillary information, in which case the constructor does nothing and need not be specified. Also, notice that `Objective` does not inherit from any other class, it needs only to implement `Evaluate` to be acceptable to the optimization framework.

The `Evaluate` method is called by the swarm algorithms. The argument, p, is the current position of a particle in the swarm, that is, a vector of possible parameter values. The first line increments `fcount`, an internal counter of the number of times `Evaluate` was called. The final value of `fcount` is displayed when *curves.py* exits.

The next line looks a little strange: it assigns the reference to the *x* data, `self.x`, to the local variable x. The following line uses Python's eval function to evaluate the function value; because `eval` uses both `x` and `p` as variable names, we need those names to exist in `Evaluate`—hence the `x = self.x`. The calculated function values are in `y`. These are the $\hat{y}$ values in Equation 4.1.

Finally, we calculate the MSE and return it to give the objective function value (or fitness value) for the supplied particle position in `p`. Notice there is no square root. I left it out to save a tiny bit of time. The smallest MSE is still the smallest even if the final square root is not applied.

When the swarm algorithm runs, it calls `Evaluate` thousands of times to map particle positions to MSE values. The swarm algorithms are utterly ignorant of *what* the objective function is measuring; all they know is that they pass a vector representing a particle position in a multidimensional space to the objective function, and it returns a scalar value where lower values are better than higher values. This makes the framework generic and applicable to a wide range of problems.

To recap, using the framework involves the following steps:

1. Determine how to map potential solutions to the problem to a position in a multidimensional space for the swarms to search.

2. Use that mapping to create an objective function class supporting at least an `Evaluate` method to accept a candidate position vector and return a scalar representing the quality of the solution it represents. The framework always minimizes, so the smaller the return value, the better the solution. To maximize, return the negative of the fitness.

3. Create a `Bounds` object to set the limits of the search space and what to do if those limits are exceeded.

4. Create an initializer (`RandomInitializer`) to supply the initial positions of the swarm particles.

5. Create an instance of the swarm class, `DE`, `PSO`, `Jaya`, `RO`, or `GA`.

6. Perform the search by calling `Optimize`, and use `Results` to return the outcome.

Our familiarity with the framework will grow with practice. For now, let's review the swarm intelligence and evolutionary algorithms to understand how they differ from each other and where the randomness is. Randomness goes deeper than the initial configuration of the swarm particles in the search space; each algorithm depends critically on randomness for its operation.

The Optimization Algorithms

There are hundreds of swarm optimization algorithms out there, but what distinguishes one from another? The short answer is the method used for searching, or how headquarters orders the agents to new locations on each iteration.

The various algorithm approaches range from the most straightforward— RO, where the agents don't communicate but wander from better position to better position independently—to sophisticated algorithms incorporating information about agents' current positions, histories, and associations into groups and neighborhoods.

In this section, I'll summarize the essential operation of the five algorithms selected for the framework. In all cases, the essential operation is the same: scatter particles throughout the search space, evaluate the quality of each particle, decide where they go next, and repeat until a best position is found or time runs out. It's the "decide where they go next" part that differentiates algorithms.

Random Optimization (RO)

Swarm particles are represented by a vector of floating-point numbers, x_i, each component of which maps to a dimension of the search space. In other words, particles are points in the search space. On each iteration, particles construct a new position some distance from their current one and ask if the new position has a better fitness value, that is, if the objective function value is lower at the new position. If so, the particle moves to the new position; otherwise, it stays put. The new candidate position is

$$x_i' = x_i + \eta x_i N(0, 1)/5$$

where η (eta) is a scale parameter ($\eta = 0.1$) and $N(0, 1)$ is a vector of samples from a normal distribution with mean 0 and standard deviation 1. If x_i' has a lower objective function value, then $x_i \leftarrow x_i'$; otherwise, the particle stays where it is for the next iteration. The particles make no use of what other particles have learned about the search space.

Jaya

Jaya, Sanskrit for "victory," is a swarm intelligence algorithm that has no adjustable parameters. Swarm algorithms depend on heuristics, so they often have adjustable parameters to improve their performance in different situations. Jaya does not. It works or it doesn't.

On each iteration, the ith particle is updated via

$$x_i \leftarrow x_i + r_1(x_{\text{best}} - |x_i|) - r_2(x_{\text{worst}} - |x_i|)$$

where x_{best} and x_{worst} are the current best and worst positions of any particle in the swarm, and r_1 and r_2 are random vectors in $[0, 1)$ per component. The vertical bars apply the absolute value to each component of the vector. In other words, Jaya moves particles toward the swarm's best position and away from the swarm's worst.

Particle Swarm Optimization (PSO)

The update equations for PSO depend on the flavor. Our framework offers two: canonical and bare-bones. The *curves.py* file uses bare-bones, hence bare=True in the PSO constructor. However, it's easier to begin with canonical PSO.

In canonical PSO, each particle (x_i) is associated with two other vectors. The first, $\hat{x}_i$, is the best position in the search space that *that* particle has found, and the second, v_i, is the particle's velocity, which controls how quickly and in which direction the particle moves through the search space.

The canonical PSO update rule is accomplished in two steps. First, the velocity is updated:

$$v_i \leftarrow \omega v_i + c_1(\hat{x}_i - x_i) + c_2(g - x_i)$$

Here, ω is the inertia factor multiplying the current velocity. It's a scalar, usually in $[0.5, 1)$, with a typical initial value of 0.9. It decreases from iteration to iteration. This slows the particle as the search progresses, in theory, because the particle is likely moving closer to the best position. The second term calculates the difference between the particle's best-known position so far, $\hat{x}_i$, and its current position, x_i. This value is multiplied, component by component, by $c_1 = c_1 U[0, 1)$, that is, a random vector in $[0, 1)$ multiplied by a scalar, c_1. We use 1.49, a typical value. The last term in the velocity update calculates the difference between the swarm's best-known position, g, and the particle's current position, and multiplies it by vector $c_2 = c_2 U[0, 1)$. Usually, $c_1 = c_2$.

Second, the particle position is updated using the newly calculated velocity:

$$x_i \leftarrow x_i + v_i$$

NOTE *If you have a physics background and are, like me, bothered by the addition of a velocity and a position, imagine a $\Delta t = 1$ multiplying $\mathbf{v}_i$ where Δt is the time step between iterations. Now the units are correct.*

Bare-bones PSO, sometimes called BBPSO, does not use a velocity vector. Instead, particle positions are updated with samples from a normal distribution. If x_i is the vector representing particle i's current position, then x_{ij} is the jth component of that vector. With that in mind, calculate

$$\bar{x} = \frac{1}{2}(g_{ij} + \hat{x}_{ij})$$
$$\sigma = |g_{ij} - \hat{x}_{ij}|$$
$$x_{ij} \sim N(\bar{x}, \sigma)$$

if $p \sim U[0, 1) < p_b$, otherwise

$$x_{ij} = \hat{x}_{ij}$$

for each component (j) of each particle (i). Here, $x_{ij} \sim N(\bar{x}, \sigma)$ means draw a sample from a normal distribution with mean $\bar{x}$ and standard deviation σ. Typically, $p_b = 0.5$, so 50 percent of the time, on average, the particle's jth component is calculated from the normal distribution, and 50 percent of the time it's simply a copy of the corresponding component of the particle's best position, $\hat{x}_i$.

Genetic Algorithm (GA)

We know from our evolution experiments that a GA involves breeding (crossover) and random mutation. The code in *GA.py* follows this pattern, but fits with the overall optimization framework. In particular, *GA.py* manipulates floating-point values by default, not integers. You can alter this behavior by subclassing `Bounds` and implementing a `Validate` method to force integer values.

The update rule for a particle, x_i, involves crossover with a randomly chosen mate, where in this case the mate is selected from the top 50 percent best performing particles—see the `top` parameter of the `GA` constructor. Additionally, the current best-particle position, that is, the fittest particle, is passed to the next generation unaltered.

Our evolution experiments bred every individual, generation to generation. Here, individuals breed only if a random value is below the `CR` probability, 0.8 by default. When an individual breeds, it is replaced by the offspring. Whether x_i breeds, there is a certain probability of a random mutation by assigning a randomly selected dimension a random value. Therefore, for any update, a particle may be replaced by its offspring, and it may undergo random mutation. The default mutation probability is 5 percent (`F` in the `GA` constructor).

As a rule of thumb, the GA seems to work best for problems that aren't mathematical (like curve fitting) and involve a higher number of dimensions to give evolution a larger "genome" to manipulate. Whether this has implications for biological evolution, I don't know; regardless, another hallmark of the GA is slow convergence. You often need an order of magnitude more iterations (or even more) to reach a solution similar to that found far more quickly by Jaya or DE. Let's turn there now.

Differential Evolution (DE)

DE was invented in 1995 by Price and Storn, the same year particle swarm optimization was invented by Kennedy and Eberhart. Like PSO, DE has stood the test of time and grown into a collection of similar approaches. DE is an evolutionary algorithm where particles are updated between iterations by a process involving crossover and mutation. However, unlike the straightforward crossover and mutation of the GA, DE replaces x_i with a new vector that is, in a sense, the offspring of *four* parents. DE isn't modeled on nature.

To update particle x_i, first select three other members of the swarm, unique and not x_i. From these three, create a donor vector:

$$v = v_1 + F(v_2 - v_3)$$

Some variants of DE require that v_1 be the best performing member of the swarm instead of a randomly selected member. Here, *F* plays the role of mutation in the GA. In this case, the default value is $F = 0.8$.

The offspring of x_i and v is created component by component (gene by gene) where, with probability *CR*, the corresponding component of v is used; otherwise, the component of x_i is retained. The default value is $CR = 0.5$, meaning the offspring of x_i retains, on average, 50 percent of its existing values (genes).

There are so many variants of DE that a nomenclature has arisen to describe them. The code in *DE.py* defaults to "DE/rand/1/bin," meaning the donor vector uses three randomly selected vectors ("rand"), a single differential ($v_2 - v_3$), and a *Bernoulli* crossover ("bin"). A Bernoulli trial is a coin flip where the probability of success is p and that of failure $1 - p$. Here, $p = CR$ is the crossover probability.

The DE class supports two additional types of selection and an additional type of crossover, should you wish to experiment with them. If one of the three vectors used to build the donor is always the current swarm best, then the label begins with "DE/best/1." Additionally, a new selection mode is supported: DE/toggle/1, which toggles between "rand" and "best" every other update. Finally, Bernoulli crossover may be replaced with GA-style crossover, meaning the DE class supports six possible differential variants from the three different selection and two crossover types. Feel free to experiment with all of them. Do you notice anything different between them, especially how quickly the swarm converges? Hint: look at all the values in the gbest element of the dictionary returned by DE's Results method combined with the giter element that tracks the iteration number for each new swarm best position (gpos) and objective function value (gbest).

The goal of this section was to fit a known function to a dataset by minimizing the MSE between the data points and the function value at those points. We were after the parameters of the function, as we already knew the form we wanted. This raises the question: What if all we have is the data and we don't know the functional form? There are several ways to answer this. One is to use a machine learning model—after all, that's what they are designed to do: learn a model (function) from a set of data. We'll do this in Chapter 5. Another is to evolve a piece of code that approximates the data. Let's give this approach a shot.

Fitting Data

Curve fitting had us searching for the parameters of a known function. In this section, all we have is the data and our goal is to evolve a piece of code approximating a function that fits the data. We still want $y = f(x)$—that is, for a given x we get an approximated y—but here $f(x)$ is Python code. Evolving code is known as *genetic programming (GP)*, and it has a long history dating back to the early 1990s. A related term is *symbolic regression*.

As the name suggests, GP often uses the GA. Our implementation, however, uses the framework from the previous section so we can select any of the swarm intelligence and evolutionary algorithms. To use a swarm, we need to find a mapping between what we want (code) and a multidimensional space where each position in the space represents a possible solution. For curve fitting, the mapping was straightforward. If there were n parameters in the function, there were n parameters in the search space where each coordinate of a specific point was, literally, the parameter value.

Here, we have to be more clever. To identify the mapping, let's think about how we want to represent the code of our function, and from there, the mapping might be somewhat easier to see.

We want a function manipulating a scalar input value, x, to arrive at a scalar output value, y. So, we need math. We'll make do with the standard arithmetic operations, plus negation, modulo, and powers.

Doing math implies mathematical expressions. Here things become murkier. Manipulating mathematical expressions is rather tricky, more tricky than we care to attack in a book like this. Traditional GP manipulates expressions using an evolutionary algorithm, complete with crossover and mutation, where crossover merges two expressions and mutation alters a term in the expression.

Fortunately, we can use a shortcut. If we have a stack and know about postfix notation, we have all we need to generate expressions and map code to a position in the search space. I'll explain, but let's make sure we're on the same page when it comes to stacks and postfix notation.

Introducing Stacks and Postfix Notation

Imagine a stack of cafeteria trays. As new trays are added to the stack, they rest on top of all the existing trays. When someone needs a tray, they take the top tray, meaning the last tray added to the stack is the first tray removed from it. *Stacks* are like cafeteria trays (though cleaner).

Consider this example. We have three numbers: 1, 2, and 5. We also have a stack that is currently empty. The first number we *push* on the stack is 1, then we push 2, and finally 5. Figure 4-3 shows what the stack looks like step by step, from left to right.

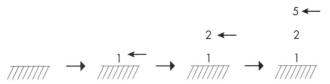

Figure 4-3: Stack manipulation

On the left, the stack is empty; then, moving right, we add 1, then 2, and finally 5. The stack is now three deep, with 1 at the bottom and 5 at the top.

Now it's time to *pop* a value off the stack. What value do we get? In a queue, we would get 1, the first value in. For a stack, we get 5, the last value pushed. Pop the stack again and we get 2, and finally 1, leaving the stack empty. Values pop off a stack in reverse order compared to how they are pushed onto the stack.

Stacks are natural structures for manipulating expressions in *postfix* form, that is, expressions where the operands come first, followed by the operation. For example, infix notation, our usual way of writing expressions, might say *a* + *b*, but in postfix notation, this becomes *a b* +. Postfix notation, also called reverse Polish notation (RPN), was developed in 1924 by Polish mathematician Jan Łukasiewicz. Postfix notation doesn't require parentheses to alter operator precedence. Instead, it builds the expression piece by piece. Combine postfix notation with a stack, and it becomes straightforward to evaluate arbitrary expressions. This is precisely what we want.

To better understand what I mean, let's translate the infix expression $y = a(b + c) - d$ to postfix notation and implement it using a stack and pseudocode statements. In postfix notation, it becomes $a\ b\ c + \times d\ -$. To evaluate it, move from left to right until you hit an operator, here +. The operands are the two variables to the left, b and c. Compute $b + c$ and replace "$b\ c$ +" with the result, t_0. The expression is now $a\ t_0 \times d\ -$. Repeat to find $\times$ with operands a and t_0. Compute the product and replace it with t_1 to get $t_1\ d\ -$. Finally, evaluate $t_1 - d$ to get the value of the expression, y.

Let's implement this process in code using a stack to hold values. Consider the following:

```
push(a)  | a
push(b)  | a b
push(c)  | a b c
add      | a t0
mul      | t1
push(d)  | t1 d
sub      | y
```

The values on the right show the stack after each instruction with $t_0 = b + c$, $t_1 = a \times t_0$, and $y = t_1 - d$. The expressions leave the answer, y, on the stack, and push places values on the stack. Binary operations like add pop two values off the stack, add them, and push the result back on the stack. Therefore, a linear sequence of statements and a stack are all we need to implement any function of x yielding y—at least, any function involving arithmetic operations, negation, and powers.

Functions written in this way become sequences of instructions with no loops. If we find a way to map the sequences to floating-point vectors, we're in business.

Mapping Code to Points

To evolve code, we need the four basic arithmetic operations: addition (add), subtraction (sub), multiplication (mul), and division (div). We also need exponentiation (pow) and will throw in modulo for good measure (mod).

Postfix notation distinguishes the subtraction operator and negation, treating the latter as a different instruction, so we need negation (neg) as well, $x \to -x$.

Finally, we need two more instructions: halt and push. If halt is executed, the code stops and ignores any subsequent instructions. The push instruction pushes x or a number (a constant) on the stack.

We have nine instructions. We want a sequence of instructions executed in order, which reminds me of a vector where each element is an instruction, and we execute the instructions one at a time from index 0 to the end of the vector.

Each instruction becomes a value, for example, add is 1 and sub is 2, so if a particle has a 2 in a particular component, then that component encodes a subtraction instruction. Particle positions are floating-point numbers, not integers, so we'll keep only the integer part of each, meaning a component with a floating-point value of 2.718 is interpreted as a 2, implying a subtraction instruction.

Table 4-1 has the mapping we'll use for instructions (the integer part of a particle position).

Table 4-1: Mapping Particle Positions to Instructions

Instruction	Number
add	1
sub	2
mul	3
div	4
mod	5
pow	6
neg	7
push(x)	8
halt	9

All that remains is to handle pushing constants on the stack. If we can't do this, we're stuck evolving expressions of only x, like $xx + x - x - x$, which will get us nowhere.

The instruction numbers begin with 1, not 0. This is intentional. Numbering this way leaves vector components in the range $[0, 1)$ available as that range is not associated with an instruction.

Let's use this range to push arbitrary numbers on the stack. When we run a search, we'll specify a lowest and highest number, like −1 and 11. Then, we'll map values in $[0, 1)$ to $[-1, 11]$. So, to push a constant of 3.1472 on the stack, we issue the instruction 0.3456 because:

$$a + f(b - a) = -1 + 0.3456(11 - {}^-1) = 3.1472$$

Handling things this way lets us specify arbitrary constant values within the given range.

For example, Table 4-2 shows a segment of code generated by *gp.py*, the program we're in the process of developing, along with the actual particle position values.

Table 4-2: An Evolved Code Sample

Instruction	Particle value
push(x)	—
push(x)	8.5251446
push(3.00482)	0.6502409
mul	3.3605457
push(7.07870)	0.8539350
push(-9.09650)	0.0451748
mod	5.0708302
add	1.3708454
halt	9.7707617
div	4.2693693
sub	2.6309877
div	4.6783009
pow	6.5429319

The task was to fit a noisy set of points representing a line. The evolved function fit the data quite nicely. The number limit in this case was −10 to 10, and I told the search (bare-bones PSO) to use 12 instructions. All evolved functions begin with x on the stack, here represented by the first push(x). Also, when the function exits, it returns the top stack value as y. Any remaining stack values are ignored.

If the particle value is ≥ 1, the integer part specifies an instruction, so $8.525 \to 8$, which is the instruction to push x, just as $3.360 \to 3$ to indicate multiplication.

Look at the second particle vector component, 0.6502409, which, with the number limits, becomes:

$$-10 + 0.6502409(10 - {}^-10) = 3.00482$$

This number is multiplied by x, that is, the second and third instructions implement $3.00482x$. The data points were generated by adding a small amount of random noise to the line $3x - 2$. The evolved function immediately implements $3.00482x$, which is quite encouraging.

The next three instructions push 7.07870 then −9.09650 on the stack before executing mod. This seems like a strange thing to do, but consider what Python does with the expression:

```
>>> 7.07870 % -9.09650
-2.0178000000000003
```

These instructions leave −2.0178 on the stack.

The next instruction is add. We add the top two stack values, which we just learned are $3.00482x$ and −2.0178. Interestingly, this is equivalent to the infix expression $3.00482x - 2.0178$, and earlier I stated that the data points were generated from $3x - 2$. The evolved code implements the expression used to create the data points in the first place.

The instruction following add is halt, which causes the function to exit with the sum on the stack. Instructions after halt are never executed.

Fabulous! We have an approach, a way to map floating-point vectors to code to implement a function. It's a bit odd, but we'll run with it and see where we get. Our next task is to create *gp.py*.

Creating gp.py

If you haven't already, read through *gp.py*. The essential framework pieces are all there, and we won't discuss every line, so it will help to be familiar with it before we begin.

The code imports all the framework components from the earlier curve fitting exercise; defines some helper functions (GetData, Number StrExpression) and the objective function class before the main code, which interprets the command line; constructs framework objects; and runs the search. Let's review Number and the objective function class here. The main code mirrors that of the curve fitting code.

The Number function transforms particle values in [0, 1) to the range specified on the command line when *gp.py* is executed. Specifically:

```
def Number(f, gmin=-20.0, gmax=20.0):
    return gmin + f*(gmax-gmin)
```

It's a direct implementation of the previous equation with gmin and gmax being the limits from the command line. These limits restrict the range of possible constants available to the evolved code; therefore, some experimentation might be necessary to find reasonable limits. For example, if you run a search and see constants at the limits, the specified range is likely too small, so double the size and try again. Remember, swarm intelligence and evolutionary algorithms are stochastic and heuristic. Parameters controlling their operation abound and must often be managed to produce good results.

A successful swarm search utilizes an objective function that drives the swarm toward good solutions, here a piece of code minimizing the MSE between the known data points and the output of the code at those data points. Therefore, our next port of call is the objective function class, Objective:

```
class Objective:
    def __init__(self, x,y, gmin=-20.0, gmax=20.0):
        self.fcount = 0
        self.x = x.copy()
        self.y = y.copy()
        self.gmin = gmin
        self.gmax = gmax

    def Evaluate(self, p):
        self.fcount += 1
```

```
y = np.zeros(len(self.x))
for i in range(len(self.x)):
    y[i] = Expression(self.x[i],p, self.gmin, self.gmax)
    if (np.isnan(y[i])):
        y[i] = 1e9
return ((y - self.y)**2).mean()
```

There are two methods, a constructor and Evaluate. The constructor keeps the data points and the number limits passed on the command line. It also initializes fcount, which tracks the number of times the objective function is evaluated.

The Evaluate method accepts a particle position (p) and passes the *x*-coordinate of the data points through it to generate output vector y. It then returns the MSE between y and self.y as the objective function value.

Not every piece of code represented by a particle position is valid. It's likely, especially early on in the search, that the randomly generated particle positions become failing blocks of code because they try impossible things like extracting values from an empty stack or dividing by zero. The NaN check in Evaluate captures such cases and ensures that a very high objective function value is returned.

The function Expression evaluates a particle position as code. It's given the *x* values, the particle position (p), and the number range; see Listing 4-1.

```
def Expression(x, expr, gmin=-20.0, gmax=20.0):
❶   def BinaryOp(s,op):
        b = s.pop()
        a = s.pop()
        if (op == 0):
            c = a + b
        elif (op == 1):
            c = a - b
        elif (op == 2):
            c = a * b
        elif (op == 3):
            c = a / b
        elif (op == 4):
            c = a % b
        elif (op == 5):
            c = a**b
        s.append(c)

    bad = 1e9
❷   s = [x]
    try:
❸       for e in expr:
            if (e < 1.0):
                s.append(Number(e, gmin=gmin, gmax=gmax))
```

```
                    else:
                        op = int(np.floor(e))
                        if (op < 7):
                            BinaryOp(s, op-1)
                        elif (op == 7):
                            s.append(-s.pop())
                        elif (op == 8):
                            s.append(x)
                        elif (op == 9):
                            break
            except:
                return bad
            try:
        ❹   return s.pop()
            except:
                return bad
```

Listing 4-1: Interpreting a particle position as code

Listing 4-1 is the heart of *gp.py*. First, there is an embedded function, BinaryOp ❶, which implements all binary operations like addition and exponentiation. The stack (s), a standard Python list, is popped twice to get the operands. Note the order: if we want $a - b$ and b is the top stack item, then the first pop returns b, not a. The second argument indexes the operation. A more compact implementation might use Python's eval function. Still, we need to be as fast as possible, so we opt for the verbose but significantly faster compound if.

The code initializes the stack with x ❷ and then begins a loop over the components of the particle position (expr) ❸. Everything is inside a try block to catch any errors. Errors return bad as the function value.

If the particle component is less than 1.0, it pushes a constant, the output of Number, on the stack. Otherwise, the integer part of the value determines the operation. If less than 7, the appropriate binary operation is performed; otherwise, the instruction is either negation, push x, or halt, which breaks out of the loop, thereby ignoring the remaining particle components. Finally, the function returns the top stack item, if there is one, as the function value ❹.

The remainder of *gp.py* is straightforward: parse the command line, create framework objects (Bounds, RandomInitializer, Objective), then, with the proper swarm object, call Optimize and Results to report how well the search went. If a final plot name is given, we generate it showing the data points and the fit.

Our curve fitting code used bare-bones PSO. This code uses both bare-bones and canonical PSO:

```
elif (alg == "PSO"):
    swarm = PSO(obj=obj, npart=npart, ndim=ndim, init=i, tol=0, max_iter=niter, bounds=b,
            rng=rng, vbounds=Bounds([-10]*ndim, [10]*ndim, enforce="clip", rng=rng),
            inertia=LinearInertia(), ring=True, neighbors=6)
```

```
elif (alg == "BARE"):
    swarm = PSO(obj=obj, npart=npart, ndim=ndim, init=i, tol=0, max_iter=niter, bounds=b,
                rng=rng, bare=True)
```

We can differentiate between the two by passing either PSO or BARE on the command line. Note that PSO uses options we haven't seen before. One is LinearInertia, which linearly decreases ω during the search from 0.9 down to 0.4. Inertia is the coefficient multiplying the previous iteration's velocity, per particle.

There are three additional options. Two are ring and neighbors, which work together. A variation of canonical PSO includes the concept of a *neighborhood*, a collection of particles that coordinate with each other. The practical effect of a neighborhood is to replace the global best position, g, with a neighborhood best. The arrangements of particles into neighborhoods is referred to as a *topology*. The PSO class supports a ring topology—the simplest. Imagine the particles forming a circle; then, for any particle, the neighbors particles to the left and right form the current particle's neighborhood. As a challenge, try to modify *PSO.py* to accommodate von Neumann neighborhoods, which is frequently the best performing topology. Careful searching online will show you what a von Neumann topology entails.

The final new option is vbounds, which sets limits on the maximum velocity per particle component, much as bounds sets spatial limits on where the swarm can move within the search space. In the velocity case, enforce is clip to keep velocity components at the limits instead of resampling along that dimension.

The number of adjustable parameters makes it sometimes tricky to set up a successful canonical PSO search, even with neighborhoods. As a result, these values are more what you'd call "guidelines" than actual rules.

Now, let's put *gp.py* through its paces to see what it can (and cannot) do for us.

Evolving Fit Functions

Let's put *gp.py* to the test with several experiments.

Fitting a Line

To run *gp.py*, use a command line like this one:

```
> python3 gp.py data/x1_2n.txt -5 5 22 20 10000 bare minstd plot.png
```

Illegal operations will likely happen early on in the search, so I recommend ignoring runtime errors by adding -W ignore. The command line uses the input file *x1_2n.txt*, which is the noisy line mentioned previously. Numbers are limited to $[-5, 5)$, and the maximum program length is 22 instructions, though halt usually appears earlier.

The swarm has 20 particles and runs for 10,000 iterations using barebones PSO and the MINSTD randomness source. The result is written to *plot.png* with the code itself displayed:

```
Minimum mean total squared error: 0.385596890  (x1_2n.txt)
    push(x)
    push(x)
    add
    push(x)
    add
    push(4.82483)
    push(2.39118)
    div
    sub
    halt
    add
    mod
    div
    sub
    sub
    sub
    div
    mul
    halt
    push(x)
    pow
    add
    div
(23 best updates, 200020 function calls, time: 257.076 seconds)
```

Note that halt appears as the ninth instruction (the initial push(x) is always present, so it's not counted).

Figure 4-4 shows the original data points along with the fit function output.

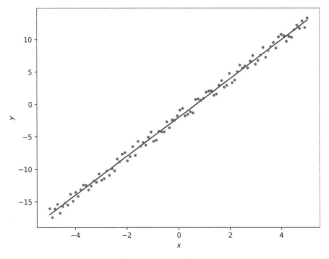

Figure 4-4: The evolved fit to a noisy line

The fit is good, which is encouraging. If we can't fit a line, we shouldn't expect to fit more complex functions.

We now have two different solutions to fitting the line, but the evolutionary path followed to arrive at a solution was quite different. The first solution evolved

$$(x)(3.00482) + (7.07870 \bmod - 9.09650) = 3.00482x - 2.0178$$

but the second produced:

$$x + x + x - (4.82483 / 2.39118) = 3x - 2.01776$$

The second solution added x to itself three times instead of multiplying by a constant. Both solutions arrived at nearly identical intercepts, not by pushing a learned value but by implementing distinct binary operations with two learned values.

The *data* directory contains several datasets, many that are the output of *gpgen.py*, which you can use to create custom datasets, noisy polynomials up to degree five. Run *gpgen.py* without arguments to learn how it works. For now, let's use a few of these datafiles to push *gp.py* to the limit.

Fitting a Quadratic

We evolved the equation of a line easily enough. What about a quadratic?

```
> python3 gp.py data/x2_2n.txt -5 5 22 20 10000 bare minstd plot.png
```

For my run, this produced:

```
Minimum mean total squared error: 0.263703051  (x2_2n.txt)
    push(x)
    push(x)
    mul
    push(-2.97844)
    sub
    push(x)
    sub
    push(x)
    sub
    halt
(98 best updates, 200020 function calls, time: 290.683 seconds)
```

Instructions after halt are ignored as they have no effect. I'll do this consistently from now on. The resulting fit is in Figure 4-5.

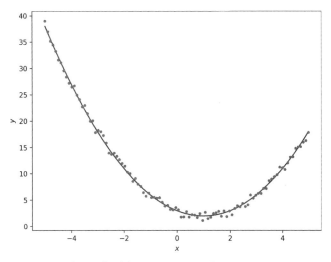

Figure 4-5: The evolved fit to a noisy quadratic

The evolved code is equivalent to:

$$(x^2 - {}^{-}2.97844) - x) - x) = x^2 - 2x + 2.97844$$

The same dataset given to NumPy's `polyfit` routine produces

$$x^2 - 2.02x + 2.99$$

giving us growing confidence in the evolutionary search.

Fitting a Quartic

The previous examples all use bare-bones PSO, which seems well suited to this task. Let's try a different dataset, a quartic, along with different algorithms. Are all of them equally effective?

Specifically, we'll fit the points in *x4_-2x3_3x2_-4x_5_50n.txt*, a noisy version of the quartic, $y = x^4 - 2x^3 + 3x^2 - 4x + 5$. The only parameter changing from run to run is the optimization algorithm. For example, here's the command line for bare-bones PSO:

```
> python3 gp.py data/x4_-2x3_3x2_-4x_5_50n.txt -25 25 22 25 15000 bare pcg64 plot.png
```

To use differential evolution, change `bare` to `DE` and run again. We'll examine the results for a single run of each algorithm.

The framework is designed for clarity, not speed. Since each particle evaluates the objective function independently of the others, opportunities for parallelization abound. Unfortunately, we take advantage of none of them, so patience is required to replicate the search for every algorithm: DE, bare-bones PSO, canonical PSO, Jaya, GA, and RO. Also, the framework does not use seed values, so your run of the code will produce different, though likely similar, output.

Figure 4-6 displays the fit for each algorithm, from DE on the upper left to RO on the lower right.

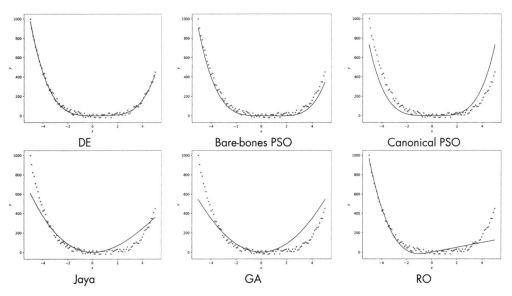

Figure 4-6: The fits for each algorithm

It's evident that not every algorithm hit the mark. Table 4-3 shows us the equivalent equations they generated.

Table 4-3: The Fit Equations Evolved by Each Algorithm

Algorithm	Equivalent equation
NumPy	$y = 1.01x^4 - 2.01x^3 + 2.76x^2 - 3.34x + 5.76$
Differential evolution	$y = x^4 - 2.19254x^3 + 3x^2$
Bare-bones PSO	$y = x^4 - 2.23835x^3$
Canonical PSO	$y = x^{4.09896}$
Jaya	$y = -x^3 + 19.36026x^2$
GA	$y = 21.78212x^2$
RO	$y = \left(\frac{-x}{1.39135}\right)\left(x \bmod 12.15209x^2 - x\right) + 24.87559x$

The first equation is the fit returned by NumPy's polyfit routine. The data was generated from a quartic, so we expect the NumPy fit to be the best, and we'll treat it as the gold standard.

Differential evolution produced the best-evolved fit function. Compare it to the NumPy fit. The DE fit recovered the first three terms of the polynomial, with coefficients in the same ballpark as the NumPy fit. Similarly, bare-bones PSO recovered the first two terms of the polynomial. Canonical PSO recovered only the first term, x^4 (or thereabouts).

Jaya produced an exciting result. The two terms fight against each other, but their sum becomes a crude approximation of the dataset. As an exercise, try plotting $-x^3$, $19.36x^2$, and their sum to see what I mean.

Both the GA and RO produced inferior output. The GA ended up with a quadratic, and whatever RO created fits only the left-hand set of dataset points, $x < -3$ or so.

These results are from single runs. We know that swarm optimization algorithms are stochastic and vary from run to run. Perhaps we're being a bit unfair, then. I ran the Jaya search five more times; here are the resulting equivalent equations:

$$y = x^4$$

$$y = x^{4.02072}$$

$$y = x^4$$

$$y = 23.10565(x^2 - x)$$

$$y = 21.49103x^2$$

The results imply that Jaya isn't converging well to a local minimum. It either captures the essential x^4 aspect of the data or stops at a quadratic.

Jaya isn't the only algorithm we might be selling short. The GA and RO produced inferior results on our initial runs. What if we increase the swarm size and churn for more iterations? Intuitively, for these algorithms, it might make sense to do this. The more particles searching on their own, the more likely we'll find a good position in the search space, so a larger swarm seems sensible for RO. For the GA, a larger population increases the size of the gene pool, so we should get better performance as well, much like the genetic drift example from Chapter 3 where the larger population was better able to adapt to the environment after a catastrophe.

Running a swarm of 125 particles for 150,000 iterations using RO produced a function calculating $21.54962x^2$, which is not encouraging as RO isn't even capturing the quartic nature of the dataset. A run of the GA with 512 particles (organisms) for 30,000 iterations requiring all organisms to breed with a member of the best performing 20 percent (top=0.2) produced Figure 4-7, which contains a graph showing the fit.

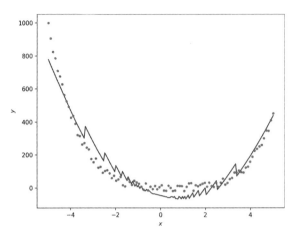

Figure 4-7: The genetic algorithm's solution for 512 organisms and 30,000 generations

The evolved set of instructions are as follows:

```
push(x)
push(-21.60479)
mul
push(21.92950)
push(-24.20913)
sub
sub
push(-24.96894)
push(-11.43579)
neg
neg
push(-21.53894)
sub
push(x)
mod
mul
push(x)
push(x)
mul
push(23.76431)
mul
add
add
```

The function is quite strange, and used all 22 possible instructions (the initial push(x) is always present), which is also quite different from the results found with other algorithms. The equivalent function is

$$y = 23.76431x^2 - 21.60479x - 46.13863 - 24.96894(10.10315 \bmod x)$$

which makes more sense in this form: a quadratic with an additional term using modulo, thereby accounting for the strange-looking oscillations on top of the general quadratic form.

Fitting a Normal Curve

The previous examples attempted to evolve a function to match a polynomial. What happens if we try to fit a noisy normal (Gaussian) curve instead? The source function, before adding random noise, was:

$$y = e^{-x^2}$$

The noisy data points are in *noisy_exp.txt*.

I ran three searches, once each for DE, bare-bones PSO, and Jaya. The searches all used 25 particles and 20,000 iterations. I restricted numbers to $[-25, 25]$, and gave the evolved functions a maximum of 22 instructions.

Figure 4-8 shows the resulting fits.

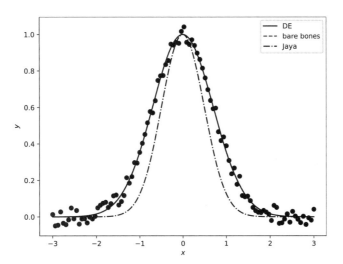

Figure 4-8: Evolving a function to fit a noisy normal curve

The DE and bare-bones PSO results are virtually identical and overlap. They fit the dataset well. As we've seen with other experiments, Jaya gets close but doesn't produce as nice a fit as the others.

So, what were the evolved functions? In this case, not only is the equivalent function illustrative, but so is the form of the code, so we'll consider both together; see Table 4-4.

Table 4-4: Comparing the Evolved Programs by Algorithm

DE	Bare-bones PSO	Jaya
push(x)	push(x)	push(x)
push(x)	push(x)	neg
push(0.35484)	push(x)	push(7.95565)
push(x)	push(2.80857)	push(x)
push(x)	push(x)	pow
mul	neg	push(x)
pow	push(x)	neg
halt	mul	pow
	pow	halt
	halt	

The equivalent function evolved is $y = 0.35484^{x^2}$ (DE), $y = 2.80857^{-x^2}$ (bare-bones PSO), and $y = 7.95565^{-x^2}$ (Jaya). Again, the function we're trying to recover from the noisy data is

$$e^{-x^2} \approx (2.71828)^{-x^2}$$

where I've approximated e with the first five digits of its decimal expansion. Written this way, it's clear that the bare-bones PSO search evolved almost exactly this function, so we should expect it to fit the data nicely.

The DE result seems odd at first. It's an exponential function, but the base is 0.35484, not e, and the exponent is x^2, not $-x^2$. However, $1/e \approx 0.36788$, meaning DE evolved the same function as bare-bones PSO since

$$\left(\frac{1}{e}\right)^{x^2} = (e^{-1})^{x^2} = e^{-x^2}$$

and 0.35484 is quite close to 0.36788. Finally, Jaya had the right idea, but didn't converge to the correct base, $7.95565 > e$.

Exercises

Swarm algorithm applications are legion. Here are a few things you may wish to explore in more detail:

- In the *curves* directory you'll find a *NIST* directory. It contains example curve fit datafiles from the National Institute of Standards and Technology (NIST), part of the United States Department of Commerce. I formatted the *.txt* versions so that they'll work with *curves.py*. The original versions end with *.dat*.

  These are challenging curve fitting test files meant to test high-performance curve fitting routines. Are any of the swarm algorithms up to the challenge? If so, which files can be fit, and which fail?

- The `Results` method of a swarm object returns a dictionary, as we saw throughout the chapter. We can track the objective function value as a function of swarm iteration by using the `gbest` and `giter` dictionary values. The first is a list of each new global best objective function value, and the second is a list marking the iteration at which that value became the global best.

  Examine the code in *plot_gbest_giter.py* to learn how to plot these values to track the swarm's learning during a search. Capture the corresponding lists for other searches using the examples in the chapter to make similar plots. Do the swarms all converge at the same rate for the same problem?

- The file *gaussian.py*, in the *micro* directory, runs a swarm search to minimize a two-dimensional function consisting of two inverted normal curves, that is, the function is $z = f(x, y)$. The file *gaussians.png* shows a 3D plot of the function with two minima, one lower than the other. Since the function being minimized has two inputs, the search space is two-dimensional, making it possible to plot the position of every particle in the swarm and track them as they move during the search.

  Run *gaussian.py* without arguments to learn how to run a search and output images of the swarm at each step. Then, page through the images to watch the swarm search. The known best position is the empty box. The swarm's current best position is the star. Change the algorithm to observe how each converges and traverses the space. Do the algorithms search the space in the same way? Do they converge on the global minimum, and, if so, at the same rate?

- The file *GWO.py*, also in the *micro* directory, implements the Grey Wolf Optimizer (GWO). GWO is a popular swarm intelligence algorithm, in theory, modeling the behavior of wolves as they hunt (I don't buy it).

  Test GWO using *gaussian.py*, then adapt *curves.py* and *gp.py* to use it as well. A quick copy-paste is all that's needed. How does GWO's performance compare to DE, bare-bones PSO, and Jaya? It's often claimed that GWO has no adjustable parameters, like Jaya. This is not strictly true. The eta parameter, which defaults to 2, can be adjusted, and this sometimes helps the search. If GWO isn't performing well, adjust eta, perhaps to 3 or 4, and try again.

- Consider the code in *MiCRO.py*, also in the *micro* directory. It implements a tongue-in-cheek swarm algorithm loosely based on grazing cattle that I call *Minimally Conscious Random Optimization*. It's meant to show how easy it is to create a "novel," "nature-inspired" swarm algorithm. The idea behind MiCRO is that the swarm is a herd of cattle, mindlessly grazing in complete ignorance of each other. With a set probability on any iteration, an animal might look up and consider another animal's position that's better than its own. If this happens, the animal jumps to the region around the better-off neighbor and continues to graze. So, the algorithm is RO with a slight probability of noticing a better performing neighbor; the swarm is *minimally conscious*.

Explore how MiCRO performs using *gaussian.py*, then invent your own swarm algorithm using the code in *RO.py* and *MiCRO.py* as guides. Does your algorithm work? Does it work well? Is it really inspired by nature, or is the nature "inspiration" an after-the-fact justification?

- The code evolved by *gp.py* is restricted to arithmetic operations plus powers and modulo. Add sine, cosine, and tangent as available operators. Each consumes one item from the stack and returns one item to the stack.

  Try to fit the *cos.txt*, *sin.txt*, and *tan.txt* datasets in the *data* directory. Can the swarm algorithms do it?

Recent work has demonstrated that GWO, along with several other popular nature-inspired algorithms, is not novel at all, but is older PSO ideas wrapped in often strained metaphors. That being the case, it's fair to wonder why I included GWO here. The emphasis in this book is on practicality and ease of application. GWO is popular and works in terms of providing solutions to problems. In that sense, it doesn't matter whether it's novel. For the larger optimization field, it's critically important to understand what is novel and what is not. I suspect, in the end, that many of the myriads of nature-inspired algorithms will prove to be alternate takes on well-known approaches. But, if GWO works, then it works, so we'll keep it in our small collection of algorithms at the risk of alienating genuine optimization researchers.

The file *nature-inspired_algorithms.pdf* lists dozens of nature- and physics-inspired swarm optimization algorithms. The list is by no means exhaustive.

Summary

This chapter introduced us to swarm intelligence and evolutionary algorithms. We used a software framework to develop two applications: one to fit data to a known functional form, that is, traditional curve fitting, and one to evolve code from scratch to implement a best fit function. We learned how to use the framework and explored each of the swarm algorithms to develop our intuition about how they work and are best applied. Each swarm algorithm critically depends on randomness, from the initial configuration of the particles to each update step that moves the particles throughout the search space.

The experiments of this chapter fit data to functions by either locating the best parameters for a known functional form or evolving the function from scratch. Both attempts were successful, though not every algorithm performed equally well.

DE proved an excellent match to these tasks, further justifying it as a go-to algorithm. However, I was surprised to see bare-bones PSO do so well. Canonical PSO wasn't as effective, but it has more parameters to tweak, so it might be made better with some experimentation (you can adjust c_1, c_2, ω, the inertia parameter change over iterations, and so on).

Jaya wasn't a complete disappointment, but it performed similarly to other places where I've used it—not particularly good or bad. It was, at times, able to recover the essence of the functional form, but not the particulars, even allowing for many iterations. As a final example, I ran a search to fit the noisy normal function one more time using Jaya and 120,000 iterations, six times as many as before. The result was a worse fit than the first run with an equivalent function of

$$y = \frac{x}{x^3 + x}$$

which looks vaguely similar to a normal curve but does not fit the data well.

RO is not, strictly speaking, a swarm algorithm because the particles don't influence each other. Still, the examples here, combined with experience in other areas, make RO worth trying in many cases. We'll encounter RO again in Chapter 5.

Curve fitting is not the GA's cup of tea. We learned in Chapter 3 that the GA is effective when simulating natural selection and genetic drift. Also, we barely explored some of its parameters, such as the fraction of top breeding organisms (top) or the particular mutation and crossover probabilities (F and CR, respectively).

We're not through with swarm algorithms. Let's leave curves and datasets behind to apply swarm techniques to other areas, like image processing and in combination with a simulation.

5

SWARM OPTIMIZATION

Swarm techniques do more than optimize mathematical functions. In this chapter, we'll use randomness to pack circles in a square, place cell towers, enhance images, and organize product placement at the grocery store. We'll apply the same collection of swarm intelligence and evolutionary algorithms as in the previous chapter.

Packing Circles in a Square

A classic math problem involves packing equal diameter circles in a square. An equivalent formulation is to locate, for a given number of points, positions in the unit square ([0, 1]) where the smallest distance between any pair of points is as large as possible. The point locations correspond to the centers of the best-packed circles. For example, where in the unit square do we put two points to maximize the distance between them? In opposite corners. In that case, the distance between the points is $\sqrt{1^2 + 1^2} = \sqrt{2}$, and it can't be any larger.

What about three points? Four points? Seventeen points? Now the answer isn't so obvious. We might approach this problem by using the elaborate algorithm detailed in Locatelli and Raber's 2002 paper, "Packing Equal

Circles in a Square: A Deterministic Global Optimization Approach," but that's not how we'll do it here. Instead, we'll use randomness in the form of a swarm search. We need to map positions in some multidimensional space to candidate solutions and then search this space for the best possible solution.

If we have n points and want to know the coordinates of n circle centers that are, for each pair, as far apart as possible while still within $[0, 1]$, we need to find n points. At first, we might believe we have an n-dimensional problem. However, the dimensionality is actually $2n$: we need both the x- and y-coordinates to specify a point. We know the search's bounds are $[0, 1]$ for each dimension. Therefore, we'll use swarms that are $2n$-dimensional vectors bounded by $[0, 1]$ where each pair of components is a point, (x, y). In other words, if we want to find five points, each particle is a 10-element vector:

$$p = (x_0, y_0;\ x_1, y_1;\ x_2, y_2;\ x_3, y_3;\ x_4, y_4)$$

To run a search, we need the dimensionality of the problem and the bounds, both of which we now have. The only remaining issue is the objective function, which tells us how good a solution each particle position represents. The problem specification lights the way: we need to maximize the smallest distance between any pair of points. If there are five points, we calculate the distance between each possible pair, find which distance is the smallest, and return the opposite. Our framework only minimizes, so to maximize, we return the negative. The largest smallest distance between pairs, when made negative, becomes the most negative number.

The Swarm Search

The code we want is in *circles.py*. Consider putting the book down and reading through the file to understand the flow. Once you've done that, we can begin with the objective function class:

```
class Objective:
    def __init__(self):
        self.fcount = 0

    def Evaluate(self, p):
        self.fcount += 1
        n = p.shape[0]//2
        xy = p.reshape((n,2))

        dmin = 10.0
        for i in range(n):
            for j in range(i+1,n):
                d = np.sqrt((xy[i,0]-xy[j,0])**2 + (xy[i,1]-xy[j,1])**2)
                if (d < dmin):
                    dmin = d
        return -dmin
```

The constructor does nothing more than initialize fcount, which counts the number of times Evaluate is called. The Evaluate method is given a position vector (p) that is immediately reshaped into a set of (x, y) pairs (xy).

The second code paragraph in Evaluate runs through each pairing of points in xy and calculates the Euclidean distance between them. If that distance is the smallest found so far, we keep it in dmin. We want to maximize the smallest distance between any pair of points, so we first find the smallest distance between any two points represented by the particle position.

The final line returns the negative of dmin. Because the framework minimizes, returning the negative of the smallest pairwise distance forces the framework to *maximize* this smallest distance—exactly what we want.

We now have everything we need to implement the search. The main body of *circles.py* follows the standard approach of pulling values off the command line and setting up the framework objects before calling Optimize to execute the search.

In code, the essential steps are:

```
rng = RE(kind=kind)
b = Bounds([0]*ndim, [1]*ndim, enforce="clip", rng=rng)
i = RandomInitializer(npart, ndim, bounds=b, rng=rng)
obj = Objective()
swarm = PSO(obj=obj, npart=npart, ndim=ndim, init=i,
            bounds=b, max_iter=niter, bare=True, rng=rng)
swarm.Optimize()
res = swarm.Results()
```

We create the desired randomness engine, followed by the bounds, initializer, and objective function instance. Notice that the objective function requires no ancillary information.

The swarm object, PSO configured for bare-bones searching, is followed by Optimize and Results. Not shown is code to report the best set of points and the distance between them before dumping all search results, including a simple plot of the point locations, into the supplied output directory.

Try running *circles.py* with no command line options to see what it expects. Now that we have it, let's use it.

The Code

Let's pack some circles. I created two shell scripts, *go_circle_results* and *go_plots*. The former runs *circles.py* for 2 through 20 circles and 7 algorithms: bare-bones PSO, canonical PSO, DE, GWO, Jaya, RO, and GA. The output is stored in the *results* directory. I recommend starting it in the evening and coming back the next morning, as the framework is designed for clarity, not speed. Run it with:

```
> sh go_circle_results
```

When *go_circle_results* finishes, execute *go_plots* to produce a series of plots showing the circle configuration each algorithm located. My results from this exercise are in Table 5-1, though yours will be somewhat different due to the stochastic nature of the swarm searches.

Table 5-1: Largest Center Distance, Known and as Found by Each Algorithm

n	Known	Bare	DE	PSO	GWO	Jaya	RO	GA
2	1.4142	1.4142	1.4142	1.4142	1.4142	1.4142	1.4142	1.4134
3	1.0353	1.0353	1.0353	1.0353	1.0353	1.0353	1.0301	1.0264
4	1.0000	1.0000	1.0000	1.0000	1.0000	1.0000	0.9998	0.9969
5	0.7071	0.7071	0.7070	0.6250	0.7025	0.7071	0.6796	0.6052
6	0.6009	0.5951	0.5953	0.5995	0.5988	0.5858	0.5723	0.5884
7	0.5359	0.5359	0.5223	0.5000	0.5072	0.5176	0.5000	0.4843
8	0.5176	0.5090	0.5045	0.5000	0.5002	0.4801	0.4661	0.4355
9	0.5000	0.5000	0.4202	0.5000	0.4798	0.5000	0.4421	0.4470
10	0.4213	0.4147	0.3697	0.4195	0.4187	0.3517	0.3788	0.3819
11	0.3980	0.3978	0.3296	0.3694	0.3895	0.3918	0.3588	0.3787
12	0.3887	0.3726	0.2989	0.3717	0.3289	0.3819	0.3496	0.3542
13	0.3660	0.3595	0.2752	0.3333	0.3277	0.2832	0.3212	0.3230
14	0.3451	0.3354	0.2537	0.3333	0.3116	0.3435	0.3037	0.3204
15	0.3372	0.3256	0.2303	0.3333	0.3278	0.2437	0.2949	0.2995
16	0.3333	0.2996	0.2269	0.2500	0.3011	0.2220	0.2760	0.2761
17	0.3060	0.2985	0.2062	0.2913	0.2952	0.1992	0.2658	0.2721
18	0.3005	0.2782	0.1927	0.2808	0.2703	0.2126	0.2516	0.2493
19	0.2900	0.2697	0.1852	0.2500	0.1905	0.1731	0.2384	0.2559
20	0.2866	0.2632	0.1789	0.2500	0.2419	0.1659	0.2200	0.2342

Table 5-1 shows the best-known distance between the points and the distance found by the swarm searches, by algorithm. These numbers will be our gold standard.

For $n < 10$, many of the distances are known from geometric arguments, as Table 5-2 shows.

Table 5-2: Known Circle Center Distances

n	Distance
2	$\sqrt{2}$
3	$\sqrt{6} - \sqrt{2}$
4	1
5	$\sqrt{2}/2$
6	$\sqrt{13}/6$
7	$4 - 2\sqrt{3}$
8	$(\sqrt{6} - \sqrt{2})/2$
9	0.5
10	0.42127954

The expressions come from Table D1 in *Unsolved Problems in Geometry* by Croft, Falconer, and Guy (Springer, 1991). The *plot_results.py* file, called by *go_plots*, uses this table to generate plots showing the packed circles. If the packing is optimal, the circles barely touch. Otherwise, the circles are separated from each other or overlap.

Examining Table 5-1 reveals 2 through 4 circles to be straightforward; every algorithm located the best arrangement. For 5 circles, bare-bones PSO, DE, and Jaya converged on the solution. We won't quibble about the one-ten-thousandth difference between the known distance and differential evolution's solution.

The swarms begin to struggle after 5 circles. For 6 circles, no swarm nails the distance, at least to four decimals, but several come pretty close. Figure 5-1 shows the output plots for each algorithm.

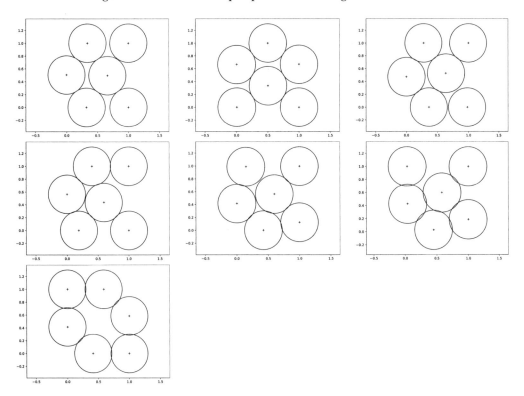

Figure 5-1: Packing 6 circles. From top left to right: PSO, GWO, DE, bare-bones PSO, GA, RO, and Jaya.

While solutions are unique regarding the distance between the circle centers, they aren't in terms of rotations. The canonical PSO, DE, and GWO results are essentially the same, only rotated by 90 degrees in some cases.

View all graphs generated by *go_plots* by the number of circles. As *n* increases, the swarms struggle more and more, but there are nice *n* values, like *n* = 9, where the swarms are more likely to arrive at the highly symmetric solution. Since we're packing a square, it makes sense that *n* values that are perfect squares—like 4, 9, and 16—lead to nicely aligned packings. However, only a few algorithms located the ideal *n* = 9 output, and none found the best *n* = 16 outcome.

Let's move on to a more practical problem.

Placing Cell Towers

Placing a cell phone tower is not an academic exercise; real utility and cost are involved. In this section we'll experiment with a (simplified) cell tower placement problem.

Our inputs are a collection of cell towers, each with a different effective range, and a mask showing where cell towers can be placed. The output is a collection of locations where the specified towers should be placed to maximize coverage.

The code we'll work with is in *cell.py*. It has the same general structure as the other swarm experiments, but is slightly more advanced because evaluating a particle position involves checking for illegal tower positions and building an image. The Evaluate method of the objective function class is more complex, but still accepts a particle position and returns a score where lower implies a better solution.

I'll lay out the plan of attack; then we'll walk through the essential parts of the code before running some experiments.

The Swarm Search

We need to translate vectors of numbers within some bounds into possible solutions. We'll work with cell towers and maps telling us where we can place them. Let's begin with representing towers and maps.

The cell towers radiate in all directions, so we'll represent them as circles where the diameter of the circle indicates the tower's strength and the center of the circle the tower's location. Not all towers are of equal strength. We specify a tower with a single floating-point number in (0, 1]; exactly how will become clear momentarily.

The maps are grayscale images. If a pixel's value is 0, that pixel is a possible tower location; if the pixel is 255, it's off-limits. I placed a collection of maps in the *maps* directory. You can make your own in any graphics program; use 255 to mark regions off-limits to towers and 0 for everything else. The maps need not be square. Note that the larger the map, the slower the search, which is why the supplied maps are rather small.

Tower sizes are fractions of half the map's largest dimension. For example, the supplied maps are 80 pixels on a side. Therefore, a tower given as 0.1 has a diameter of $0.1 \times 40 = 4$ pixels, while a 0.6 tower's diameter is $0.6 \times 40 = 24$ pixels. Towers are stored in a text file, one number per line, with the number of lines indicating the number of towers. Examine the files in the *towers* directory to see what I mean.

Particle positions represent tower locations. If there are n towers, we need $2n$-dimensional particles, as we did for packing circles. Every two elements of a particle position are the center location for a tower. The order in which towers are specified maps, one-to-one, to pairs of particle elements. For example, *towers0* has six lines for six towers: 0.1, 0.2, 0.3, 0.4, 0.5, 0.6. Therefore, a search using *towers0* involves 12-dimensional particles

$$(x_0, y_0; \ x_1, y_1; \ x_2, y_2; \ x_3, y_3; \ x_4, y_4; \ x_5, y_5)$$

where (x_0, y_0) is the location of the 0.1 tower, (x_1, y_1) is the location of the 0.2 tower, and so on.

Figure 5-2 shows the default maps.

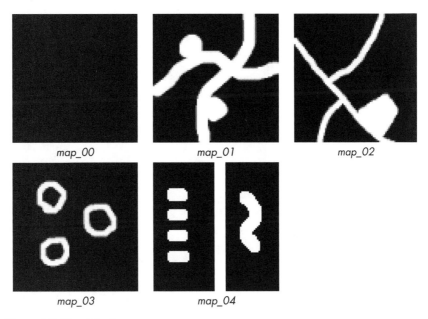

map_00 map_01 map_02

map_03 map_04

Figure 5-2: The default maps

The first map is blank, with no forbidden regions. Any marked area on the other maps is off-limits to a tower. Think of these as roads, parking lots, lakes, and so on.

We have towers and maps. We know how to represent tower locations and sizes. How do we put towers, maps, and locations together to get a score? We want to cover the map as much as possible by placing towers in allowed locations. Therefore, we want to minimize the number of allowed map pixels that aren't covered by a tower; we want the number of zero pixels after placing towers to be as small as possible. This sounds like a job for the objective function.

For a given particle position, the objective function needs to determine whether any proposed cell tower centers are in an illegal location. If even one tower is, the objective function rejects the entire configuration by immediately returning a score of 1.0, the largest possible, implying that none of the map is covered.

If all proposed tower locations are allowed, it's time to calculate the coverage. The number of uncovered pixels divided by the number of pixels in the map is a value in [0, 1], where 0 means all pixels are covered. The lower this value, the better the coverage.

The approach I chose begins with an empty image the same size as the map image. We add towers to the image by adding each pixel covered by the tower to any current pixel value.

Adding pixels in this way accomplishes two things: first, any pixel of the initially empty image that is still 0 after adding all the pixels covered by towers is uncovered; second, adding pixels tower by tower builds a comprehensible image. We'll be able to see each tower and its covered region clearly, including areas where towers overlap.

To recap, for a given particle position, we:

1. Convert the tower coordinates to a set of points as we did earlier for packing circles.

2. Return 1.0 as the score if any tower centers land on illegal regions of the map.

3. Add each tower to an initially empty image array, including all covered pixels, if all tower centers are allowed.

4. Return the count of uncovered pixels divided by the total number of pixels as the score.

These steps map particle positions to solutions, thereby generating a single number representing the quality of the solution.

Figure 5-3 shows an input map on the left and output generated after a canonical particle swarm search using the six towers in *towers0* on the right.

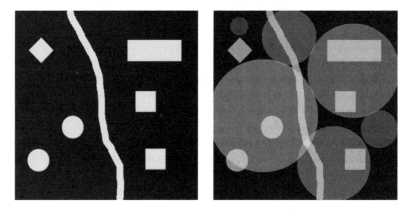

Figure 5-3: An input mask (left) and resulting tower placement (right)

The towers overlap only slightly, and no centers are in masked areas. Examine the files in the *example* directory to view these images in more detail.

Let's review the essential parts of *cell.py* to understand how the steps were translated into code.

The Code

The code in *cell.py* is relatively complex. Spend some quality time with the file before continuing.

The most important piece of code is the objective function class and friends; see Listing 5-1.

```
class Objective:
    def __init__(self, image, towers, radius):
        self.image = image.copy()
        self.R, self.C = image.shape
❶       self.radii = (towers*radius).astype("int32")
        self.fcount = 0

    def Collisions(self, xy):
        n = 0
        for i in range(xy.shape[0]):
            x,y = xy[i]
            if (self.image[x,y] != 0):
                n += 1
        return n

    def Evaluate(self, p):
        self.fcount += 1
        n = p.shape[0]//2
❷       xy = np.floor(p).astype("uint32").reshape((n,2))
        if (self.Collisions(xy) != 0):
            return 1.0
        empty = np.zeros((self.R, self.C))
❸       cover = CoverageMap(empty, xy, self.radii)
        zeros = len(np.where(cover == 0)[0])
        uncovered = zeros / (self.R*self.C)
        return uncovered
```

Listing 5-1: The objective function class

The constructor stores the map (image), towers vector (towers), and radius, which is one-half the largest dimension of the map image. This sets the largest possible cell tower range; for example, if the tower's range is 1, then the circle representing the tower has a radius of radius pixels, which is half the height or width of the map image, whichever is larger. Internally, radii is a vector of tower radii in pixels ❶.

The Evaluate method first reshapes the particle position vector into (x, y) points, as we did for *circles.py*. In this case, we want pixel coordinates, so floor ensures that points are integer valued ❷.

The Collisions method first checks whether any proposed cell tower centers are in a forbidden region. This is a simple query against the map image. If the center pixel isn't 0, count it as a collision. If any collisions happen, return a score of 1.0, implying all pixels are uncovered.

Assuming no collisions, it's time to place the towers and calculate the score. An empty image the same size as the map image is created and passed to CoverageMap along with the tower centers (xy) and radii ❸. The return value, cover, is an image similar to the right side of Figure 5-3, but without proper scaling to $[0, 255]$—it is a floating-point array. If an element of cover is 0, that element isn't in the range of any tower, so we count it with NumPy's where function and divide that count by the number of pixels in the map to calculate the score.

The CoverageMap method is not of the Objective class because it's used elsewhere in *cell.py*. It is, however, critical to the success of the code, so let's walk through it in some detail (Listing 5-2).

```
def CoverageMap(image, xy, radii):
    im = image.copy()
    R,C = im.shape
❶   for k in range(len(radii)):
        x,y = xy[k]
❷       for i in range(x-radii[k],x+radii[k]):
            for j in range(y-radii[k],y+radii[k]):
                if ((i-x)**2 + (j-y)**2) <= (radii[k]*radii[k]):
                    if i < 0 or j < 0:
                        continue
                    if i >= R or j >= C:
                        continue
                    im[i,j] += 0.5*(k+1)/len(radii)
    imax = im.max()
❸   for k in range(len(radii)):
        x,y = xy[k]
        im[x,y] = 1.4*imax
    return im
```

Listing 5-2: Generating the coverage map for a set of tower positions

The CoverageMap method accepts the map image, tower center locations, and tower radii as input. Its goal is to fill in im. Passing an empty image to CoverageMap seems odd at present, but later calls to the function pass the map image itself.

Towers are applied in turn with center at (x, y) ❶. The (inefficient) double loop ❷ examines every pixel in the map that could be inside the range of the current tower. The body of the inner loop asks whether the current pixel, (i, j), is within the disk of the current tower (the if statement). If so, and the (i, j) pixel is within the space of the image, the current pixel value is incremented according to

$$0.5(k + 1)/n$$

where n is the number of towers. This equation increments the pixel value (im is a floating-point array) by an amount specific to each tower. The result leads to the right side of Figure 5-3, where each disk is a different intensity and overlaps are visible.

After all towers are added, a final loop over adds each tower's center point ❸. The value of the center point is always 1.4 times the intensity of the maximum pixel value to make the center points visible (best viewed on a computer screen). Since im is a floating-point array, it isn't restricted to $[0, 255]$. Scaling comes later in the code when writing output images to disk.

The CoverageMap method returns a two-dimensional array where any remaining zero values are pixels not covered by any tower. The number of zero values scaled by the number of pixels is the final score for the given tower locations.

The main body of *cell.py* follows *circles.py* in form: parse the command line, create framework objects, and do the search. However, instead of calling Optimize, the search is performed by repeated calls to Step so the current best score can be displayed per iteration.

The search is configured as in Listing 5-3.

```
rng = RE(kind=kind)
x,y = map_image.shape
lower = [0,0]*len(towers)
upper = [x,y]*len(towers)
b = Bounds(lower, upper, enforce="resample", rng=rng)
ndim = 2*len(towers)
w = x if (x>y) else y
radius = w//2
i = RandomInitializer(npart, ndim, bounds=b, rng=rng)
obj = Objective(map_image, towers, radius)
swarm = DE(obj=obj, npart=npart, ndim=ndim, init=i, bounds=b,
           max_iter=niter, tol=1e-9, rng=rng)
```

Listing 5-3: Configuring the search

Here radius sets the maximum radius for any tower.

The search itself is a loop (Listing 5-4).

```
k = 0
swarm.Initialize()
while (not swarm.Done()):
    swarm.Step()
    res = swarm.Results()
    t = "   %5d: gbest = %0.8f" % (k,res["gbest"][-1])
    print(t, flush=True)
    k += 1
res = swarm.Results()
```

Listing 5-4: Running the search

The Initialize method configures the swarm, Done returns True when the search is complete (all iterations done or tolerance met), and Step performs one update of the swarm (it acts like headquarters).

Every iteration calls the Results method to report on the current best value—the fraction of image pixels not covered by a tower. After the loop exits, the final call to Results returns the best set of tower locations. Additional code captures the per-iteration output and generates the output map for the swarm's best configuration. See the frames option on the command line.

Finally, the coverage map is generated and stored in the output directory, as shown in Listing 5-5.

```
p = res["gpos"][-1]
n = p.shape[0]//2
xy = p.astype("uint32").reshape((n,2))
radii = (towers*radius).astype("int32")
cover = CoverageMap(map_image, xy, radii)
c2 = (cover/cover.max())**(0.5)
c2 = c2/c2.max()
img = Image.fromarray((255*c2).astype("uint8"))
img.save(outdir+"/coverage.png")
```

Listing 5-5: Generating the coverage map

The swarm best position (p) is converted to a set of (x, y) points, which, along with the corresponding radii for the towers, are passed to CoverageMap along with the input map itself (map_image). Unlike the Evaluate method in the objective function, the map with masked regions is passed rather than an empty image.

The resulting coverage map (cover) is passed through a square root function to squash intensities before converting to a grayscale image (img) and writing to the output directory.

By necessity, we skipped code in *cell.py*, but a careful reading of the file will make those sections clear. Now let's see what *cell.py* can do.

Running *cell.py* without arguments shows us how to use it:

```
> python3 cell.py
cell <map> <towers> <npart> <niter> <alg> <kind> <outdir> [frames]

   <map>     -  map image (.png)
   <towers>  -  text file w/towers and ranges
   <npart>   -  number of swarm particles
   <niter>   -  number of swarm iterations
   <alg>     -  DE|RO|PSO|BARE|GWO|JAYA|GA
   <kind>    -  randomness source
   <outdir>  -  output directory (overwritten)
   frames    -  'frames' ==> output frame per iteration
```

Let's run the code using *map_01* and *towers0*. For example:

```
> python3 cell.py maps/map_01.png towers/towers0 20 100 ga pcg64 test frames
```

We're using the GA with 20 particles for 100 iterations and dumping the output to a directory called *test*. The frames keyword outputs the current best tower placement per iteration so we can trace the evolution of the swarm visually.

Notice that 100 iterations is not many compared to the 10,000 or more we used with *circles.py*. All the manipulation to generate the coverage map takes time, so running for 10,000 iterations is out of the question. Fortunately, we don't usually need more than a few hundred iterations.

As *cell.py* runs, it dumps the current swarm best score along with a summary when the search ends. The output directory contains this text in *README.txt* along with the original map image (*map.png*) and the final coverage map (*coverage.png*). The Python `pickle` file (*.pkl*) contains the swarm objects, should you wish to explore the evolution in more detail. The output directory also contains a *frames* directory holding images representing the swarm's best configuration by iteration. Page through these files to watch the swarm evolve.

My run produced a final coverage value of 0.358, meaning about 36 percent of the map wasn't covered by a tower, as seen in Figure 5-4.

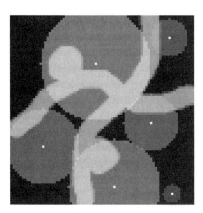

Figure 5-4: Placing towers

All six towers in *towers0* are visible in the output. The towers overlap only slightly, which is a good sign. Each tower center avoids masked regions.

To experiment with *cell.py*, run the shell scripts *go_tower_results*, *go_towers*, and *go_towers0*. The first applies *towers0* to all sample maps using all algorithms. The second applies all tower files and algorithms to *map_02*. Finally, the last one applies each tower file to *map_00* using only bare-bones PSO to illustrate how the swarm places towers when there are no obstacles.

Run *go_results* and *go_towers*, then run *make_results_plot.py* and *make _towers_plot.py* to produce an image file containing all the results with each row showing the output of a different swarm algorithm.

The final script, *go_towers0*, produces the output seen in Figure 5-5.

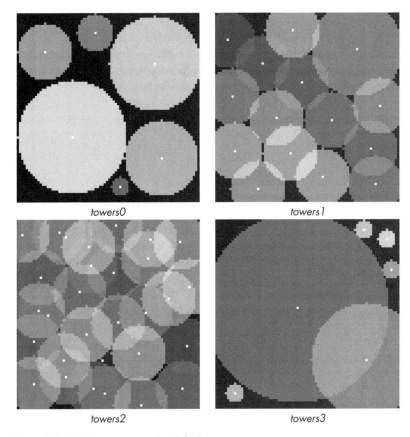

towers0

towers1

towers2

towers3

Figure 5-5: Placing towers on the default maps

It runs bare-bones PSO over the blank map for each tower file. Note that *towers0* isn't able to completely cover the map, but the resulting positions do not overlap, meaning bare-bones PSO found an optimal configuration (one of many). The output for *towers3* is much the same with the small towers not overlapping the larger ones.

Contrast these results with those for *towers1* and *towers2*. It's not immediately clear whether *towers1* is capable of completely covering the map, but *towers2* certainly is—yet tiny parts of the map remained uncovered. I suspect running for more than 300 iterations would take care of this. Does it? Is there a difference in runtime between using any map with masked regions and the blank map? If so, why might that be?

Feel free to experiment with different custom maps and numbers and sizes of towers to find whether there's an optimal or more utilitarian mix of tower sizes. What works best, many small towers or a few larger ones?

Let's use randomness to implement a "make it pretty" image filter.

Enhancing Images

When I view images I've taken with my phone in its gallery, I'm offered the option to remaster the picture; I call it the "make it pretty" filter. In this section, we'll use a swarm search to implement a "make it pretty" filter for grayscale images.

Our filter is based on one that has consistently appeared in academic literature. It's a good example of the tendency to take an existing paper, slightly alter the technique, and publish it as new. A quick review of the literature turned up eight papers all implementing this approach with only a tweak of the optimization algorithm: PSO, Firefly, Cuckoo search, DE, PSO, Cuckoo, Cuckoo, and DE, respectively. While our implementation is yet another in this illustrious line of research, I'm claiming the excuse of pedagogy and do not appeal to novelty or applicability.

Disclaimers aside, the "make it pretty" filter applies a local image enhancement function to an input grayscale image to make it look nicer. If that sentence is as clear as mud, have no worries—I'll explain.

We intend to apply a function to each pixel of the input image to produce a new output pixel. Let's apply the function

$$g'_{ij} = \frac{kG}{\sigma + b} \left(g_{ij} - m\mu\right) + \mu^a \tag{5.1}$$

to the pixel at row i and column j; that is, g_{ij}. We need to find a, b, m, and k to make the image look as nice as possible.

G is the original image's *mean intensity*, or the value found by adding all the pixel intensities and dividing by the number of pixels. The μ and σ variables are the mean (μ) and standard deviation (σ) of a 3×3 region around the current pixel, g_{ij}. Imagine a tic-tac-toe (naughts and crosses) board as in Figure 5-6.

$g_{i-1,j-1}$	$g_{i-1,j}$	$g_{i-1,j+1}$
$g_{i,j-1}$	g_{ij}	$g_{i,j+1}$
$g_{i+1,j-1}$	$g_{i+1,j}$	$g_{i+1,j+1}$

Figure 5-6: Pixel offsets

The 3×3 region slides over the image visiting each pixel, calculates μ and σ, and then uses Equation 5.1 to create a new pixel value, g'_{ij}. Note that g'_{ij} is the value of the ij pixel in the output image; it does not update the original image pixel, g_{ij}.

The function for updating image pixels has four parameters we need to find, along with other values that depend on the original image and how we apply the function. However, the function only tells us how to update the image for a given a, b, m, and k; it doesn't say anything about how nice the output image will appear to a human observer. For that, we need an objective function.

Researchers claim that the following function captures something of what makes an image look pleasant to humans:

$$F = \log(\log(I)) \left(\frac{\text{edgels}}{rc}\right) h$$

We'll use F as our objective function. The higher F is, the better the image will look to a human observer—at least, that's the theory. I is the sum of the pixel intensities of an edge-detected version of the input image. The "edgels" variable is the number of edges in the edge-detected version of the image above a threshold, here 20. Finally, r and c are the image dimensions (rows and columns) and h is the entropy of the image:

$$h = -\sum_i p_i \log_2 p_i$$

Here p_i is the probability of pixel intensity in bin i of a 64-bin histogram. Entropy, in this sense, refers to the information content of the image.

This exercise seems superficially similar to curve fitting. We have a function with parameters to be optimized, but the algorithm behind the application of the function has many more steps. However, as we'll soon learn, the extra effort means little to our swarm algorithms. They still provide floating-point vectors to the objective function and expect a scalar quality measure in return. The swarm is blissfully unaware of what it's optimizing.

The Enhancement Function

We have a four-parameter optimization problem: for a given image, we want to find the best a, b, m, and k values, all of which are floating-point numbers. From a framework perspective, the setup is straightforward once we decide on bounds for the parameters. All the cool code will be in the objective function class. The full program is in *enhance.py*.

We start with Listing 5-6, the objective function class, which implements both the elaborate image enhancement and F functions given previously.

```
class Objective:
    def __init__(self, img):
        self.img = img.copy()
        self.fcount = 0

    def F(self, dst):
        r,c = dst.shape
        Is = Image.fromarray(dst).filter(ImageFilter.FIND_EDGES)
        Is = np.array(Is)
        edgels = len(np.where(Is.ravel() > 20)[0])
        h = np.histogram(dst, bins=64)[0]
        p = h / h.sum()
        i = np.where(p != 0)[0]
        ent = -(p[i]*np.log2(p[i])).sum()
```

```
        F = np.log(np.log(Is.sum())))*(edgels/(r*c))*ent
        return F

    def Evaluate(self, p):
        self.fcount += 1
        a,b,m,k = p
        dst = ApplyEnhancement(self.img, a,b,m,k)
        return -self.F(dst)
```

Listing 5-6: The image enhancement objective function class

The class constructor stores a copy of the original image. The Evaluate method extracts *a*, *b*, *m*, and *k* from the supplied particle position, passing them and the original image to ApplyEnhancement to return a new image, dst. We pass the new image to the F method to calculate the score. Since we want to maximize *F*, we return the negative.

I'll explain ApplyEnhancement momentarily; let's focus on F for now. The method is linear and makes heavy use of powerful functions supplied by NumPy and PIL (Image and ImageFilter). Walking through, line by line, makes sense in this case.

First, we extract the image dimensions (r, c). In the next line, we apply an edge detection filter to the image producing Is from dst. The output of an edge detector looks like Figure 5-7.

Figure 5-7: An edge detector in action

We recast the PIL image, Is, as a NumPy array before using where to count the number of edge pixels greater than 20. We chose 20 because it's an empirically selected threshold that works well. The count is stored in edgels.

The only part of *F* yet to determine is the entropy, *h*, here given as ent. To get this, we first need the histogram of the image using 64 bins, conveniently acquired in a single line of code courtesy of NumPy's histogram function. Scaling the histogram by the sum of all bins converts from counts per bin to an estimate of the probability per bin, p.

With p in hand, we calculate the entropy by summing the probabilities multiplied by the log, base 2, of the probabilities (ent). The penultimate line in F is a direct analog of the equation for *F*, the value of which is returned.

Now for `ApplyEnhancement` in Listing 5-7.

```
def ApplyEnhancement(g, a,b,c,k):
    def stats(g,i,j):
        rlo = max(i-1,0); rhi = min(i+1,g.shape[0])
        clo = max(j-1,0); chi = min(j+1,g.shape[1])
        v = g[rlo:rhi,clo:chi].ravel()
        if len(v) < 3:
            return v[0],1.0
        return v.mean(), v.std(ddof=1)

    rows,cols = g.shape
    dst = np.zeros((rows,cols))
    G = g.mean()
    for i in range(rows):
        for j in range(cols):
            m,s = stats(g,i,j)
            dst[i,j] = ((k*G)/(s+b))*(g[i,j]-c*m)+m**a
    dmin = dst.min()
    dmax = dst.max()
    return (255*(dst - dmin) / (dmax - dmin)).astype("uint8")
```

Listing 5-7: Applying a set of parameters to an image

We enhance the original image by applying Equation 5.1. The output image (dst) is constructed, pixel by pixel, using the local 3×3 region mean (m) and standard deviation (s) in conjunction with the arguments a, b, c, and k. Note that c is m in Equation 5.1.

The helper function, stats, defines the 3×3 region around (i,j), accounting for the image index limits. It then returns the mean and standard deviation. The if catches the edge case where too few pixels exist for a meaningful standard deviation calculation. Notice the ddof keyword in the call to std. By default, NumPy calculates the biased estimator of the variance by dividing by the number of values instead of the unbiased estimate found by dividing by one less than the number of values. Many statistics packages use the unbiased estimator by default. In most cases, especially if there are < 20 values in the dataset, we want the unbiased estimator, so we set ddof=1. Recall that the standard deviation is the square root of the variance.

All that remains is to configure the search, as shown in Listing 5-8.

```
orig = np.array(Image.open(src).convert("L"))
img = orig / 256.0
ndim = 4
rng = RE(kind=kind)
b = Bounds([0.0,1.0,0.0,0.5], [1.5,22,1.0,1.5], enforce="resample", rng=rng)
i = RandomInitializer(npart, ndim, bounds=b, rng=rng)
```

```
obj = Objective(img)
swarm = GWO(obj=obj, npart=npart, ndim=ndim, init=i, bounds=b, max_iter=niter, rng=rng)
```

Listing 5-8: Configuring the search

The configuration follows our framework: randomness source (rng), bounds (b), initializer (i), objective function (obj), and a swarm object, here GWO. We force the input image (orig) to grayscale (convert) and scale it by 256 to be in the range $[0, 1)$ (img). It's not uncommon to manipulate images in this range instead of $[0, 255]$. After manipulation, the image is scaled to $[0, 255]$ and converted to an integer type before writing it to disk.

The search is four dimensional (ndim), so there are four bounds. The bounds vary by dimension. The limits $a \in [0, 1.5]$, $b \in [1.0, 22]$, $m \in [0, 1]$, and $k \in [0.5, 1.5]$ are based on values used in the literature. As we'll see, they appear to work well, but try experimenting with them, especially if you notice output values near the limits. We'll soon learn how to find these values after a search.

Running the search is as simple as calling Optimize on the swarm object, but we want to track the F score as we go, so we'll loop manually instead (Listing 5-9).

```
k = 0
swarm.Initialize()
while (not swarm.Done()):
    swarm.Step()
    res = swarm.Results()
    t = "    %5d: gbest = %0.8f" % (k,res["gbest"][-1])
    print(t, flush=True)
    s += t+"\n"
    k += 1
res = swarm.Results()
pickle.dump(res, open(outdir+"/results.pkl","wb"))
a,b,m,k = res["gpos"][-1]
dst = ApplyEnhancement(img, a,b,m,k)
Image.fromarray(dst).save(outdir+"/enhanced.png")
Image.fromarray(orig).save(outdir+"/original.png")
```

Listing 5-9: Running the search

The search ends after all specified iterations. We then dump the final results (res) to the output directory via pickle. Use the gpos key to return the final set of parameters. To conclude, the image is enhanced with the best set and written to the output directory along with the original image for comparison purposes.

Does *enhance.py* work? Let's find out.

The Code

Run *enhance.py* without arguments to learn what it expects on the command line:

```
> python3 enhance.py

enhance <src> <npart> <niter> <alg> <kind> <output>

   <src>    - source grayscale image
   <npart>  - number of particles
   <niter>  - number of iterations
   <alg>    - BARE,RO,DE,PSO,JAYA,GWO,GA
   <kind>   - randomness source
   <output> - output directory (overwritten)
```

We need to supply the original image, swarm size, iterations, algorithm type, randomness source, and output directory name.

The *images* directory contains a set of 128×128-pixel grayscale images that we'll use for our experiments. The search isn't particularly fast, given the sequential nature of the framework and the extensive image manipulations each call to the objective function entails, so smaller images work best. The program will work with larger images, which need not be square; all that's required is patience.

Give this command line a try:

```
> python3 enhance.py images/barbara.png 10 60 gwo minstd babs
  0: gbest = -4.77187094
  1: gbest = -4.80063898
  2: gbest = -5.09058855
  3: gbest = -5.09058855
--snip--
```

The output shows the current swarm best F score by iteration. The value is negative because we want to maximize F. The command line specified GWO with a swarm of 10 particles, 60 iterations, and an output directory named *babs*.

On my system, the search finishes with this output:

```
59: gbest = -6.09797382

Search results: GWO, 10 particles, 60 iterations

Optimization minimum -6.09797382 (time = 216.419)
(14 best updates, 610 function evaluations)
```

Therefore, the best set of parameters led to $F = 6.09797$ after 14 swarm best updates. The *babs* output directory contains

```
enhanced.png
original.png
README.txt
results.pkl
```

giving us the enhanced image, the original image, the pickled results, and a README file containing all output generated during the search.

The enhanced image should look sharper with better contrast than the original. Unfortunately, a printed version will likely not show the differences clearly. Nonetheless, Figure 5-8 shows both images, with the original on the left and the enhanced version on the right.

Figure 5-8: The original and enhanced images

Look, in particular, at the books in the bookcase. They are better defined and show improved contrast.

To extract the enhancement parameters, load the *results.pkl* file

```
> python3
>>> import numpy as np; import pickle
>>> res = pickle.load(open("babs/results.pkl","rb"))
>>> res["gpos"][-1]
array([0.01867829, 1.00785356, 0.45469097, 1.1731131 ])
```

which tells us that $a = 0.01867$, $b = 1.0078$, $m = 0.45469$, and $k = 1.17311$. We haven't used pickle before; it requires a file object (the output of open) and must use binary mode ("rb").

There are nine images in the *images* directory. We'll run various swarm and evolutionary algorithms against these images, and then collect the resulting output to produce composite images showing the original and the enhanced versions so we might rate each algorithm's performance. To that end, I created two Python scripts: *process_images.py* and *merge_images.py*.

Run *process_images.py* first. I recommend starting it in the evening and returning in the morning. The script processes every image in *images* using each swarm algorithm. The swarm consists of 10 particles and runs for 75 iterations in all cases.

When *process_images.py* finishes, use *merge_images.py* to produce composite images showing the results, which are in the *output* directory. For example

```
> python3 merge_images.py zelda zelda_results.png
```

creates *zelda_results.png*, as in Figure 5-9. Canonical PSO looks like the winner here.

Figure 5-9: A composite image. From top left: original, bare-bones PSO, DE, GA, GWO, Jaya, canonical PSO, RO.

Table 5-3 lists the *F* scores and parameters for each algorithm.

Table 5-3: *F* Scores and Parameters for Each Algorithm

	F	a	b	m	k
GWO	4.59328	0.00030	2.21893	0.71926	0.99232
Canonical PSO	5.07767	0.00585	1.50612	0.39545	1.46734
Jaya	4.31448	1.21667	1.00493	0.76448	1.47750
DE	4.29245	1.24982	1.00076	0.76535	1.46355
Bare-bones PSO	4.27962	1.09779	1.01205	0.89588	1.49625
GA	4.10057	1.14040	1.64773	0.12196	1.19704
RO	4.01708	1.08478	5.07904	0.53249	0.97552

The subjectively best-looking image is also the image with the largest *F* score—a good sign. The GWO image, which has low contrast but is quite sharp, has the second-highest *F* score. The GWO parameters are also quite different from the others. It's possible that the enhancement function parameter space has a fairly complex structure, and there are multiple local minima. For the Zelda image, the GWO algorithm seems to have landed apart from the others. Does this happen with the other images? Which algorithm seems to work best overall?

The program *F.py* applies a specific set of parameters to an image. What happens if we apply a from the GWO result and b, m, and k from the canonical PSO result?

```
> python3 F.py images/zelda.png zelda2.png 0.00030 1.50612 0.39545 1.46734
F = 5.28999404
```

The new F score is higher still, and the output file, *zelda2.png*, looks even better than the canonical PSO result.

There is much more to explore here. I'll provide a few suggestions in "Exercises" on page 169, including a way to (possibly) enhance color images. For the time being, let's move on to another experiment, which combines optimization with simulation to maximize profit by figuring out how to best arrange the products in a grocery store.

Arranging a Grocery Store

Have you noticed that grocery stores usually put milk in the back, as far from the entrance as possible? Or that candy is located right at the front, near the checkouts? The placement of products in a grocery store is not an accident; it's intentional to maximize profit. Many people stop by the store to pick up necessities, like milk, and often pick up something else if they run across it, like candy. Grocery stores arrange the products to maximize such occurrences and increase revenue.

This section attempts to replicate such an arrangement of products to validate or refute common store practice. This experiment combines optimization with simulation. We'll optimize product placement in the store using a set of simulated shoppers to evaluate the arrangement. Our goal is an arrangement of products, our objective function score is revenue over a day, and the function itself is a simulation of several hundred shoppers. Randomness is everywhere, from swarm initialization and position updates to the collection of shoppers and their habits. To jump in, read through *store.py*.

The Environment

Let's define our operating environment. Actual stores are essentially two dimensional; there is a layout over some floor space. Our framework uses position vectors, one-dimensional entities. We'll make the store one-dimensional as well, so a position vector can be a store layout with each element a product. Shoppers will enter the store on the left (index 0) and progress through the store to the right, like in Figure 5-10.

Figure 5-10: Shopping at a one-dimensional grocery store
(Illustration by Joseph Kneusel)

People typically go to the store to pick up a specific product; we'll call this the *target product*. The shoppers will also have impulse products that they'll buy if they encounter them before the target product.

For example, Figure 5-10 shows two shoppers who are thinking of a target product (signified by the question mark) and an impulse product (the exclamation point).

The shopper on the left is searching for the diamond product, but will purchase the circle product if they see it. Since they encounter the diamond before the circle, they buy only the diamond product and leave the store.

The shopper on the right is searching for the triangle product, but they'll buy the square product if they come across it. They find the square while searching for the triangle and purchase both.

The shopping simulation requires products, located in the *products.pkl* file. The file contains three lists, each of 24 elements: counts, names, and prices, in that order. The data comes from an actual collection of products purchased over some period. The products are stored in decreasing purchase frequency, so the item most often purchased is first, and the least often purchased is last.

We convert the counts to a purchase probability by dividing each by the sum of all the counts. We'll use the purchase probability in our objective function.

If there are 24 products in the store, we have 24-dimensional position vectors. We're searching for the best ordering of products to maximize daily revenue. We'll visit the simulation part in more detail later, but for the moment, let's focus on the product order and how we'll represent it in a swarm.

At first, we might think to make the position vectors discrete values in $[0, 23]$, where each number is a product, an index into the list of products read from *products.pkl*. However, I implemented an alternate approach using position vectors in $[0, 1)$.

Ultimately, we want a vector that places products in a particular order, some permutation of the vector $\{0, 1, 2, 3, \ldots, 23\}$. The trick is to abstract this permutation so we can still use continuous floating-point values in $[0, 1)$. Instead of using the product numbers directly, we pass each position vector to NumPy's argsort function, which returns the order of indices that would sort the vector. For a vector of 24 elements, the output of argsort is a permutation of the numbers 0 through 23.

That this approach works—and we'll see that it does—is impressive. We're asking the swarms to generate a collection of floating-point numbers, [0, 1), that are useful only when the additional operation of determining their sort order has happened. The reason it works likely has to do with the fact that using integer values requires some kind of truncation or rounding of a floating-point number where the implemented approach uses the numbers as they are. If changing a particular element of the particle position from 0.304 to 0.288 alters the sort order of the entire vector to a more profitable configuration, the swarm makes use of that change, whereas truncation might call both numbers 0.

Here are the steps we need to implement:

1. Initialize the swarm with 24-element position vectors in [0, 1).

2. Initialize a collection of randomly generated shoppers.

3. Run our usual swarm search where each position vector is evaluated by passing shoppers through the store using the sort order of the current vector as the arrangement of products. Then, tally the amount of money each shopper spends. They'll always find their target product, but may not find their impulse products. Finally, return the negative of the total spent by the shoppers, as we want to maximize daily revenue.

4. Let the swarm algorithm update positions as usual until all iterations have run.

5. Report the sort order of the best position found as the "ideal" ordering of the products.

The following two sections detail how to implement shoppers and how the objective function works. With those in mind, we'll be ready to go shopping and see if our simulation agrees with grocery industry experts.

The Shoppers

A shopper is an instance of the *Shopper* class, as shown in Listing 5-10.

```
class Shopper:
    def __init__(self, fi, pv, rng):
        self.item_values = pv
      ❶ self.target = Select(fi,rng)
      ❷ self.impulse = np.argsort(rng.random(len(fi)))[:3]
        while (self.target in self.impulse):
            self.impulse = np.argsort(rng.random(len(fi)))[:3]

    def GoShopping(self, products):
        spent = 0.0
        for p in products:
```

```
    if (p == self.target):
        spent += self.item_values[p]
    ❸ break
    if (p in self.impulse):
        spent += self.item_values[p]
return spent
```

Listing 5-10: The Shopper *class*

The constructor configures the shopper by selecting the target (target) and impulse (impulse) products. It also keeps a copy of the product prices (item_values) for when it's time to go shopping.

A call to Select returns the target product, which is an index into the product list ❶. The Select method takes advantage of the fact that fi is the probability of a product being purchased in decreasing order (Listing 5-11).

```
def Select(fi, rng):
    t = rng.random()
    c = 0.0
    for i in range(len(fi)):
        c += fi[i]
        if (c >= t):
            return i
```

Listing 5-11: Selecting a product according to its purchase frequency

We select a random value, $[0, 1)$ (t). We then add successive probabilities for each product to c until equaling or exceeding t. When that happens, as it must because $t < 1$ and the sum of all product probabilities is 1.0, the current product index is returned (i).

Let's see an example to clarify how Select works. Imagine there are five products, each selected with probability:

$$0.5, 0.3, 0.1, 0.07, 0.03$$

This means product 0 is purchased about 50 percent of the time while product 4 is purchased only 3 percent of the time. The sum is 1.0, or 100 percent. Now, pick a random value in $t \in [0, 1)$. This will be less than 0.5 half of the time, meaning Select will return index 0. The sum of the first two product probabilities is $0.5 + 0.3 = 0.8$. But half the time $t < 0.5$, so the difference between 0.5 and 0.8 is the fraction of the time $0.5 > t \leq 0.8$: 30 percent of the time. Similarly, 10 percent of the time $0.8 > t \leq 0.9$, 7 percent of the time $0.9 > t \leq 0.97$, and 3 percent of the time $0.97 > t \leq 1.0$. Therefore, the index returned by Select reflects the true purchase probability for the item.

Figure 5-10 shows a single impulse purchase product. In reality, the simulation selects three unique impulse products that aren't the target product ❷. The call to argsort returns a permutation of the product indices, so keeping the first three ensures unique products. The while loop repeats the process, if necessary, to ensure that the target isn't one of the impulse purchases.

When evaluating a particle position, we call the GoShopping method. It's passed a list of products, the sort order for the current particle. It then walks through the list, checking if the current product is the target or one of the impulse buys. If it is either, the method adds the price to spent to indicate that the shopper purchased the item. If the item is the target, the loop exits, and any unencountered impulse products are ignored ❸. The method then returns the total spent.

The Shopper class represents a single shopper. The Objective class manages a collection of shoppers.

The Objective Function

The Objective class evaluates a single particle position, or a configuration of products, as shown in Listing 5-12.

```
class Objective:
    def __init__(self, nshoppers, pci, pv, rng):
        self.nshoppers = nshoppers
        self.fcount = 0
        self.shoppers = []
        for i in range(nshoppers):
            shopper = Shopper(pci, pv, rng)
            self.shoppers.append(shopper)

    def Evaluate(self, p):
        self.fcount += 1
❶      order = np.argsort(p)
        revenue = 0.0
        for i in range(self.nshoppers):
            revenue += self.shoppers[i].GoShopping(order)
        return -revenue
```

Listing 5-12: The Objective function class

The constructor builds a list of randomly initialized shoppers, meaning we use the same collection of shoppers for the entire simulation. Here, pci is the probability of each product being selected, most probable first, and pv is the associated price.

The Evaluate method receives a single particle position; however, we aren't interested in p's values, only the order in which they need to be moved to sort them ❶. This is the product order GoShopping uses to determine how much money the shopper spends. To get the total revenue, then, each shopper is asked to go shopping while tallying the amount of money spent (revenue). The objective function value is the negative of this amount (to maximize).

The remainder of *store.py* loads the products and then parses the command line:

```
products = pickle.load(open("products.pkl","rb"))
nshoppers = int(sys.argv[1])
npart = int(sys.argv[2])
niter = int(sys.argv[3])
alg = sys.argv[4].upper()
kind = sys.argv[5]
```

The code then creates the list of purchase probabilities (pci)

```
ci = products[0]   # product counts
ni = products[1]   # product names
pv = products[2]   # product values
pci = ci / ci.sum()   # probability of being purchased
N = len(ci)           # number of products
```

before initializing the swarm and running the search:

```
ndim = len(ci)
rng = RE(kind=kind)
b = Bounds([0]*ndim, [1]*ndim, enforce="resample", rng=rng)
i = RandomInitializer(npart, ndim, bounds=b, rng=rng)
obj = Objective(nshoppers, pci, pv, rng)
swarm = Jaya(obj=obj, npart=npart, ndim=ndim, init=i, max_iter=niter, bounds=b, rng=rng)
swarm.Optimize()
res = swarm.Results()
```

The remainder of the file generates a report showing how successful the search was.

The Shopping Simulation

Enough prep; let's run and see what output we get:

```
> python3 store.py 250 20 200 pso mt19937
Maximum daily revenue $1114.28 (time 38.440 seconds)
(25 best updates, 4020 function evaluations)

Product order:
        cream cheese  ( 2.3%) ($1.57)
             berries  ( 1.9%) ($1.98)
      misc. beverages ( 1.6%) ($2.37)
               candy  ( 1.7%) ($2.23)
             chicken  ( 2.4%) ($1.49)
                beef  ( 3.0%) ($1.35)
             dessert  ( 2.1%) ($1.76)
              onions  ( 1.8%) ($2.10)
              coffee  ( 3.3%) ($1.23)
```

```
       salty snack  ( 2.1%) ($1.66)
            apples  ( 1.9%) ($1.87)
            butter  ( 3.1%) ($1.29)
         chocolate  ( 2.8%) ($1.42)
        frankfurter ( 3.4%) ($1.19)
    root vegetables ( 6.2%) ($1.01)
      shopping bags ( 5.6%) ($1.02)
        canned beer ( 4.4%) ($1.08)
        brown bread ( 3.7%) ($1.15)
       bottled beer ( 4.6%) ($1.06)
 fruit/vegetable juice ( 4.1%) ($1.11)
            pastry  ( 5.1%) ($1.04)
            yogurt  ( 7.9%) ($1.01)
         rolls/buns (10.5%) ($1.00)
         whole milk (14.5%) ($1.00)


milk rank = 23
candy rank = 3

Upper half median probability of being selected =  2.1
                        median product value = 1.71
Lower half median probability of being selected =  4.8
                        median product value = 1.05
```

The code expects the number of shoppers to simulate (250), the number of particles (20) and iterations (200), the algorithm (pso), and a randomness source (mt19937). The output prints to the screen.

First, we're told this run generated $1,114.28 as the maximum daily revenue. The product order is as given, where the first product is at the front of the store, here cream cheese (strangely). The order provides the product name, the probability of purchasing it, and the price.

Conventional wisdom says to put the milk at the back of the store and the candy at the front. In this case, milk ended up as product 23, which is at the back of the store, while candy was product 3, very close to the front. This run followed conventional wisdom—a good sign.

The remainder of the output gives the median purchase probability for products in the first half of the store along with the median price for those products, and then again for products in the second half of the store. If the swarm is ordering the store along the lines we expect it to, then lower probability items that cost more will appear in the front part of the store (the top half of the product list). At the same time, higher probability items that generally cost less will appear toward the back of the store. This is precisely what we see in the output: items toward the back are more likely to be purchased and, in general, cost less.

Overall, the search produced reasonable output validating conventional wisdom.

The script *go_store* runs 10 searches for each algorithm capturing the results in the *output* directory. Run it with

```
> sh go_store
```

then follow with *process_results.py*:

```
> python3 process_results.py
```

This should produce a *results* directory that contains NumPy files (*.npy*) holding the milk and candy rankings for each algorithm and run, along with the best revenue for each run by algorithm. Also included is a plot tracking the milk and candy rankings across runs for each algorithm; see Figure 5-11.

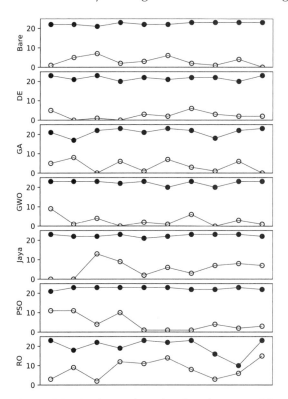

Figure 5-11: Product rankings by algorithm across 10 runs, in which solid circles are milk and open circles are candy

In the figure, solid circles represent where milk ended up in the rankings, and open circles are for candy. Almost all algorithms were able to put milk near the very back of the store, with both PSO variants and Jaya perhaps the most consistent (for this single run of *go_store*). An obvious exception is RO. While it managed to put milk behind candy on every run, at times the placement wasn't particularly great. For example, in run 9, milk and candy were almost next to each other.

The *process_results.py* code also outputs a summary. For example:

```
Mean revenue by algorithm:     t-test, best vs rest:
    Bare: $1148.59 ( 5.09)        Bare vs   DE: 0.05296
      DE: $1127.86 ( 8.62)        Bare vs   GA: 0.00017
      GA: $1112.06 ( 5.81)        Bare vs  GWO: 0.36133
     GWO: $1141.01 ( 6.30)        Bare vs Jaya: 0.04042
    Jaya: $1130.88 ( 6.20)        Bare vs  PSO: 0.01244
     PSO: $1125.73 ( 6.47)        Bare vs   RO: 0.00000
      RO: $1023.75 (10.03)
```

The first part shows the mean revenue across the 10 runs by algorithm. The standard error of the mean (standard deviation divided by the square root of the number of samples, here 10) is in parentheses. Bare-bones PSO is the winner, averaging a daily revenue of $1,148.59 compared with RO's meager performance of $1,023.75.

The right-hand part of the output requires some explanation. I wanted to compare the revenue across runs by algorithm. The code locates the highest performing algorithm, bare-bones PSO in this case, and runs a t-test with it against the others. A *t-test* is a hypothesis test that asks whether two datasets are plausibly from the same data generating process. The value shown is the p-value, the probability of the observed difference in the means and standard deviations of the two datasets (or greater), given they are from the same data generating process. If the p-value is high, then the two datasets are likely from the same data generating process, meaning the test's null hypothesis is likely valid. In this case, the bare-bones PSO result is not meaningfully different from the GWO result, because the p-value is 0.36.

The smaller the p-value, the more likely that the results aren't from the same data generating process. For RO and GA, the p-values are very low, giving us confidence that the bare-bones PSO result is better. However, the other p-values are also low. So, is bare-bones PSO head and shoulders above all the others for this task, or was it just lucky this time around?

To find out, I ran *go_store* five more times and accumulated the output of *process_results.py* in the *results_per_run.txt* file. The best-performing algorithm varied across runs, but there were trends. For three of the six runs, Jaya was the top performer; for two, it was bare-bones PSO; and for one, GWO. The GA and RO results were always the worst. Looking at the p-values from the t-tests, Jaya, bare-bones PSO, and GWO are good algorithms to use for this task, with likely no meaningful difference between Jaya and bare-bones PSO.

The products in *products.pkl* are arranged in decreasing order of purchase probability but increasing order of price, meaning the least likely product to be purchased is the most expensive and vice versa. Therefore, we might expect to maximize profit by arranging the products so the least likely but most expensive is first and the most likely, least costly is last, at the back of the store. The order generated by a run of *store.py* is the swarm's attempt to meet this ideal.

Run *product_order.py*, passing it an output directory (*output* created by *go_store*) followed by another directory name, like *orders*. You'll generate a collection of plots, one for each algorithm, showing the mean value of each product slot, from 0 through 23, for all 10 runs of each algorithm. Also plotted is the curve reflecting the ideal product ordering, the reverse of the product order in *products.pkl*.

Figure 5-12 shows the mean product cost by position in the store over 10 runs of each algorithm compared to the ideal ordering (smooth curve).

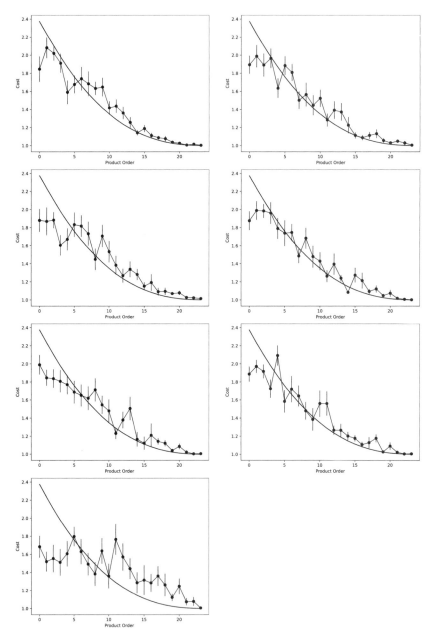

Figure 5-12: Comparing the mean swarm value by product order to the ideal. From top left to right: bare-bones PSO, DE, GA, GWO, Jaya, canonical PSO, RO.

First, note that in no way did the swarm search seek to match the ideal sequence. Instead, any match is an emergent effect of the swarms' attempts to maximize daily profit.

Second, all algorithms except RO were effective at matching the order of the cheapest products. We also see this in Figure 5-11 by the consistency of milk placement compared to candy. Most of the variation between algorithms is in the order of the first few products. We also see this in the larger error bars for products near the front of the store compared to those at the back.

The placement of products near the front of the store might be more difficult because those products are the least often purchased. Both barebones PSO and Jaya did reasonably well with more expensive products, but one could argue that GWO more closely matched the ideal curve. This tells us that the differences between algorithms are subtle, at least when viewed in this fashion, though GA was nearly as poor at matching the early product order as RO.

Exercises

Had enough of swarm algorithms? Me neither. Here's more to explore and contemplate:

- *Circles.py* used `enforce="clip"` to pack circles. Change this to `enforce="resample"`. If the results are suddenly different, why?

- Packing circles in a square is a similar problem to packing spheres in a cube. Modify a copy of *circles.py* to pack spheres in a cube, making a 2D problem a 3D problem. Run your code and compare it to the numbers in *sphere_dmin.png*. If you get stuck, take a peek at *spheres.py*.

- Placing cell towers on an empty map returns (primarily) non-overlapping towers. What sort of output do you get from *cell.py* if the map is particularly busy with relatively few allowed locations? You can make your own map or give *map_busy.png* in the *maps* directory a try. Can the swarm algorithms find places for the towers?

- The *enhance.py* file manipulates grayscale images. Modify it to enhance RGB images instead. A crude approach is in the file *process_rgb_images.py* in the *rgb* directory. This directory also contains some RGB images (*original*). Does *process_rgb_images.py* consistently produce good results? Why? Implement a new version that doesn't enhance channels independently, but instead seeks a set of parameters that work best across all channels, perhaps by summing *F* per channel.

- Modify *enhance.py* to use `ddof=0` instead of `ddof=1`—to use the biased variance and not the unbiased. Do you notice any difference in the results?

- The grocery store simulation used the same collection of shoppers for each iteration of the swarm. What happens to the results if the collection of shoppers is regenerated before each iteration? Do you expect this to make a difference?

- The simulation results presented in the grocery store simulation used 250 shoppers. What happens to my claim that Jaya and bare-bones PSO are well suited to the task if there are only five shoppers? Ten? Fifty?

Summary

This chapter continued our exploration of swarm optimization algorithms. We learned how to pack circles in squares, place cell towers while avoiding restricted locations, implement a "make it pretty" filter, and mix optimization and simulation to develop a product placement plan for a grocery store.

We did all this using the same collection of algorithms. We didn't change a single swarm intelligence or evolutionary algorithm to adapt it to the problem. Instead, casting the problem in the proper form enabled the direct application of the algorithms. This is a powerful ability that's widely applicable. Many processes in the real world are, in the end, optimization problems, meaning swarm algorithms likely have a role to play. They are general-purpose algorithms, as are many machine learning algorithms to which we now turn.

The experiments in this chapter and the previous one introduced a powerful approach to general optimization problems. If we can cast the problem as finding the best position in a multidimensional space where each point represents a possible solution, then swarm intelligence and evolutionary algorithms are likely applicable. I can't emphasize enough the usefulness of this concept.

We used a simple framework supporting a handful of standard swarm algorithms, of which there are hundreds to choose from—though not all are created equal. We designed the framework to be easy to use and pedagogical instead of performant. The framework acts as a stepping stone to more sophisticated tools, should you often return to swarm algorithms. If that's the case, then consider exploring more advanced toolkits like these:

inspyred   *https://pythonhosted.org/inspyred*

pyswarms   *https://github.com/ljvmiranda921/pyswarms*

DEAP   *https://github.com/DEAP/deap*

These toolkits support multiple swarm algorithms, both swarm intelligence and evolutionary, and are designed for performance. Toolkits for other languages include:

Java   *https://cs.gmu.edu/~eclab/projects/ecj*

C++   *https://eodev.sourceforge.net*

Swarm algorithms will make another appearance in Chapter 7, but for now, we'll explore randomness in the world of artificial intelligence.

6

MACHINE LEARNING

The goal of machine learning is to train models to generate correct outputs when given previously unseen inputs. This is usually done by repeatedly presenting the model with a collection of known inputs and outputs until the model succeeds in properly assigning outputs to inputs, or *learns*.

In this chapter, we'll explore randomness in machine learning by building two datasets for histology slides and images of handwritten digits. As we'll learn, randomness is essential to building suitable machine learning datasets.

Next, we'll explore randomness in neural networks—the driving force behind the AI revolution. We'll restrict ourselves to traditional neural network architectures; randomness is just as important, if not more so, when working with advanced models.

After neural networks come *extreme learning machines*, simple neural networks that fundamentally depend on randomness. Unlike their grown-up cousins, extreme learning machines don't require extensive training, but instead rely on the power of randomness to do most of the learning for them.

Random forests close out the chapter, and are also critically dependent on randomness for their success.

I'll point out where randomness appears as we move through the chapter. Randomness is central to the success of machine learning, from your favorite smart speaker to the self-driving car you may (someday soon) be riding in.

Datasets

In machine learning, we train models from sample data. Therefore, we must build datasets before beginning our explorations. Randomness plays a critical role in this process.

We'll build two datasets. The first consists of measurements of cells from histology slides that hopefully enable the model to learn whether the tissue sample is benign (class 0) or malignant (class 1).

The second dataset consists of 28×28-pixel grayscale images of handwritten digits: 1, 4, 7, and 9. The images aren't stored in the usual format; rather, they're unraveled into vectors where the first row is followed by the second, and so on, to map the 28×28 pixels to 784-element vectors.

Histology Slide Data

Machine learning etiquette dictates that training a model requires a minimum of two datasets. The first is the *training set*, a collection of pairs, (x, y), where x are input vectors and y are the corresponding output labels. The second is a *test set*, of the same kind as the training set, but it's not used until training is complete. The model's performance on the test set decides how well it's learned.

The *raw* directory holds the *bc_data.npy* and *bc_labels.npy* files. The first is a dataset that contains a two-dimensional NumPy array of 569 rows and 30 columns. Each row is a *sample*, and each column a *feature*. The 30 elements of a sample represent 10 measurements of three different cells on the histology slide. The second file contains the label, 0 for benign and 1 for malignant. There's a one-to-one correspondence between the rows of the data and the labels. Therefore, row 0 of *bc_data.npy* represents features from a benign sample, while row 2 has features from a malignant sample because the first element of the vector in *bc_labels.npy* is a 0, and the third is a 1.

We'll build two datasets from the 569 samples, using a 70/30 split, meaning 70 percent of the samples are for training (398) and the remaining 30 percent for testing (171).

As machine learning models are notoriously slow to learn, we should be concerned that 398 samples are not enough to condition the model. We want more data, but don't have any more; what are we to do?

Randomness comes to our aid. We can *augment* the data by creating fake samples that, plausibly, come from the same source as our training data. We'll apply random alterations to the existing data—enough to make it different, but not so much that the labels are no longer accurate. Data augmentation is a powerful part of modern machine learning that helps models learn not to pay too much attention to the particulars of the training set, but instead seek more general characteristics that differentiate between the classes.

Before we augment the training samples, we need to standardize the data. Many machine learning models have difficulty with features that have different ranges. For example, one feature might sit in the range $[0, 2]$ while another uses $[-30,000, 30,000]$. To bring both features into the same relative range, we subtract the mean value of each feature and then divide by the standard deviation of the features. After this transformation, each feature has a mean value close to zero and a standard deviation of one.

We have standardized features that we've split into two disjoint groups, one for training and one for testing. We're now ready to augment the training data by employing *principal component analysis (PCA)*. If we were able to plot the data in 30 dimensions, we'd see that it's spread out in some directions more than others. PCA finds these directions and, in effect, rotates the 30-dimensional coordinate system so that the first coordinate aligns with the direction where there's the most variation in the data, then the second coordinate with the next, and so on. This means that later coordinates are less important in representing the data (though perhaps not in distinguishing between classes). We'll take advantage of this decreasing importance in directions to randomly alter the coordinate directions, producing training data similar to the original but not identical. By making small changes, we can be (reasonably) confident that the new data represents an instance of the original class.

The code we need is in *build_bc_data.py*. Let's walk through the important bits, beginning with loading the raw data and separating it into train and test (Listing 6-1).

```
np.random.seed(8675309)
x = np.load("raw/bc_data.npy")
y = np.load("raw/bc_labels.npy")
❶ x = (x - x.mean(axis=0)) / x.std(ddof=1,axis=0)
i = np.argsort(np.random.random(len(y)))
x = x[i]
y = y[i]
n = int(0.7*len(y))
xtrn = x[:n]
ytrn = y[:n]
xtst = x[n:]
ytst = y[n:]
```

Listing 6-1: Splitting the raw histology data

First, we fix NumPy's pseudorandom number seed so the same dataset is built each time the code is run. Generally, altering NumPy's seed this way is not a good idea, as it affects *all* code using NumPy, even inside other modules (like the scikit-learn modules we'll use later in the chapter). However, in this case, we'll take the risk.

Next, we load the raw data and labels before standardizing ❶. We want the mean value of each feature. The columns of x are the features, requiring the use of axis=0. This keyword applies the function, mean, across the rows of x, thereby delivering a 30-element vector where each element is the mean of the corresponding column of x.

We subtract this mean from each sample, or each row of x. With NumPy's broadcast rules, we can do this automatically, with no looping required. NumPy is smart enough to see that we're attempting to subtract a 30-element vector from a 2D array where the second dimension is 30, so it performs the subtraction and repeats it for each row.

Next, we divide the mean-subtracted data by the standard deviation of each feature. Again, axis=0 lets us apply the function across the rows. Because of NumPy's broadcast rules, the division is applied down the rows of x to produce the final, standardized dataset.

The next three lines randomize the dataset by assigning i to a random permutation of the numbers 0 through 568. NumPy's argsort function doesn't sort a vector, but instead returns the sequence of indices that would sort it. The following two lines apply this permutation to both x and y to scramble the data and the labels in sync.

The final five lines split the raw dataset into training sets (xtrn, ytrn) and testing sets (xtst, ytst). Note that we split the dataset before augmenting; if we split it after, augmented versions of a sample will likely end up in the test set, thereby making the model seem better than it is.

We're now in a position to augment the data. First, we need to learn the principal components of the raw data, then we build a new training set where each original sample is kept, along with nine augmented versions of that sample.

In Listing 6-2, scikit-learn supplies PCA.

```
from sklearn import decomposition
pca = decomposition.PCA(n_components=xtrn.shape[1])
pca.fit(x)
```

Listing 6-2: Using PCA to learn the principal components

The PCA class follows scikit-learn's standard approach of defining an instance of a class before calling fit on that instance. We set the number of components to the number of features in the dataset (30).

We'll use the trained pca object in a loop to build a new training set of augmented samples, as shown in Listing 6-3.

```
start = 24
nsets = 10
nsamp = xtrn.shape[0]
newx = np.zeros((nsets*nsamp, xtrn.shape[1]))
newy = np.zeros(nsets*nsamp, dtype="uint8")

for i in range(nsets):
    if (i == 0):
        newx[0:nsamp,:] = xtrn
        newy[0:nsamp] = ytrn
```

```
else:
    newx[(i*nsamp):(i*nsamp+nsamp),:] = generateData(pca, xtrn, start)
    newy[(i*nsamp):(i*nsamp+nsamp)] = ytrn
```

Listing 6-3: Augmenting the samples

The new training data is in newx and newy. Each existing training sample will be accompanied by nine augmented versions of it, so the new training set will have 3,980 samples instead of a mere 398.

The loop constructs the dataset in blocks of 398 samples. The first pass stores the original data, and subsequent passes call generateData to return a new, augmented version of the samples in the original dataset. The order of the new samples is identical to the original, meaning the labels are in the same order as well.

In Listing 6-4, the function generateData applies the PCA transformation and alters the least important coordinate directions beginning with start (24).

```
def generateData(pca, x, start):
    original = pca.components_.copy()
    ncomp = pca.components_.shape[0]
    a = pca.transform(x)
    for i in range(start, ncomp):
        pca.components_[i,:] += np.random.normal(scale=0.1, size=ncomp)
    b = pca.inverse_transform(a)
    pca.components_ = original.copy()
    return b
```

Listing 6-4: Applying PCA

PCA is a reversible transformation. The generateData function alters the PCA components by adding a small, normally distributed value to each one beginning with component 24 (out of 30). When the inverse transform uses the altered components, the resulting values (b), a block of 398 samples, are no longer identical to the original. These augmented versions build the next block of the new dataset.

The new training dataset is in newx and newy, but the order of the samples is not random because it was built in blocks. Therefore, we perform a final randomization before writing the train and test datasets to disk (Listing 6-5).

```
i = np.argsort(np.random.random(nsets*nsamp))
newx = newx[i]
newy = newy[i]
np.save("datasets/bc_train_data.npy", newx)
np.save("datasets/bc_train_labels.npy", newy)
np.save("datasets/bc_test_data.npy", xtst)
np.save("datasets/bc_test_labels.npy", ytst)
```

Listing 6-5: Storing the augmented datasets

The histology training and test datasets are now ready for use. We applied randomness multiple times to scramble the order of the data and to alter the principal components to build augmented versions of the training data.

Handwritten Digits

One of the first great successes of the deep learning revolution involved correctly identifying objects in images. While the dataset we'll build isn't very advanced comparably, it is part of MNIST, a larger, workhorse dataset commonly used in machine learning. We'll build a dataset consisting of handwritten 1s, 4s, 7s, and 9s. I selected these four digits because even humans often confuse one for another, so we might expect a machine learning model to do the same (time will tell).

Figure 6-1 shows samples of each digit type.

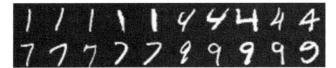

Figure 6-1: Sample digits, where 1s and 7s are often confused, as are 4s and 9s

The images are 28×28-pixel grayscale, with each pixel an integer in [0, 255]. We'll work with the digits as 784-element vectors (28 × 28 = 784). I collected the digits and their labels in the files *mnist_data.npy* and *mnist_labels.npy*, respectively.

The raw digits dataset is relatively small, with 100 samples of each digit; we split the raw data to have 50 train and 50 test samples. We'll augment each image multiple times to expand the size of the training set.

We used PCA to augment the histology data, but because we're working with images here, we'll apply basic image processing transformations to randomly produce slightly altered versions of each training image. In particular, we'll rotate each image in the range [−3, 3] degrees; about 10 percent of the time, we'll zoom the image to [0.8, 1.2] times its original size while maintaining a final size of 28×28 pixels by cropping or embedding a smaller image inside a blank 28×28 image.

The code we need is in *build_mnist_dataset.py*. While it mirrors the code that built the histology dataset, differences include splitting the raw data 50/50 between train and test, storing the unaugmented training data, and augmenting the training data 20 times instead of 10 times, resulting in an augmented training set of 4,200 samples (200 original plus 20 more for each original sample); see Listing 6-6.

```
newx = []
newy = []
for i in range(len(ytrn)):
    newx.append(xtrn[i])
```

```
          newy.append(ytrn[i])
❶ for j in range(20):
            newx.append(augment(xtrn[i]))
            newy.append(ytrn[i])
    xtrn = np.array(newx)
    ytrn = np.array(newy)
❷ i = np.argsort(np.random.random(len(ytrn)))
    xtrn = xtrn[i]
    ytrn = ytrn[i]
```

Listing 6-6: Augmenting the digit images

The original 200 training samples (xtrn, ytrn) are examined, one by one. First, we add the original image to the augmented output (newx, newy). Then, we add 20 augmented versions of the same image by making repeated calls to augment ❶. After adding the augmented images, we scramble the entire training set again to mix the order of the images ❷.

Listing 6-7 shows the code to augment images.

```
from scipy.ndimage import rotate, zoom
def augment(x):
    im = x.reshape((28,28))
    if (np.random.random() < 0.5):
        angle = -3 + 6*np.random.random()
        im = rotate(im, angle, reshape=False)
    if (np.random.random() < 0.1):
        f = 0.8 + 0.4*np.random.random()
        t = zoom(im, f)
        if (t.shape[0] < 28):
            im = np.zeros((28,28), dtype="uint8")
            c = (28-t.shape[0])//2
            im[c:(c+t.shape[0]),c:(c+t.shape[0])] = t
        if (t.shape[0] > 28):
            c = (t.shape[0]-28)//2
            im = t[c:(c+28),c:(c+28)]
    return im.ravel()
```

Listing 6-7: Augmenting an image

The input image (x), a NumPy vector, is first reshaped into a 28×28-element two-dimensional array, im. Next, two if statements ask whether a random value is less than 0.5 or 0.1. The first, executed 50 percent of the time, applies a random rotation of the image by some value in the range [−3, 3] degrees. Notice the use of rotate from scipy.ndimage. The reshape=False keyword forces rotate to return an output array that's the same size as the input array.

The second if statement, executed 10 percent of the time, uses zoom to magnify the image by a random scale factor in [0.8, 1.2], meaning the zoomed image is anywhere from 80 to 120 percent the size of the original.

The code after the call to zoom ensures the output image is still 28×28 pixels by either embedding the smaller image in a blank 28×28 image or selecting the central 28×28 pixels if magnifying beyond 100 percent. The newly augmented image is returned after unraveling it and transformed back into a 784-element vector.

The code in *build_mnist_dataset.py* stores the smaller, unaugmented training set and the augmented training set. The file *mnist_test.py* uses sklearn's `MLPCLassifier`—which we'll learn about soon—to train 40 models using the training sets, keeping the overall accuracy of each. The models use default values and a single hidden layer of 100 nodes. The mean accuracy for the unaugmented dataset was 87.3 percent, while augmented training data led to a mean overall accuracy of 90.3 percent, a statistically significant difference, thereby delivering evidence that the random augmentation process aids training models. It may seem tedious to spend this section detailing how the datasets are constructed, but it's hard to overemphasize the importance randomness plays in the process. Dataset construction is so crucial to modern machine learning that competitions are held where the models are fixed and good dataset construction is required to produce winning results (search for "Data-Centric AI Competition").

Now that we have our datasets, let's put them to the test.

Neural Networks

A *neural network* is a set of nodes that, layer by layer, successively transforms an input to an output. Network nodes accept multiple inputs and produce a single output, an operation sufficiently similar to the operation of a biological neuron that the name "neural network" has persisted. Neural networks are not artificial brains; rather, they are feed-forward, directed, acyclic *graphs*, a data structure commonly used in computer science.

The operation of the network transforms an input to an output; in other words, neural networks are a type of function, $y = f(x; \theta)$. Training the neural network means locating θ, the set of parameters causing the network to perform as desired.

Anatomy Analysis

If a neural network is a set of nodes working in layers, then it makes sense to begin with a node. See Figure 6-2.

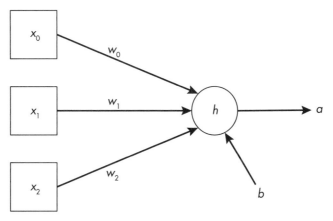

Figure 6-2: A neural network node

Data flows from left to right. The inputs (x), either the input to the network or the output of a previous network layer, are multiplied by *weights* (w) and summed along with a *bias* (b) before passing the total to an *activation* function (h) to produce the output, a. In symbols:

$$a = h(w_0x_0 + w_1x_1 + w_2x_2 + b)$$

The activation function is nonlinear, or some function beyond x to the first power. Modern networks often use *rectified linear unit (ReLU)* activation functions, which output the input unless the input is less than zero, in which case the output is zero, $h = \max(0, x)$.

A network layer consists of a collection of nodes that, collectively, receive the output of the previous layer as their inputs and produce new output, one per node, passed to the next layer. Each node in a layer receives each input; in graph terms, the network is *fully connected*.

Data flows through the network, layer by layer, to the output. In this way, the network maps input x to output y. The output is often a vector, but can be a scalar, y.

Moving data through a neural network is most easily accomplished by representing the weights between layers as a matrix and the biases as a vector. This representation automatically applies each input to each node and activation function, thereby reducing the entire layer operation to a matrix multiplied by a vector plus another vector passed to a vector version of the activation function. Training means learning a set of matrices for the weights, one matrix per layer, and a set of bias vectors, one per layer.

The weights and biases, which live on the connections between the nodes of the layers, are the parameters of θ; these are the things the network learns during training. This is analogous to fitting a curve, but in this case, the function is determined by the architecture of the network. Unlike curve fitting, training a neural network usually doesn't involve reducing the error on the training set to zero. Instead, the goal is to coax the network to learn weights and biases that make the network generally applicable to new inputs. After all, the entire point of training the model is to use it with new, unknown inputs.

A detailed discussion of neural network training is beyond our present scope, as our goal is to understand randomness's role in the process. If you're interested in learning more about neural networks, I recommend my book *Practical Deep Learning: A Python-Based Introduction* (2021), also available from No Starch Press. Remember that the training process produces a specific set of weights and biases that tailor the network to the problem at hand.

Randomness

Randomness is critically important at the beginning of the training process, in the selection of the initial weights and biases.

Training a neural network follows this general algorithm:

1. Randomly initialize the weights and biases of the network.

2. Pass a randomly selected subset of the training data through the network.

3. Use a measure of the error between the network's output and the desired output to update the weights and biases.

4. Repeat from step 2 until training is done (however decided).

In this section, we're concerned with step 1. Steps 2 and 3 involve a *loss function*, a measure of the mistakes the network has made, and a two-step process to update the weights and biases: *backpropagation* and *gradient descent*. The former uses the chain rule from differential calculus to determine how each weight and bias value contributes to the error, and the latter uses the gradient formed from those measures to alter the weights and biases to minimize the loss function.

Initialization

It was only in 2010–2012 that deep neural networks (with many layers) exploded on the scene. One of the factors contributing to this was the realization that previous algorithms for initializing the weights and biases of models were relatively poor; better options exist.

We will explore these options by employing the MLPClassifier class from sklearn to implement a traditional neural network. *MLP* stands for *multilayer perceptron*, with "perceptron" being an old name for a neural network (I recommend searching for Frank Rosenblatt and his Perceptron machine—fascinating research that was not sufficiently appreciated in its time).

Initializing the Network

The MLPClassifier class includes an internal method, _init_coef, which is responsible for assigning the initial weights and biases. We will subclass MLPClassifier and override this method, allowing us to alter the initialization approach while still taking advantage of everything else the MLPClassifier has to offer. Take a look at Listing 6-8.

```
from sklearn.neural_network import MLPClassifier

def normal(rng, mu=0, sigma=1):
    if (normal.state):
        normal.state = False
        return sigma*normal.z2 + mu
    else:
        u1,u2 = rng.random(2)
        m = np.sqrt(-2.0*np.log(u1))
        z1 = m*np.cos(2*np.pi*u2)
        normal.z2 = m*np.sin(2*np.pi*u2)
        normal.state = True
        return sigma*z1 + mu
normal.state = False


class Classifier(MLPClassifier):
    def _init_coef(self, fan_in, fan_out, dtype):
❶       def normvec(fan_in, fan_out):
            vec = np.zeros(fan_in*fan_out)
            for i in range(fan_in*fan_out):
                vec[i] = normal(self.rng)
            return vec.reshape((fan_in,fan_out))

        if (self.init_scheme == 0):
❷           return super(Classifier, self)._init_coef(fan_in, fan_out, dtype)
        elif (self.init_scheme == 1):
❸           vec = self.rng.random(fan_in*fan_out).reshape((fan_in,fan_out))
            weights = 0.01*(vec-0.5)
            biases = np.zeros(fan_out)
        elif (self.init_scheme == 2):
❹           weights = 0.005*normvec(fan_in, fan_out)
            biases = np.zeros(fan_out)
```

```
elif (self.init_scheme == 3):
❺ weights = normvec(fan_in, fan_out)*np.sqrt(2.0/fan_in)
    biases = np.zeros(fan_out)

return weights.astype(dtype, copy=False), biases.astype(dtype,
    copy=False)
```

Listing 6-8: Overriding the initialization method

The subclass overrides scikit-learn's approach ❷ with three other approaches. The method returns a weight matrix and a bias vector for a layer that has fan_in inputs and fan_out outputs. The dtype parameter specifies the data type, typically 32- or 64-bit floating point.

By default, scikit-learn uses *Glorot initialization*. We get it by calling the superclass version of the method ❷. Glorot initialization depends on the number of inputs and outputs and is one of the initialization approaches leading to improved model performance. At least, that's the claim. We'll put it to the test.

Another modern initialization approach is *He initialization* ❺, which is suitable for networks using ReLU activation functions. He initialization depends on a matrix of samples from a normal distribution with a mean of 0 and a standard deviation of 1. We get that here via the embedded normvec function ❶, which permits us to use our RE class. We'll see precisely how momentarily.

We initialize classical neural networks through the use of small uniformly distributed ❸ or normally distributed ❹ random values. The first draws random samples from a uniform distribution, $[-0.005, 0.005]$. The second uses normally distributed samples scaled by 0.005. Both approaches are intuitively reasonable, but, as we'll see, they are not ideal.

The normal function returns a normally distributed sample with a mean value of mu and a standard deviation of 1. A normal distribution selects samples symmetrically around the mean; return to Figure 1-1 on page 4 as an example. NumPy provides a function to select samples from a normal distribution, but we want to use our RE class, so we need to define a function that creates normally distributed samples from the uniformly distributed samples RE returns. We can use the Box-Muller transformation introduced in Chapter 1:

$$z_1 = \sqrt{-2 \log u_1} \cos(2\pi u_2)$$
$$z_2 = \sqrt{-2 \log u_1} \sin(2\pi u_2)$$

Here u_1 and u_2 are uniformly distributed samples in $[0, 1)$, precisely what RE returns (how convenient!).

The code implementing normal requires explanation. Note the indentation on the final line: it's not part of normal, but refers to it immediately after its definition.

The Box-Muller transformation equations use a pair of uniform samples to produce a pair of normally distributed samples. While we can ask RE objects to return a pair of uniform samples, we want normal to return only one normally distributed sample when called. We can generate two samples and throw one away, or carefully try to preserve both. This way, each call to normal returns a single sample and generates new samples only when required. Accomplishing this requires normal to preserve state between calls. The Python way to do this is to use a class, but that seems a bit overkill. Instead, we'll make use of the fact that Python functions are objects; we can add new attributes (member variables) to objects at will. We'll define normal as a Python function and then immediately add a state variable to it initialized to False—the final line in Listing 6-8.

When we call normal, the state variable has a value, False. This value persists between calls because it's an attribute of the normal function, which uses the value of state to decide whether to generate two new samples and return the first caching the second, or to return the second. If state is True, return the second sample, z2, after multiplying it by the desired standard deviation (sigma) and adding the mean value (mu). Then set state to False. If state is False, get two uniform samples from rng and use them to calculate two normal samples, z1 and z2. Cache z2 and return z1, properly scaled and offset.

To use Classifier we create a neural network using scikit-learn and add two new attributes: init_scheme to specify the desired initialization scheme, and rng, an instance of RE, to allow access to the different randomness sources we've developed throughout the book.

Experimenting with Initialization

Let's experiment with Classifier and neural network initialization. Along the way, we'll set up a scikit-learn training session.

The code we need is in *init_test.py*. It loads the digits dataset created previously and trains 12 models using each initialization scheme. Training multiple models for each is essential because initialization is random; for example, we might get a bad dice roll for one initialization scheme when it's actually a good approach. Averaging the results from multiple models lets us apply statistical tests.

Listing 6-9 shows the beginning of the main part of *init_test.py*.

```
import numpy as np
from Classifier import *
from RE import *
from scipy.stats import ttest_ind, mannwhitneyu

xtrn = np.load("../../data/datasets/mnist_train_data.npy")/256.0
ytrn = np.load("../../data/datasets/mnist_train_labels.npy")
xtst = np.load("../../data/datasets/mnist_test_data.npy")/256.0
ytst = np.load("../../data/datasets/mnist_test_labels.npy")

N = 12
init0 = []
```

```
for i in range(N):
    init0.append(Run(0, xtrn,ytrn,xtst,ytst))
init0 = np.array(init0)
```

Listing 6-9: Setting up the initialization experiment

The imports make the `Classifier` class available, along with `RE` and two statistical tests: the t-test and the *Mann-Whitney U* test, a nonparametric version of the t-test. *Nonparametric tests* make no assumptions about the data and are harder to satisfy. I often consider both together to account for the possibility that the t-test results are invalid because the data isn't normally distributed—a fundamental assumption of the t-test.

Next, we load the digits dataset, first the training (xtrn) followed by the test (xtst). We divide the data by 256 to map it from $[0, 255]$ to $[0, 1]$.

The following code paragraph accumulates the output of `Run` for initialization scheme 0. The return value is the overall accuracy of a model using scikit-learn's Glorot initialization. Similar code captures the accuracy for the other four initialization schemes.

We then transform the results for all schemes into means and standard errors before printing, as in Listing 6-10.

```
m0,s0 = init0.mean(), init0.std(ddof=1)/np.sqrt(N)
m1,s1 = init1.mean(), init1.std(ddof=1)/np.sqrt(N)
m2,s2 = init2.mean(), init2.std(ddof=1)/np.sqrt(N)
m3,s3 = init3.mean(), init3.std(ddof=1)/np.sqrt(N)
```

Listing 6-10: Reporting the results

Finally, we run the statistical tests to compare one initialization scheme against another. Theoretically, we might expect initialization scheme 3 (He) to be the best performing, so we compare it (init3) with each of the others and report the corresponding p-values (the "u" is the p-value for the Mann-Whitney U test); see Listing 6-11.

```
_,p = ttest_ind(init3,init0)
_,u = mannwhitneyu(init3,init0)
print("init3 vs init0: p=%0.8f, u=%0.8f" % (p,u))
_,p = ttest_ind(init3,init2)
_,u = mannwhitneyu(init3,init2)
print("init3 vs init2: p=%0.8f, u=%0.8f" % (p,u))
_,p = ttest_ind(init3,init1)
_,u = mannwhitneyu(init3,init1)
print("init3 vs init1: p=%0.8f, u=%0.8f" % (p,u))
```

Listing 6-11: Running the statistical tests

The statistical tests, `ttest_ind` and `mannwhitneyu`, first return their respective test statistics and then their associated p-value. The p-value tells us the probability that we'll measure the difference in the means between the two sets of accuracies, given the two sets are from the same distribution. The smaller the p-value, the less likely the two sets are from the same distribution, thereby giving us confidence that the difference observed is real.

Training the Network

The code to configure and train a neural network is in the Run function, as in Listing 6-12.

```
def Run(init_scheme, xtrn,ytrn,xtst,ytst):
    clf = Classifier(hidden_layer_sizes=(100,50), max_iter=4000)
    clf.init_scheme = init_scheme
    clf.rng = RE()
    clf.fit(xtrn,ytrn)
    pred = clf.predict(xtst)
    _,acc = Confusion(pred,ytst)
    return acc
```

Listing 6-12: Training a neural network

Here's where we begin to appreciate the power of scikit-learn. The arguments to Run include which initialization scheme to use, [0, 3], followed by the train and test datasets. First, we create the neural network by creating an instance of Classifier, our subclass of scikit-learn's MLPClassifier. By default, the neural network uses a ReLU activation function, which is what we want. It also trains until training no longer improves things or max_iter passes through the entire training set. Each pass through is known as an *epoch*.

The hidden_layer_sizes keyword defines the architecture of the model. We know the input has 784 elements and the output 4 because there are four classes, the digits 1, 4, 7, and 9. The layers between the input and output are hidden; we're specifying two: the first with 100 nodes and the second with 50.

Before training, we need to set the initialization scheme (init_scheme) and define rng, an instance of RE, using the default of PCG64 and floating-point values in [0, 1).

Train the neural network with clf.fit(xtrn,ytrn), in which the fit method accepts the input vectors (xtrn) and the associated labels (ytrn). When the method returns, the model is trained and has found values for all the weights and biases (θ).

Evaluating the Network

When given the test data (xtst), predict generates a set of predicted class labels, pred—a vector of the same length as ytst, the known class labels. We compare the two to evaluate the model by building a *confusion matrix*. The rows of the confusion matrix represent the known, true class labels. The columns are the model's assigned labels. The matrix elements are counts of the number of times each possible pairing of the true label and assigned label happened. A perfect classifier will always assign correct labels, meaning the confusion matrix will be diagonal. Any counts off the main diagonal represent errors.

The Confusion function generates a confusion matrix, and overall accuracy, from a set of known and predicted labels; see Listing 6-13.

```
def Confusion(y,p):
    cm = np.zeros((4,4), dtype="uint16")
    for i in range(len(p)):
        cm[y[i],p[i]] += 1
    acc = np.diag(cm).sum() / cm.sum()
    return cm, acc
```

Listing 6-13: Creating a confusion matrix

The confusion matrix (cm) is a 4×4 matrix because there are four classes. The most important line updates cm by indexing first by the known class label (y) and then by the assigned class label (p). Adding one counts each time that particular pairing of the true class label and assigned label occurs.

The overall accuracy is either the number of times the known and assigned labels matched or the sum along the main diagonal divided by the sum of all matrix elements.

Running *init_test.py* takes a few minutes. My run produced:

```
init0: 0.92667 +/- 0.00167
init1: 0.89958 +/- 0.00311
init2: 0.90083 +/- 0.00183
init3: 0.92500 +/- 0.00213

init3 vs init0: p=0.54429253, u=0.55049935
init3 vs init2: p=0.00000002, u=0.00004054
init3 vs init1: p=0.00000088, u=0.00006765
```

The first block displays the mean accuracy over the 12 models for each initialization scheme (mean ± SE). Glorot and He initialization averaged 92.7 and 92.5 percent, respectively. The older uniform and normal initialization strategies achieved 90.0 and 90.1 percent, respectively. These are large differences, but are they statistically significant? For that, look at the second block of results.

Comparing He initialization (init3) with Glorot (init0) returns p-values of about 0.5 for both the t-test and Mann-Whitney U. Such p-values strongly indicate that there is no difference between either approach.

Now, look at the p-values comparing He initialization with the older methods. They are virtually zero, implying the difference in mean accuracy is very likely real—He and Glorot initialization both lead to significantly better performing models. Modern deep learning is vindicated. While there was never any doubt, it's beneficial to confirm directly rather than take it purely on faith.

Extreme Learning Machines

An *extreme learning machine* is a simple, single, hidden-layer neural network. What distinguishes an extreme learning machine from a traditional neural network is the source of the weights and biases.

To map an input vector through the first hidden layer of a neural network involves a weight matrix, W, and a bias vector, b

$$z = h(Wx + b)$$

where x and z are vectors, and h is the activation function that accepts a vector input and produces a vector output.

In a traditional neural network, W and b are learned in the training process. In an extreme learning machine, W and b are generated randomly, with no regard for the training data. Random matrices are (sometimes) capable of mapping inputs, like the digit images as vectors, to a new space where it's easier to separate the classes.

If we use the equation to map all training data, the output of the hidden layer (z) becomes a matrix (Z) with as many rows as there are training samples and as many columns as there are nodes in the hidden layer. In other words, the random W matrix and b bias vector have produced a new version of the training data, Z.

The random weight matrix and bias vector link the input and the hidden layer. To finish building the neural network, we need a weight matrix between the hidden layer's output and the model's output; there is no bias vector in this case. Learning happens here, but we need to take a detour to see it.

The training dataset consists of input samples and associated class labels, $[0, 3]$. Many machine learning models don't use integer class labels, but transform the labels into *one-hot vectors*. For example, if there are four classes, the one-hot vector has four elements, all 0 except for the element corresponding to the class label, which is 1. If the class label is 2, then the one-hot vector is $\{0, 0, 1, 0\}$. Likewise, if the class label is 0, the one-hot vector is $\{1, 0, 0, 0\}$. To finish building the extreme learning machine, we need the training set class labels in this form. I assume that the labels are so transformed for the remainder of this section.

As one-hot vectors, the class labels become a matrix, Y, with as many rows as there are training samples and as many columns as there are classes. A linear mapping from the output of the hidden layer, Z, to the known labels (as one-hot vectors), Y, uses a matrix, B:

$$Y = BZ$$

We want to find B as the second weight matrix.

Because we know Z by pushing the training data through the hidden layer and Y, the known labels, we generate B via

$$YZ^{-1} = B$$

where Z^{-1} is the inverse of the matrix Z. Multiplying by the inverse of a matrix is akin to multiplying by the reciprocal of a scalar value; they cancel each other. To find the inverse of Z, we need the Moore-Penrose pseudoinverse, which NumPy provides in its linear algebra module, linalg.

We now have all we need to build the extreme learning machine:

1. Select a weight matrix, W, and bias vector, b, at random.

2. Pass the training data through the first hidden layer, $Z = h(WX + b)$, where X and Z are now matrices, one row for each training sample.

3. Calculate the output weight matrix, $B = YZ^{-1}$, using the pseudo-inverse of Z, the output of the hidden layer.

4. Use W, b, and B as the weights and biases of the extreme learning machine.

This is rather abstract. Let's make it concrete in code.

Implementation

The file *elm.py* implements an extreme learning machine and applies it to the digits dataset. Listing 6-14 shows the train function.

```
def train(xtrn, ytrn, hidden=100):
    inp = xtrn.shape[1]
    m = xtrn.min()
    d = xtrn.max() - m
    w0 = d*rng.random(inp*hidden).reshape((inp,hidden)) + m
    b0 = d*rng.random(hidden) + m
    z = activation(np.dot(xtrn,w0) + b0)
    zinv = np.linalg.pinv(z)
    w1 = np.dot(zinv, ytrn)
    return (w0,b0,w1)
```

Listing 6-14: Defining an extreme learning machine

The first line sets inp to the number of features in the training data, here 784. The next two lines define the smallest training feature value (m) and the difference between that and the largest (d).

The following two lines generate the random weight matrix and bias vector, Wi (w0) and b (b0), by randomly sampling in the range of the training data, $[m, m + d]$. Again, W and b are completely random, but fixed once selected.

The final piece of the extreme learning machine is B (w1), which is found by passing the training data through the hidden layer

```
z = activation(np.dot(xtrn,w0) + b0)
```

where activation is the selected activation function ($h()$).

Finally, to define B, we multiply YZ^{-1}:

```
zinv = np.linalg.pinv(z)
w1 = np.dot(zinv, ytrn)
```

The return value for the function is the defined and trained extreme learning machine: (w0,b0,w1). Let's take it for a test drive.

Testing

The file *elm.py*, from which we extracted train, trains an extreme learning machine to classify the digits dataset using a user-supplied number of nodes in the hidden layer. For example:

```
> python3 elm.py 200 mt19937
[[49  0  1  0]
 [ 1 42  0  9]
 [ 2  4 39  4]
 [ 1  5  6 37]]

accuracy = 0.835000
```

We use the Mersenne Twister pseudorandom generator to build a machine with 200 hidden layer nodes. The output shows the confusion matrix and the overall accuracy, 83.5 percent. The confusion matrix is almost diagonal, which is a good sign.

Let's scrutinize the confusion matrix. The rows represent the true class labels from top to bottom, ones, fours, sevens, and nines. The columns, from left to right, are the model's assigned class label (we'll see how soon). Looking at the top row, of the 50 ones in the test dataset, the model called a one a one 49 times, but once called it a seven.

The model had the most difficulty with nines, the final row of the confusion matrix. The model got it right 37 out of 49 times, but it called it a seven 6 times and a four 5 times. Only once did the model confuse a nine for a one.

The train function creates the extreme learning machine. To use it, we need predict, as shown in Listing 6-15.

```
def predict(xtst, model):
    w0,bias,w1 = model
    z = activation(np.dot(xtst,w0) + bias)
    return np.dot(z,w1)
```

Listing 6-15: Prediction with an extreme learning machine

We pull the weight matrices and bias vector from the supplied model, and then pass the test data through the hidden layer and, finally, the output layer. The activation variable is set to the specific activation function currently in use—by default, the ReLU. We'll experiment with different activation functions in the next section.

The output of predict is a two-dimensional matrix with as many rows as are in xtst (200) and as many columns as there are classes (4). For example, the output of predict for the first test sample (xtst[0]) is

$$0.19551782, 0.90971894, 0.05398019, -0.06542743$$

and the first known test label, as a one-hot vector, is:

$$0, 1, 0, 0$$

Notice the largest value in the model's output vector is at index 1, as is the largest for the one-hot label. In other words, the first test sample belongs to class 1 (four), and the model successfully predicted class 1 as the most likely class. The last output value is negative: the model is not outputting a probability, but a decision function value where the largest is the most likely class label, even if we don't have a true probability associated with that value.

Therefore, to construct a confusion matrix, we will need code like Listing 6-16.

```
def confusion(prob,ytst):
    nc = ytst.shape[1]
    cm = np.zeros((nc,nc), dtype="uint16")
    for i in range(len(prob)):
        n = np.argmax(ytst[i])
        m = np.argmax(prob[i])
        cm[n,m] += 1
    acc = np.diag(cm).sum() / cm.sum()
    return cm,acc
```

Listing 6-16: Building a confusion matrix for an extreme learning machine

The `confusion` function returns the confusion matrix and overall accuracy, but instead of directly using the values in `ytst` and `prob`, we apply NumPy's `argmax` function to return the index of the largest value in the four-element vectors.

Listing 6-17 shows the remainder of *elm.py*, which loads the digit datasets, scaling them by 256, and then trains and tests the extreme learning machine with three lines of code.

```
model = train(xtrn, ytrn, nodes)
prob = predict(xtst,model)
cm,acc = confusion(prob,ytst)
```

Listing 6-17: Training and testing an extreme learning machine

The `nodes` parameter is the number of nodes in the hidden layer as read from the command line.

How sensitive is the extreme learning machine to the number of nodes in the hidden layer? That's a good question.

By default, the code in *elm.py* uses a ReLU activation function. However, the file defines several other activation functions. In this section, we'll explore each in combination with differing numbers of hidden layer nodes to find whether there's a best combination.

The ReLU activation function uses NumPy's `maximize` function, which returns the largest of the two arguments, element by element:

```
def relu(x):
    return np.maximum(x,0)
```

We're not constrained to use only the ReLU. Classically, neural networks made heavy use of the sigmoid and hyperbolic tangent functions:

```
def sigmoid(x):
    return 1 / (1 + np.exp(-0.01*x))
def tanh(x):
    return np.tanh(0.01*x)
```

Both of these functions return S-shaped curves. I added a factor of 0.01 to scale x in the sigmoid and hyperbolic tangent. This is not usually done, but it was necessary here to prevent overflow errors.

For fun, I defined several other activation functions:

```
def cube(x):
    return x**3
def absolute(x):
    return np.abs(x)
def recip(x):
    return 1/x
def identity(x):
    return x
```

Let's test the activation functions as the number of nodes in the hidden layer is varied from 10 to 400. The code is in *elm_test.py*. It makes use of the train, predict, and confusion functions. The main loop looks like Listing 6-18.

```
acts = [relu, sigmoid, tanh, cube, absolute, recip, identity]
nodes = [10,20,30,40,50,100,150,200,250,300,350,400]
N = 50
acc = np.zeros((len(acts),len(nodes),N))

for i,act in enumerate(acts):
    for j,n in enumerate(nodes):
        for k in range(N):
            activation = act
            model = train(xtrn, ytrn, n)
            prob = predict(xtst, model)
            _,a = confusion(prob, ytst)
            acc[i,j,k] = a

np.save("elm_test_results.npy", acc)
```

Listing 6-18: Testing activation functions and hidden layer sizes

Fifty extreme learning machines are trained for each combination of activation function and hidden layer size, tracking the overall accuracy (acc). Notice the assignment of act to activation. Functions may be freely assigned to variables in Python and then used when the variable is referenced (as in train).

Run *elm_test.py* by passing it a randomness source (I used MT19937). When it finishes, run *elm_test_results.py* to parse the output and generate a plot like that of Figure 6-3 showing the mean accuracy by hidden layer size and activation function. Error bars are present but small.

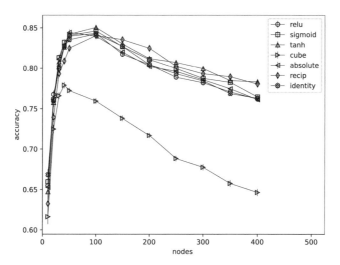

Figure 6-3: Extreme learning machine performance as a function of the hidden layer size and activation function

Figure 6-3's most obvious statement is that $y = x^3$ is a lousy activation function, as it always results in inferior models compared to the others. Another interesting observation is that the activation functions follow the same shape: a rapid increase in model accuracy as the number of nodes in the hidden layer increases, followed by a maximum and a slower decline. The difference between activation functions is slight, especially near the maximum of 100 hidden nodes; however, the hyperbolic tangent came out on top in this run. In fact, tanh wins for most runs, so it's fair to say that for this particular dataset, an extreme learning machine using tanh and 100 nodes in the hidden layer is the way to go.

The identity activation function, $f(x) = x$, eschews nonlinearity; all the performance comes from the linear top layer mapping the hidden layer output to the predictions per class.

Reckless Swarm Optimizations

An extreme learning machine's attraction is the incredible speed with which a model is trained compared to all the calculations necessary to train a traditional neural network of the same size. Sure, the first model might not be that good, but trying a few times in a row and keeping the best-performing model seems a reasonable thing to do. I can imagine a scenario where an autonomous system might want to train a model quickly in response to rapidly changing input data.

However, the random part of the extreme learning machine—that is, the weight matrix and bias vector from the input to the hidden layer—made me wonder: Might we be able to learn the weight matrix and bias vector via a swarm optimization instead? Would this work, and if so, might it be any better than the random versions? My thought is clearly missing the point that extreme learning machines are meant to use random weights and biases, but I want to know if swarm optimization might prove helpful, even if vastly more computationally intensive than traditional neural network training.

The true challenge is the dimensionality of the problem. Our inputs are 784-element vectors. We learned that 100 nodes in the hidden layer seems a good thing to have, so the total dimensionality of the weight matrix and bias vector is:

$$784 \times 100 + 100 = 78,500$$

We'll be asking the swarms to search a space of 78,500 dimensions to come up with a good position, one that leads to a model with the highest accuracy. That's a tall order.

The code I experimented with is in *elm_swarm.py*. I won't walk through it, but you'll see it follows similar optimization code from earlier chapters. The objective function uses each particle position as a weight matrix and bias vector, and then learns the output weight matrix and evaluates the model on the test set to produce an overall accuracy. Therefore, each call to the Evaluate method results in a trained and tested model.

To run the code, use a command line like this:

```
> python3 elm_swarm.py 100 tanh 20 60 de de.pkl
```

Here, differential evolution (de) searches for a model with tanh activation functions and 100 hidden layer nodes. The swarm has 20 particles and runs for 60 iterations before reporting the final confusion matrix and accuracy. The best model is stored in the file *de.pkl*. The current best accuracy is shown on each iteration, so you can watch the swarm learn. Run *elm_swarm.py* without arguments to see all options.

For example, the previous command produced the following output:

```
 0: 0.90000 (mean swarm distance 111.844796164)
 1: 0.90000 (mean swarm distance 111.253958636)
 2: 0.90000 (mean swarm distance 110.932668482)
 3: 0.90000 (mean swarm distance 110.743740374)
 4: 0.90000 (mean swarm distance 110.701071153)
--snip--
57: 0.93000 (mean swarm distance 106.803938775)
58: 0.93000 (mean swarm distance 106.803938775)
59: 0.93000 (mean swarm distance 106.803938775)
[[50  0  0  0]
 [ 0 52  0  0]
```

```
[ 1  2 41  5]
 [ 1  3  2 43]]
final accuracy = 0.930000 (DE-tanh-100, 20:60, 1220 models
  evaluated, 4 best updates)
```

The per-iteration best accuracy is shown along with the mean distance between the particles in the swarm. If the swarm is converging, this distance shrinks during the search, as it does here. The distance between two swarm particles uses the formula to find the distance between two points in two-dimensional or three-dimensional space

$$d = \sqrt{(x_1 - x_0)^2 + (y_1 - y_0)^2} \text{ or } d = \sqrt{(x_1 - x_0)^2 + (y_1 - y_0)^2 + (z_1 - z_0)^2}$$

but extended to 78,500 dimensions. The result is still a scalar.

After 60 iterations, the swarm located a weight and bias vector leading to a 93 percent overall accuracy on the held-out test set. There were four swarm best updates. A second run, using GWO for 600 iterations, found a model with 94 percent overall accuracy. In that run, the swarm collapsed from an initial inter-particle distance of around 90 to less than 1 by the time it reached iteration 600. There were eight swarm best updates.

I conducted runs of *elm_swarm.py* for all algorithms, five runs for each. Table 6-1 displays the resulting average accuracies.

Table 6-1: Average Model Accuracies for Each Optimization Algorithm

Algorithm	Accuracy (mean ± SE)
GWO	0.9260 ± 0.0087
Differential evolution	0.9200 ± 0.0016
Bare-bones PSO	0.9190 ± 0.0019
Jaya	0.9180 ± 0.0034
GA	0.9170 ± 0.0058
RO	0.9170 ± 0.0025
PSO	0.9160 ± 0.0024

The results are not statistically significantly different, but the ranking from best to worst is typical (except PSO). The GWO standard error of the mean is larger because one search found a model with an accuracy of 95.5 percent, the highest I encountered.

How many runs of *elm.py* do we need, on average, to find a model that meets or exceeds a given accuracy? The file *elm_brute.py* generates extreme learning machine after extreme learning machine, for up to a given maximum number of iterations, attempting to find a model that meets or exceeds the specified test set accuracy.

Structurally, *elm_brute.py* is a tweak to *elm.py* to do the model creation and testing in a loop and then report the performance if successful or note that it wasn't able to meet the accuracy threshold. Running *elm_brute.py* for different thresholds, 10 runs for each with a maximum of 2,000 iterations, produced Table 6-2.

Table 6-2: Number of Models Tried to Achieve a Given Accuracy

Target accuracy	Mean	Min	Max	Successes
0.70	1	1	1	10
0.75	1	1	1	10
0.80	1	1	1	10
0.85	2.2	1	6	10
0.90	147.2	15	270	10
0.91	504.2	87	1,174	9
0.915	858.4	69	1,710	9
0.92	788.3	74	1,325	4

The first column shows the target accuracy. Any model meeting or exceeding this accuracy is considered a success. The mean number of models needed to find one at or above the threshold comes next, followed by the minimum and maximum numbers. Finally, the number of successful runs at that threshold, out of 10, is shown.

Targets at or below 85 percent are easy to find, averaging a little more than two searches. At the 90 percent threshold, however, there is a sudden jump requiring the creation of about 150 models on average. The minimum of 15 and the maximum of 270 implies a long tail to the distribution of the number of models tested.

Above 90 percent, the mean number of models increases again, rather dramatically, including long tails of up to 1,710 for 91.5 percent. The number of successful searches also goes down, implying that 2,000 iterations were an insufficient maximum in most cases to find a model with 92 percent or greater accuracy.

We now have two different approaches. The first blindly tries to find a suitable model by randomly assigning weights and biases and then testing over and over—a brute force approach. The second uses a principled swarm search to locate the weights and biases, and is more successful. For example, the previous swarm approach used 1,220 candidate models to find one with an accuracy of 93 percent, and a run of *elm_brute.py* needed 21,680 candidates to find a model with an accuracy of 93.5 percent.

It's impressive that the swarm techniques can find good models while searching such a high-dimensional space. The approach isn't practical without seriously reengineering the code to be orders of magnitude faster, but that we achieved any level of success is fascinating.

Extreme learning machines are a prime example of randomness in action. Their structure invites experimentation, so please experiment. If you discover something interesting, let me know. In the meantime, let's move on to our last example of randomness in machine learning.

Random Forests

A *random forest* is a collection (or ensemble) of *decision trees*. We'll define these terms shortly. The ideas behind random forests were developed in the 1990s and brought together by Breiman in his appropriately named 2001 paper, "Random Forests." Hence they have some history behind them. Decision trees themselves are even older, dating from the early 1960s. Let's begin there.

Decision Trees

A decision tree is a machine learning model consisting of a series of yes or no questions asked not of a person, but of a feature vector. The sequence of answers for that feature vector moves through the tree from the root node to a leaf, a terminal node. We then assign the class label associated with the leaf to the input feature vector.

Decision trees have the benefit of interpretability—by their very operation, they explain themselves. Neural networks can't easily explain themselves, an issue that's given birth to *Explainable AI (XAI)*, a subfield of modern deep learning.

Decision trees are best understood with an example, for which we'll use a tiny dataset of two measurements of three species of iris flowers. The dataset is two-dimensional; there are two features with 150 samples total. We'll use the first 100 for training the decision tree and the remaining 50 for testing. As the dataset has only two dimensions, we can plot its features by class; see Figure 6-4.

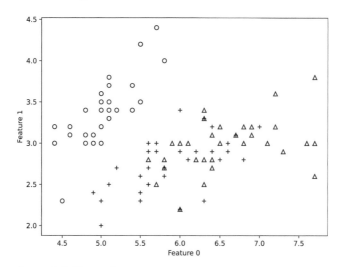

Figure 6-4: The iris features

In Figure 6-4, circles represent class 0, squares class 1, and diamonds class 2. Class 0 is well separated from the other two classes, which overlap considerably. Therefore, we might expect the decision tree classifier to do well with class 0 and be most often confused by class 1 and class 2.

Let's build a decision tree for this dataset. The code I'm using, which also generates Figure 6-4, is in the file *iris_tree.py*. It uses scikit-learn's `DecisionTreeClassifier` class and limits the depth of the tree to three. As with most scikit-learn classes, the `fit` method uses the training data and the `predict` method uses the test data. The output of *iris_tree.py* is:

```
[[18  0  0]
 [ 1  4 11]
 [ 0  1 15]]
0.74
```

This is a confusion matrix and has an overall accuracy of 74 percent. Class 0 was perfectly classified, 18 out of 18, while the decision tree labeled most of class 1 as class 2.

Figure 6-5 shows what the tree looks like.

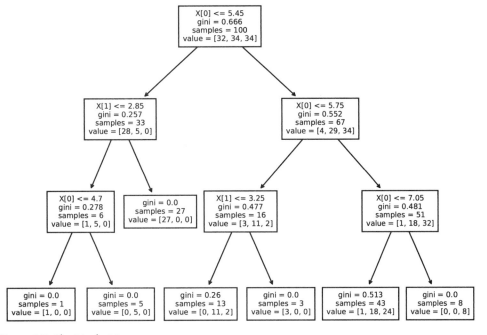

Figure 6-5: The iris decision tree

The tree's root is at the top. Each box is a node, and the first line of each box contains a question—in this case, "Is $x_0 \leq 5.45$?" or is feature 0 less than or equal to 5.45? If the answer is yes, move to the left; otherwise, move to the right. Then consider the question in that node. Continue this process until you reach a *leaf*, or a node with no children.

The value part of a node indicates the number of training samples of each class present at that node. For example, the leftmost leaf node has value=[1,0,0], meaning only one member of class 0 is present at this node. Therefore, any path leading to this node assigns class 0 to the input feature vector. Likewise, the leaf immediately to the right is labeled [0,5,0], so feature vectors leading to this node are assigned to class 1. Finally, the leaf second from the right is labeled [1,18,24], meaning one class 0 training sample landed in this node, as did 24 samples of class 2. The majority rules when more than one class is represented, so this node assigns feature vectors to class 2.

There are two other lines in each node: samples and the Gini index. The former is the sum of the values vector, the number of training samples present at that node. The decision tree algorithm uses the Gini index to split nodes; we won't go into details here, but you can check out the scikit-learn documentation to learn more.

Classically, decision trees are deterministic; the same dataset generates the same decision tree. Scikit-learn modifies this behavior somewhat, but by fixing the pseudorandom seed, which I do recklessly in *iris_tree.py*, we can generate a repeatable tree. We'll assume going forward that decision trees are entirely deterministic.

Decision trees explain themselves. For example, an input feature vector of {5.6, 3.3} will traverse the path in Figure 6-6.

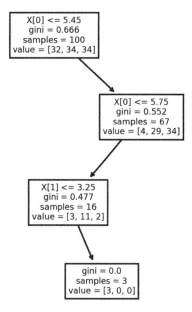

Figure 6-6: A path through the decision tree

This is an example of class 0 because feature 0 is between 5.45 and 5.75, and feature 1 is greater than 3.25.

The deterministic nature of decision trees led to the development of random forests. The decision tree for a particular dataset either works well or doesn't; they tend to *overfit* the data, being too sensitive to the particulars of the training set and not sensitive enough to the general characteristics of the type of data the model might encounter—at least not as sensitive as we'd like. In other words, they might do well on the training set and less well on everything else.

Let's learn how we can curtail the decision tree's penchant for overfitting by introducing randomness.

Additional Randomness

Three techniques turn a solitary decision tree into a forest of decision trees, or a random forest. The first is *bagging*, which generates many new datasets by sampling the original dataset with replacement (called *bootstrapping*). The second uses random subsets of the available features for each new, bootstrapped dataset, and the third is *ensembling*, the creation of multiple models that somehow vote or otherwise combine their output. Let's review these techniques before putting them together to grow random forests.

Creating Datasets with Bagging

I have a dataset with 100 samples, feature 0 from the iris dataset. The values are measurements from a parent distribution, a population from which this particular collection of values is a sample. The sample has a mean value of 5.85, which we can take as an *estimate* of the population mean, but not necessarily the actual value. Over what range of values do we expect, with 95 percent confidence, to find the actual population mean?

We have only one set of 100 values, but we'll use bootstrapping to generate another set, taken randomly from the first and allowing for the possibility of selecting the same sample more than once. This is known as "sampling with replacement." We now have two collections of samples, both plausibly from the parent population. We can repeat this process many times to generate many sets, each of which has a mean value. Then, with the collection of mean values, we use NumPy's quantile function to estimate the 95 percent confidence range.

The file *bootstrap.py* implements this process. It loads the iris training set, keeping only feature 0, and then generates 10,000 bootstrap datasets, keeping the mean value of each. To get the 95 percent confidence interval, we need to know the 2.5 percent and 97.5 percent quantile values. A quantile provides the percentiles for a dataset. The 50th percentile is the median, the middle value once the data has been sorted. The 25th percentile means that 25 percent of the data is below this value, while the 75th percentile means that 75 percent of the data is under this value.

To get the 95 percent confidence range, we want the values where 95 percent of the data is between them, implying we need to exclude 5 percent of the data: the bottom and top 2.5 percents. For that, we need the 2.5th and 97.5th percentiles, both of which we find courtesy of NumPy. Let's review the code, shown in Listing 6-19.

```
import numpy as np
from RE import *
```

```
def bootstrap(x):
    n = RE(mode="int", low=0, high=len(x)).random(len(x))
    return x[n]

x = np.load("iris_train_data.npy")[:,0]

means = [x.mean()]
for i in range(10000):
    y = bootstrap(x)
    means.append(y.mean())
means = np.array(means)

L = np.quantile(means, 0.025)
U = np.quantile(means, 0.975)

print("mean from single measurement %0.4f" % x.mean())
print("population mean 95%% confidence interval [%0.4f, %0.4f]" % (L,U))
```

Listing 6-19: Bootstrapping confidence intervals

First, load the iris training data, keeping feature 0 (x). Then generate 10,000 bootstrap versions of x tracking the mean of each. Use quantile to find the lower (L) and upper (U) confidence interval bounds before reporting them.

The bootstrap function creates n, a vector of the same length as x where each value is an integer in the range 0 through len(x)-1. These are indices into x where it's possible the same index appears more than once—a sample with replacement.

Each run of *bootstrap.py* produces a slightly different range:

```
> python3 bootstrap.py
mean from single measurement 5.8500
population mean 95% confidence interval [5.6940, 6.0090]
```

This output indicates that the true population mean is, with 95 percent confidence, between 5.694 and 6.009. Without the bootstrap process, we couldn't know this range from a single set of measurements. If we made assumptions about the shape of the distribution, that it's normal or follows a t-distribution, then we could make an estimate; by bootstrapping, we don't need to assume normally distributed data.

Bootstrapping is a helpful technique to get confidence intervals when building random forests because each bootstrapped dataset is a plausible collection of measurements. In that sense, the bootstrapped datasets are newly acquired datasets for training a model. Bagging is the process of using bootstrapped datasets to train multiple decision trees.

Combining Models with Ensembling

Bagging helps because each decision tree is trained on a slightly different dataset; anything causing overfitting in one training set is hopefully compensated for in another.

We'll make an ensemble of decision trees with bagging, in which we train multiple models using bootstrapped datasets and average their predictions to see if it improves results over ordinary decision trees. We'll use the histology dataset from the beginning of this chapter. The file *bagging.py* trains a user-supplied number of decision trees, each with a different bootstrapped version of the histology dataset. We'll then apply each model to the histology test data and average the resulting predictions to produce an ensemble output. Averaging model output is one approach to ensembling; we'll use another, voting, later in this chapter.

Consider the relevant code in Listing 6-20.

```
def Bootstrap(xtrn, ytrn):
    n = RE(mode="int", low=0, high=len(xtrn)).random(len(xtrn))
    return xtrn[n], ytrn[n]


xtrn = np.load("../data/datasets/bc_train_data.npy")
ytrn = np.load("../data/datasets/bc_train_labels.npy")
xtst = np.load("../data/datasets/bc_test_data.npy")
ytst = np.load("../data/datasets/bc_test_labels.npy")


trees = []
for i in range(N):
    tr = DecisionTreeClassifier()
    if (bag):
        x,y = Bootstrap(xtrn,ytrn)
        tr.fit(x,y)
    else:
        tr.fit(xtrn,ytrn)
    trees.append(tr)


preds = []
for i in range(N):
    preds.append(trees[i].predict(xtst))
preds = np.array(preds)
pred = np.floor(preds.mean(axis=0) + 0.5).astype("uint8")
cm, acc = Confusion(pred, ytst)
```

Listing 6-20: Using bagging to build an ensemble of decision trees

The code defines a `Bootstrap` function, loads the histology train and test datasets, creates and trains multiple decision trees using bootstrapped training sets, and makes predictions on the test data before averaging the results to create a final, aggregate set of predictions.

The first for loop creates a decision tree object (that is, an instance of DecisionTreeClassifier). If bag is true, it trains the tree using a bootstrapped version of the training data. If bag is false, it uses the entire dataset each time (no bagging). The Bootstrap function needs to select both the feature vectors and the proper label.

The next loop creates preds, a list of predictions for each decision tree. Each tree was trained on a different bootstrapped dataset, so if bag is true, the predictions will have slightly different errors. Turning preds into a NumPy array lets us average down the rows to get a single vector that's the average of all tree outputs for each test sample (the columns). We want to assign a class label, 0 or 1, to the average, so we add 0.5 and then use floor to round to the nearest integer (0 or 1).

Finally, a call to Confusion builds the confusion matrix and overall accuracy, which later code (not shown) displays.

One run of *bagging.py*, using a collection of 60 trees and bagging, returned:

```
> python3 bagging.py 60 1
Bagging with 60 decision trees:
[[101   3]
 [  9  58]]
overall accuracy 0.9298
first six accuracies: 0.9357 0.9064 0.9357 0.8830 0.9123 0.9123
```

The overall ensemble accuracy was 92.98 percent. The accuracies of the first six decision trees are also shown. Your runs will produce different output.

Running the code again with no bagging returns:

```
> python3 bagging.py 60 0
Bagging with 60 decision trees:
[[99  5]
 [13 54]]
overall accuracy 0.8947
first six accuracies: 0.9006 0.9006 0.8947 0.8947 0.8713 0.8947
```

This is a significantly worse outcome.

I ran *bagging.py* 30 times, first with bagging and again without. The respective mean accuracies were 92.73 percent and 89.77 percent, with a p-value of less than 10^{-10} using the Mann-Whitney U test. Bagging has a dramatic effect on the quality of the models.

Bootstrapped training sets and ensembling make up two-thirds of a random forest. Now let's add the remaining third: random feature sets.

Using Random Feature Sets

The final ingredient in a random forest is to use random subsets of the available features. Traditionally, the number of features we'll use is the square root of the available features. For example, the histology dataset has 30 features, so we'll use 5, selected randomly, whenever we want a bootstrapped dataset.

The formula is as follows:

1. Select a bootstrapped dataset using five randomly selected features.
2. Train a decision tree using this dataset. Keep it and the specific set of features selected (for testing).
3. Repeat step 1 and step 2 for each tree in the forest.
4. Apply each tree to the test data using only the features that the tree expects.
5. Average the results across the test set to get the final predictions from the random forest.

In code, adding random feature selection is only a minor tweak to the bagging example; take a look at *forest.py*. Listing 6-21 shows the relevant changes from *bagging.py*.

```
def Bootstrap(xtrn, ytrn):
    n = RE(mode="int", low=0, high=len(xtrn)).random(len(xtrn))
    nf = xtrn.shape[1]
    m = np.argsort(RE().random(nf))[:int(np.sqrt(nf))]
    return xtrn[n][:,m], ytrn[n], m

trees = []
for i in range(N):
    tr = DecisionTreeClassifier()
    x,y,m = Bootstrap(xtrn,ytrn)
    tr.fit(x,y)
    trees.append((tr,m))

preds = []
for i in range(N):
    tr,m = trees[i]
    preds.append(tr.predict(xtst[:,m]))
preds = np.array(preds)
```

Listing 6-21: Implementing a random forest

First, we must modify Bootstrap to select not only a random sampling of the training set (n) but also a random set of features (nf leading to m). The particular features extracted must be returned so we can use them at test time.

The second paragraph trains the trees as before, but drags m along for the ride. Finally, at test time, we apply each tree to xtst after keeping only the proper subset of features. The output of *forest.py* is identical to that of *bagging.py*, minus the first six individual tree accuracies.

I ran the code 30 times using 60 trees, as before, to get the mean accuracies in Table 6-3. I've included previous mean accuracies to show the improvement as each phase of the random forest is added.

Table 6-3: Mean Accuracy by Model Type

Option	Mean accuracy
No bagging	89.47%
Bagging, ensembling	92.98%
Bagging, ensembling, random features	95.25%

All random forest steps combined led to a significant improvement over a simple decision tree.

As you might expect, scikit-learn also supports random forests via the RandomForestClassifier class:

```
from sklearn.ensemble import RandomForestClassifier
```

The RandomForestClassifier class supports all three tricks plus additional tricks; see the scikit-learn documentation. Training 30 instances of RandomForestClassifier using 60 trees and all defaults results in a mean accuracy of 96.55 percent, which is even better than *forest.py*.

Models Combined with Voting

Before we leave this section, let's investigate how the size of the forest affects performance. For this experiment, we'll switch to the MNIST digits dataset. Also, instead of averaging each model's output, we'll vote to assign class labels.

The code we want is in *forest_mnist.py*. Built on *forest.py*, it loops over different sized forests to determine the mean accuracy for 20 models with that many trees. The output is a plot, but before we examine it, let's review how to implement voting:

```
preds = []
for i in range(N):
    tr,m = trees[i]
    preds.append(tr.predict(xtst[:,m]))
preds = np.array(preds)

pred = []
for i in range(preds.shape[1]):
    pred.append(np.argmax(np.bincount(preds[:,i])))
pred = np.array(pred)

cm, a = Confusion(pred, ytst)
acc.append(a)
```

The first code paragraph captures the predictions for each tree in the forest. The second creates pred, a vector that holds the most commonly selected class label across each model for each test sample. To get the winner for test sample i, we first use bincount to count how often each label appears for all models (the rows of `preds`) and then use `argmax` to return the index of the class label with the highest count. Ties go to the first occurrence of the highest value, that is, the lower index. Breaking ties this way might introduce a slight bias toward lower class labels, but we can live with that—consider it a systematic error as all forest sizes are likewise affected.

Figure 6-7 illustrates how the accuracy changes with forest size.

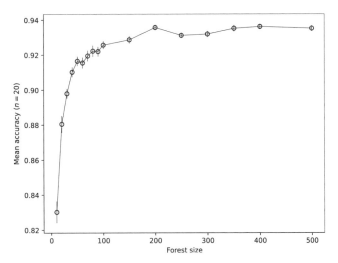

Figure 6-7: Mean accuracy on the digits dataset as a function of forest size

At first, increasing the number of trees helps, but eventually, saturation sets in, producing diminishing returns as the forest grows.

Exercises

Machine learning is a vast and critically important field. Here are some exercises related to the topics of this chapter to help you enhance your machine learning expertise and intuition:

- We augmented the digits dataset rather conservatively, with small rotations and zooming. Other image processing options might help improve the performance of the models in this chapter. Experiment with them by adding other options to the `augment` function in *build_mnist_dataset.py*. The `Image` and `ImageFilter` classes in Python's `PIL` module might be helpful. Use `Image.fromarray` to convert a NumPy array of `dtype` uint8 to a PIL image. To go the other way, pass the `Image` object to `np.array`.

- The `Classifier` subclass of `MLPClassifier`, used by *init_test.py*, defines multiple approaches to initializing a neural network. Add new ones to see how they affect results. What happens if all weight matrices are initially zero or some constant value? What about when the bias vectors are zero? Consider experimenting with the beta distribution (np.random.beta), as adjusting its two parameters can generate samples with a wide range of shapes.

- The extreme learning machine uses a randomly generated weight matrix and bias vector to map the input to the first hidden layer. The selection method we used in this chapter is commonly found in the literature. What happens if you alter this method and select random values according to a nonuniform distribution? Consider np.random.normal (or the RE-based version) along with the beta distribution. Is there much of an effect?

- Create a two-layer extreme learning machine where w0, w1, b0, and b1 are randomly selected. The final weight matrix, now w2, will be learned as before from the output of the second hidden layer. Compare the performance to the single-layer version.

- You'll find files in the *datasets* directory beginning with *mnist_14x14*. They contain 14×14-pixel versions of all MNIST digits, $[0, 9]$. Try them in place of the four-digit version used throughout the chapter. How well do the various models perform?

- Modify *elm_brute.py* to track the number of all tested models over many runs for the same accuracy. Then use np.histogram and Matplotlib to plot histograms for a fixed accuracy (like 0.92). What shape do they have? Does the shape make sense to you?

- We largely ignored scikit-learn's RandomForestClassifier in favor of our homegrown version. Explore scikit-learn's approach in more detail by reading through the documentation page for the class and experimenting with the different options. Consider using both the histology and digits datasets.

- Run *rf_vs_mlp.py* followed by *rf_vs_mlp_results.py*. (Some patience is required.) Consider the output, which demonstrates how scaling the input feature vectors affects model performance. Which type of model is sensitive to the relative ranges of the features, the neural network or the random forest? Why might that be? Think about what a neural network is trying to do and compare that to the individual decision trees of the random forest. Why might one type of model care about the feature ranges while the other might not?

Summary

This chapter explored the importance of randomness in machine learning when building datasets, both in terms of ordering the samples during training and in augmenting existing samples with plausible new ones to enlarge the type of data from which models learn.

Next, we considered the initialization of neural networks. We subclassed scikit-learn's MLPClassifier to override its initialization method, allowing us to add alternate initialization approaches. We then examined their effect on model performance.

Following this, we explored extreme learning machines, a subtype of neural networks that employs randomness as one of its essential components. We learned how these machines perform on our datasets, then considered the effect of hidden layer size and activation function. We concluded the section by replacing the random weight matrix and bias vector with ones learned by a swarm optimization exercise. We discovered that swarm algorithms could generate models with performance beyond what most extreme learning machines provide (for the same architecture).

Lastly, we experimented with random forests, a collection of decision trees. We learned what decision trees are and how to build a random forest from collections of decision trees employing bagging, ensemble voting, and random feature selection.

The next chapter takes a break from the practical to enhance our lives through generative art.

7

ART

In this chapter, we'll explore generative art—images created by algorithms that use randomness. We'll begin with "random art," a catchall phrase for three different approaches to producing artistic images. Alternative methods exist, but these will give you a taste of what's possible. We'll also learn about fractals, which were extremely popular a few decades ago when personal computers were finally powerful enough to do interesting things with graphics.

We'll generate many color images in this chapter, though this book is in black and white. I suggest running the code to see them properly.

Creating Random Art

We'll begin with three examples. The first mimics a program that amazed early Apple II personal computer enthusiasts. It also introduces us to Python's turtle graphics package.

The second implements a random walk, effectively a simulation of Brownian motion, the wanderings of particles in a fluid when viewed through a microscope. It's a simple process that, with appropriate color tables, leads to pleasant images suitable for T-shirts and coffee mugs.

The last example warps a two-dimensional grid of points using randomly selected functions, rotations, and color tables to generate unique images, likewise suitable for print.

Moiré Patterns

Moiré patterns occur in digital images when the discrete world of the computer screen meets what should be a continuous representation, that is, when drawing a line that can't be accurately represented by the grid imposed by the display. Figure 7-1 shows an example.

Figure 7-1: "Brian's Theme" as rendered by an Apple II computer

Figure 7-1 is 1979's "Brian's Theme," a simple BASIC program created by Brian Howard; it could be found on the Apple II DOS 3.3 system master diskette from 1983. The program selected a random point, from which it drew line after line to the edge of the screen, stepping each time by a randomly selected value. The crude resolution of the 280×160-pixel Apple II display, along with the funky way points were colored when plotting, produced a moiré pattern.

Let's create our own moiré patterns. The code we want is in *moire.py*. It introduces us to Python's turtle graphics module (turtle), part of the standard Python distribution.

NOTE *The phrase* turtle graphics *comes from the Logo programming language, which used a graphical "turtle" to drag a pen across the screen to teach programming concepts to children.*

Simple commands like FD 10 (forward 10), RT 90 (right turn 90 degrees), and PD (pen down) made it possible to create complex drawings with relatively little programming knowledge; imagine an old-school spirograph game, but on the computer.

We want the turtle to draw straight lines from a randomly selected point near the center to the edges, stepping by some distance along the edge each time. The result is something like Figure 7-2.

Figure 7-2: "Brian's Theme" redux

Figure 7-2 is nothing more than straight line after straight line; the pattern emerges from the moiré effect.

First, we import the necessary modules and ask the turtle to configure the display:

```
import time
import turtle as tu
from RE import *

tu.speed(0)
tu.ht()
tu.getscreen().setup(500,500)
tu.getscreen().bgcolor('black')

x = np.linspace(-200,200,400)
y = np.linspace(-200,200,400)
```

We configure the turtle, here tu, to go as fast as it can and then hide itself (ht for "hide turtle"). This way, we see only what it draws.

The drawing window is adjusted to 500×500 pixels with a black background. The *x* and *y* dimensions follow using NumPy's linspace to generate

400 point vectors spanning −200 to 200. The turtle's screen places the origin in the center of the display window.

We'll draw a moiré pattern, wait a bit, then clear the screen and draw another—interactive visual art at its finest. We need a loop:

```
while (True):
    tu.clear()
    X,Y = RE(mode='int', low=-100, high=100).random(2)
    step = RE(mode='int', low=2, high=9).random()
    r,g,b = RE(mode='int', low=1, high=256).random(3)
    color = "#%02x%02x%02x" % (r,g,b)
    tu.color(color)

    for i in range(0,400,step):
        Line(X,Y, x[i],y[0], color)
        Line(X,Y, x[0],y[i], color)
        Line(X,Y, x[i],y[-1], color)
        Line(X,Y, x[-1],y[i], color)

    time.sleep(4)
```

To start a drawing, we clear the screen and select the origin point (X, Y) and the step size. We specify color as red, green, and blue using HTML notation; for example, #FF0000 is bright red, #E0B0FF is mauve, #A0522D is sienna, and so on, mixing red, green, and blue in proportion.

A simple loop then uses Line to draw the pattern from the center point to each of the four screen edges:

```
def Line(x0,y0,x1,y1,color):
    tu.color('white')
    tu.pu()
    tu.goto(x0,y0)
    tu.pd()
    tu.goto(x1,y1)
    tu.color(color)
    tu.goto(x0,y0)
    tu.pu()
```

The Line method first draws the requested line in white, then backs over it using the randomly selected color to produce a flashing effect.

Given the speed of modern computers, you might expect the moiré pattern to flash on the screen. However, Python's turtle is similar to its namesake in that it draws the pattern at its leisure, essentially matching the speed of the Apple II code we're mimicking.

We'll return to turtle graphics later in the chapter. For now, it's time to take a walk.

Random Walks

A *random walk* algorithm follows the "two steps forward, one step backward" model in two dimensions: beginning at the origin, repeatedly take steps in random directions. To turn this trivial algorithm into art, we track the sequence of steps to create an image where each step is assigned a color from one of Matplotlib's color tables.

A color table—also called a color map or lookup table—is a list of colors in red, green, and blue format. Most color tables, Matplotlib's included, have 256 entries. For example, if some value of interest is 129, or assignable to 129, then the associated color is stored at index 129 of the color table currently in use.

If the desired index is 129 and the color table is `viridis`, then the resulting RGB color value is:

```
>>> from matplotlib import cm
>>> cmap = cm.get_cmap("viridis")
>>> cmap(129)
(0.126453, 0.570633, 0.549841, 1.0)
```

This code snippet demonstrates how to access Matplotlib's color tables, but the output might not be as expected. Why are there four values, and why are they not in the range [0, 255] like turtle graphics' hexadecimal color values? RGB color values are often mapped to [0, 1], or as fractions of the maximum possible value of 1.0. To find the corresponding byte value, multiply by 255 and keep the integer part. The color in the previous code is #20918C in hexadecimal.

This explains the first three values, but the fourth component is the alpha value, representing the color's transparency. An alpha value of 1.0 is opaque, 0.0 is transparent, and 0.5 blends the color and the color of the pixel behind it (similar to stacked graphics planes). For our purposes, we'll allow for transparent backgrounds (what this means will become apparent when examining the code).

Matplotlib comes with 84 predefined color tables, called *maps*; see *color_map_names.txt* for a list. Our random walk code supports all of them in any combination.

While random walks "repeatedly take steps in random directions," I didn't specify the allowed set of directions. From a point, (x, y), there are either four or eight possible directions one might go, as shown in Figure 7-3.

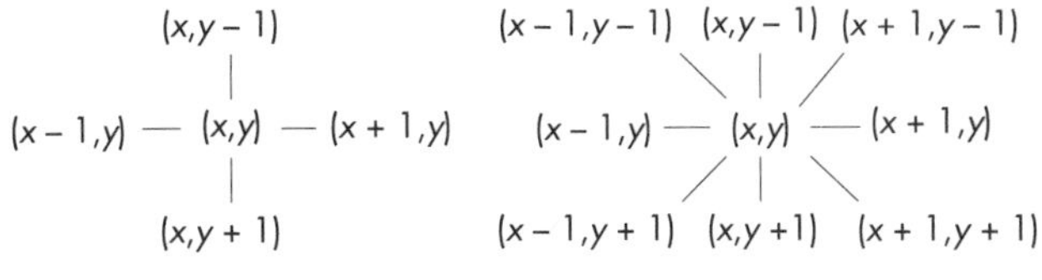

$$(x,y-1)$$
$$|$$
$$(x-1,y) \,—\, (x,y) \,—\, (x+1,y)$$
$$|$$
$$(x,y+1)$$

$$(x-1,y-1) \quad (x,y-1) \quad (x+1,y-1)$$
$$\diagdown \quad | \quad \diagup$$
$$(x-1,y) \,—\, (x,y) \,—\, (x+1,y)$$
$$\diagup \quad | \quad \diagdown$$
$$(x-1,y+1) \quad (x,y+1) \quad (x+1,y+1)$$

Figure 7-3: The four (left) or eight (right) directions a random walker can move

Random walkers on the left (called *4-connected*) are restricted to the cardinal directions: north, south, east, or west. Those on the right (*8-connected*) have the option to walk diagonally. The code in *walker.py* supports both options.

Notice how the offsets from (x, y) are marked. If you're used to plotting with Cartesian coordinates, as in math class, you might be a bit thrown by the signs. Ordinarily, we expect to plot in the first quadrant with x and y both positive and increasing to the right and up, respectively; this is not how most computers handle things. Instead, they put the origin at the upper-left corner of the screen so y increases as you move down and x increases as you move to the right. Mathematically, we're plotting in the fourth quadrant and ignoring the sign of the y-axis.

Let's run *walker.py* and see what it gives us:

```
> python3 walker.py 4 1000000 Reds,Oranges,Reds,Oranges none portrait tshirt.png mt19937 8675309
```

The command produces an output image according to the given extension, here a PNG file; see Figure 7-4.

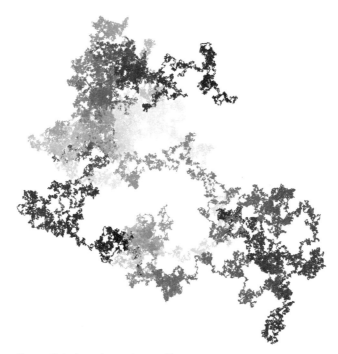

Figure 7-4: Sample random walker output

The first argument tells the code to take a 4-connected walk. The second is the number of steps to take *per specified color table*. The color tables come next, separated by commas and no spaces. Here, I've repeatedly used Matplotlib's Reds and Oranges. With 1,000,000 steps per color table and four color tables, the image in Figure 7-4 represents a stroll of 4 million random steps, each generating a single pixel.

After the color tables, we add the background color or the word none for a transparent background. We can specify colors using HTML hexadecimal format without the leading # so that 000000 is a black background and FFFFFF is white.

The following argument is the orientation for the output image. The walk will naturally be more extensive in one direction or the other. This argument orients the output as either portrait so the longest dimension of the output image is in the *y* direction, or landscape to use the *x* dimension.

The final two arguments are the randomness source and seed value for the source. Both are optional.

NOTE   *I called the output file* tshirt.png *because I used this example to make an actual T-shirt. There are multiple online services where you can upload an image and order a shirt. If you want to make your own, I recommend many millions of steps, a transparent background, and even encapsulated PostScript output format (use a* .eps *file extension on the command line).*

Let's take a walk through *walker.py*. I'll skip the part that imports the usual modules and parses the command line. Here's the code:

```
if (len(sys.argv) == 8):
    kind = sys.argv[7]
    rng = RE(kind=kind, mode="int", low=0, high=mode)
elif (len(sys.argv) == 9):
    kind = sys.argv[7]
    seed = int(sys.argv[8])
    rng = RE(kind=kind, mode="int", low=0, high=mode, seed=seed)
else:
    rng = RE(mode="int", low=0, high=mode)
```

We begin by configuring the randomness source. In this case, we choose random integers in the range $[0, 3]$ if 4-connected and $[0, 7]$ if 8-connected. We'll use the values as indices into a list of *x* and *y* offsets and add the offsets to the current point to take a step. The three cases in the code handle specifying a randomness source, a source with a seed value, or defaults.

Next comes the walk. The main loop tracks points in the lists X and Y, with colors in C. At this level, the loop is over the specified color tables:

```
X = []; Y = []; C = []
for cname in cnames:
    x,y,c = Walk(steps,cname,mode)
    X = X + x
    Y = Y + y
    C = C + c
```

The Walk function simulates a complete walk of steps steps for the current color table (cname) and mode:

```
def Walk(steps, cname, mode):
    try:
        cmap = cm.get_cmap(cname)
```

```
except:
    cmap = cm.get_cmap("inferno")
if (mode == 8):
    offset = [[0,-1],[1,-1],[1,0],[1,1],[0,1],[-1,1],[-1,0],[-1,-1]]
else:
    offset = [[0,-1],[1,0],[0,1],[-1,0]]
X = [0]
Y = [0]
C = [cmap(0)]
for i in range(steps):
    m = rng.random()
    X.append(X[-1] + offset[m][0])
    Y.append(Y[-1] + offset[m][1])
❶   c = cmap(int(256*i/steps))
    C.append((c[0],c[1],c[2]))
return X,Y,C
```

First, we get the color table (cmap), and then we define offset, a list of x and y offset values for 4-connected and 8-connected walks. The walk itself begins at $(0, 0)$ using the color at index 0 (C).

The loop over steps simulates the walk. We add a randomly selected offset to the last position for both x and y. Consider the code to select the color ❶. The loop index, i, ranges over [0, steps). Dividing by steps produces a fraction, [0, 1). Multiplying by 256 selects an index of the color table. Therefore, a full random walk steps through the entire color table once. There are other options to experiment with; see "Exercises" on page 238.

For now, there is no output image, only a collection of points (x, y) and associated colors in C. To make an image, we need CreateOutputImage:

```
def CreateOutputImage(X,Y,C, background):
    x = np.array(X)
    y = np.array(Y)
    xmin = x.min(); xmax = x.max()
    dx = xmax - xmin
    ymin = y.min(); ymax = y.max()
    dy = ymax - ymin
    img = np.zeros((dy,dx,4), dtype="uint8")

    if (background is not None) and (background != "none"):
        try:
            r = int(background[:2],16)
            g = int(background[2:4],16)
            b = int(background[4:],16)
            a = 255
        except:
            r,g,b,a = 0,0,0,0
    else:
        r,g,b,a = 0,0,0,0
```

```
img[:,:,0] = r; img[:,:,1] = g
img[:,:,2] = b; img[:,:,3] = a

for i in range(len(x)):
    xx = int((dx-1)*(x[i] - xmin) / dx)
    yy = int((dy-1)*(y[i] - ymin) / dy)
    c = C[i]
    img[yy,xx,0] = int(255*c[0])
    img[yy,xx,1] = int(255*c[1])
    img[yy,xx,2] = int(255*c[2])
    img[yy,xx,3] = 255
return img
```

The function is in three parts. The first makes NumPy vectors of the x and y points and determines the extent in each direction (dx, dy) to specify the output image, img. The output image uses 4 as the number of channels, not 3 as we've used previously; because we want to support a transparent background, we need to explicitly specify the alpha channel.

Next, we set the entire image to the transparent background color. Use 0 for the alpha channel for a transparent background and 255 for one that is completely opaque.

We loop over the points, mapping each point to a pixel of the image with explicit red, green, and blue color values. Review the assignments to xx and yy to follow how they map raw points to valid image coordinates.

Notice how img is indexed. The code assigns the point (x, y) to the image, but the image is indexed as (yy,xx). Images are indexed by row and then by column, meaning the y-coordinate—using the origin at the upper-left corner convention—represents the row, and the x the column.

We're almost done. The CreateOutputImage function returns the image as specified by the actual points of the walk. To produce the final output, we reorient the image to respect the portrait or landscape orientation:

```
img = CreateOutputImage(X,Y,C,background)
rows, cols, _ = img.shape
if (orient == "portrait"):
    if (rows < cols):
        img = img.transpose([1,0,2])
else:
    if (rows > cols):
        img = img.transpose([1,0,2])
Image.fromarray(img).save(oname)
```

The transpose method rearranges the columns of a NumPy array as specified. We swap rows and columns as needed, but ensure the number of channels remains in place. Read through *walker.py* to follow the overall flow of the code.

I've included several example outputs in the chapter files to inspire you to create your own. I especially like *example1.png*; it reminds me of the sci-fi sets used in *Doctor Who* episodes from the 1970s. The *example5.png* file uses *hotbits.bin* when that file was 1,130,496 bytes. The cyclic pattern is due to the number of requested points, which exceeds the number of random samples extracted from the file, thereby repeating. The *example5.txt* file contains the text of the command. If you create any compelling examples, please share them. I'll make a small gallery on the book's GitHub page.

Let's change gears and use randomness in combination with determinism to produce images constructed from warpings of the *xy*-plane.

A Grid

What happens if you take an evenly spaced grid of points in two dimensions, apply a function that maps each point to a new point, then plot the points? In other words, what happens when you warp a grid of points? In this section, we'll discover the answer and use it to build random works of abstract digital art.

Figure 7-5 illustrates the basic process.

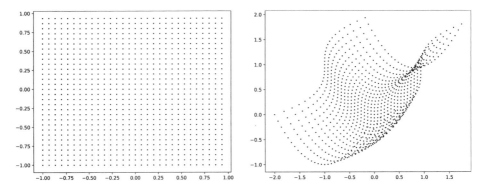

Figure 7-5: A grid of points (left) and the same grid warped (right)

On the left of Figure 7-5, we have a grid of 30×30 points. We apply a function to each point, (x, y), to generate a new point, (x', y'), plotted on the right of Figure 7-5.

The function is:

$$(x', y') \leftarrow (y^3 + x, x^2 + y)$$

For input point (x, y), the output *x*-coordinate is $y^3 + x$ and the *y*-coordinate is $x^2 + y$.

Likewise, the function $(x, y) \rightarrow (yx^2, xy^2)$ changes the uniform grid of points into Figure 7-6.

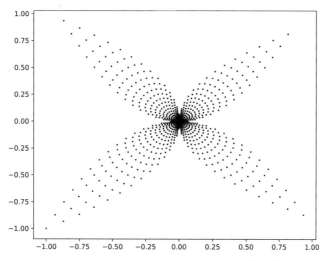

Figure 7-6: A uniform grid warped by $(x, y) \rightarrow (yx^2, xy^2)$

We'll generate a set of warping functions, select one randomly, and apply it to the uniform grid of points. Then, we repeat this process for a specified number of cycles, keeping all output points as we go.

To add color, we make each warping function return a color table index along with the new point location. Therefore, each warping function maps an input point to a new output point and color table index. Each cycle selects a new color table at random.

To add more randomness, we'll shift the collection of points by adding randomly selected x and y offsets. Because we're adventurous types, we'll also rotate the points by a random angle about the origin.

I've defined a collection of five warping functions, two of which we've already seen:

$$(x, y) \rightarrow (x^3 + y, \; y^2 + x, \; (xy + 1)/2)$$
$$\rightarrow (y^3 + x, \; x^2 + y, \; (xy + 1)/2)$$
$$\rightarrow (yx^2, \; xy^2, \; (x + 1)/2)$$
$$\rightarrow (xy^2, \; yx^2, \; (y + 1)/2)$$
$$\rightarrow (xe^y, \; ye^x, \; (xy + 1)/2)$$

The input grid is always in $[-1, 1]$. Each function returns the new point and color table index found by keeping the integer part after multiplying the third return value by 255. The cumulative effect of the random function applications, rotations, offsets, and color tables produces the desired output image.

The code is in *warp.py*. Run it like so

```
> python3 warp.py 300 11 example.png minstd 6502
```

to produce Figure 7-7.

Figure 7-7: Warping in action

The first argument to *warp.py* is the number of points along each grid dimension. The second is the number of cycles, or the number of times we apply a randomly selected function to the grid. Finally, the last three arguments are the output image name, randomness source, and seed, if desired.

First are the warp functions. For example:

```
def f0(a,b):
    x,y = a**3 + b, b**2 + a
    c = int(255*(a*b+1)/2)
    return x,y,c
```

This implements the first warping function in the collection. The remaining four functions are similar. We store the warp functions in a list so they can be selected at random:

```
funcs = [f0,f1,f2,f3,f4]
```

The main body of *warp.py* parses the command line and loads the color table names from *color_map_names.txt*. It then defines empty lists to hold all the (x, y) points and associated RGB colors, X, Y, and C, respectively.

The grid is specified as an npoints long vector from −1 to +1:

```
v = -1 + 2*np.arange(npoints)/npoints
```

All the action is in the loops over cycles and grid points:

```
for k in range(cycles):
    n = int(len(cnames)*rng.random())
    cmap = cm.get_cmap(cnames[n])
    n = int(len(funcs)*rng.random())
    fn = funcs[n]
    xoff,yoff = rng.random(2)-0.5
    theta = np.pi*rng.random()
    for i in range(len(v)):
        for j in range(len(v)):
            n,m,c = fn(v[i],v[j])
            x = n*np.cos(theta) - m*np.sin(theta)
            y = n*np.sin(theta) + m*np.cos(theta)
            X.append(x+xoff)
            Y.append(y+yoff)
            C.append(cmap(c))
```

The outermost loop over k handles the cycles. Each cycle randomly selects a color map (cmap) and a warp function (fn). It also selects random x and y offsets (xoff, yoff) and a rotation angle (theta).

The two inner loops, over i and j, walk through v in both the x and y directions; they visit every point in the 2D grid. The selected function is applied to each point, n,m,c = fn(v[i],v[j]), to return a new point (n, m) and color table index (c). To rotate a point about the origin, we multiply it by a rotation matrix

$$\begin{bmatrix} x \\ y \end{bmatrix} = \begin{bmatrix} \cos\theta & -\sin\theta \\ \sin\theta & \cos\theta \end{bmatrix} \begin{bmatrix} n \\ m \end{bmatrix}$$

which, following the rules for matrix multiplication, becomes:

$$x = n\cos\theta - m\sin\theta$$
$$y = n\sin\theta + m\cos\theta$$

This matches the previous code.

Rotation complete, the final step is to add the x and y offsets. These shift the points in the x and y directions to keep them from overlapping.

Every warped, rotated, and offset point is appended to the list of points and colors. Notice that the warping function returns a color table index, but C holds the tuple returned by cmap for the current color table.

With all cycles complete, we produce the plot and dump it to disk. Matplotlib obliges:

```
plt.scatter(X,Y, marker=',', s=0.6, c=C)
plt.axis('off')
plt.tight_layout(pad=0, h_pad=0, w_pad=0)
plt.savefig(oname, dpi=300)
plt.show()
```

Here `oname` is the output image name; it's taken from the command line argument.

Unlike Matplotlib's `plot` command, `scatter` accepts a list of per-point colors, which is why we constructed `C` for each point. The `tight_layout` command removes extraneous space, which is handy after turning the axes off.

Run *warp.py* a few times to experiment with the code. If you make the grid too fine, with the first argument above 300 or thereabouts, you may run out of memory. While it isn't necessary to specify a pseudorandom generator and seed value, doing so lets you re-create the output. The file *warp_factor_9.py* uses a single pseudorandom generator with a user-supplied global seed to create a specified number of warp images. I ran the command

```
> python3 warp_factor_9.py 3141592 100 warpings
```

to create 100 images in the *warpings* directory. It took about 15 minutes to run and produced some attractive output. Matplotlib's cyclic color tables, like `flag`, make stunning displays.

While *warp.py* makes pretty pictures, the randomness isn't particularly profound; it selects from among a set of options, relying on combinations of possible options to produce novelty.

The following section introduces us to fractals: a world of math, emergent behavior, and randomness.

Fun with Fractals

Fractals are mathematical objects constructed from smaller copies of themselves; they are *self-similar*. Mathematical fractals are our focus, but approximate fractals are a common sight in nature—for example, the branching of a tree, the airways in your lungs, and the fronds of a fern.

In this section, we'll explore randomness as a means of generating fractal images. There is a rich literature on fractals for art and computer graphics. We'll accomplish the barest of introductions.

First, we'll play the chaos game to build simple fractal images with Python's turtle. Then, we'll learn about a more sophisticated approach to generating fractal images, iterated function systems (IFS). Finally, we'll put everything together to build *ifs.py*.

The Chaos Game

Let's play the *chaos game*, so named by mathematician and fractal explorer Michael Barnsley. It works like this:

1. Pick three points, (x_0, y_0), (x_1, y_1), and (x_2, y_2), to form the vertices of a triangle.

2. Set $x = x_0$ and $y = y_0$.

3. Pick a vertex of the triangle at random, (x_n, y_n) for n in $[0, 2]$.

4. Update using $x \leftarrow 0.5(x + x_n)$ and $y \leftarrow 0.5(y + y_n)$.

5. Plot (x, y).

6. Repeat from step 3.

This algorithm plots points in the xy-plane. Let's find out whether the set of points eventually covers the plane, or there's a pattern of some kind. Run *sierpinski.py*:

```
> python3 sierpinski.py
```

A small window should appear. The Python turtle is also there, jumping around the window seemingly at random. Each time it moves, it leaves a small dot behind. The dot color corresponds to which triangle vertex was selected in step 3.

Let the program run for some length of time to draw whatever it is it wants to draw. When you have waited as long as you can for the turtle to do its thing, press a key and examine the resulting image. It should look similar to Figure 7-8.

Figure 7-8: The chaos game in action

The image is a fractal known as the Sierpiński triangle after Wacław Sierpiński, a Polish mathematician who explored such shapes in 1915.

Fractals are constructed from copies of themselves. For example, the main triangle is built from three smaller triangles, each constructed from three still smaller triangles, and so on forever. Therefore, the game didn't fill the xy-plane but only a subset of points on it. Such things are called "fractal" because their dimension is a noninteger.

The fractal is more than a line, dimension 1, and less than a plane, dimension 2. There are multiple ways to define a fractal dimension, but perhaps the most common is the Hausdorff dimension. For the Sierpiński triangle, the Hausdorff dimension is $\log 3 \,/\, \log 2 = 1.5849625\ldots$, greater than a line but less than a plane.

Let's look at *sierpinski.py*, shown in Listing 7-1.

```
X = [-200,0,200];  Y = [-200,200,-200]
x = X[0]; y = Y[0]
colors = ['#E7FFAC','#ACE7FF','#97A2FF']
tu.color(colors[0])
rng = RE(mode='int', low=0, high=3)

done = False
def Done():
    global done
    done = True
tu.onkeypress(Done)
tu.listen()

while (not done):
    n = rng.random()
    x = 0.5*(x + X[n])
    y = 0.5*(y + Y[n])
    tu.color(colors[n])
    tu.goto(x,y)
    tu.pd()
    tu.dot(1)
    tu.pu()

tu.ht()
tu.done()
```

Listing 7-1: Uncovering the Sierpiński triangle

The first code paragraph configures the turtle, some colors, and the pseudorandom generator (rng). It also defines two lists holding the triangle's vertices, X and Y. The other important bit is assigning x and y to the first vertex. Any random initial value will do, but a few iterations of the game might be necessary to land on points within the fractal.

The second paragraph is merely for convenience. It defines a global, done, and an event handler for turtle graphics to call when we press a key. The handler's sole task is to set global done to True so the while loop exits. The turtle's onkeypress and listen methods tell the turtle that it should listen for keypresses in the drawing window and perform some action when it hears one.

All the cool stuff is in the third paragraph—the while loop that runs until a keypress. It selects a vertex at random, n, and updates x and y before placing a properly colored dot at that point. The process repeats, placing point after point.

The chaos game uses the vertices of the triangle and randomness to uncover the *attractor* of the fractal, or the set of points that make up the fractal.

The chaos game works for n-sided polygons ($n > 2$) as well, as the code in *polygon.py* demonstrates. I won't walk through it here; it's a glorified version of *sierpinski.py* accepting n on the command line, followed, optionally, by a randomness source and a seed value. Figure 7-9 shows the results for $n = 5$ and $n = 9$.

Figure 7-9: The chaos game with five-sided (left) and nine-sided polygons (right)

Read through *polygon.py* to understand what it's doing.

Iterated Function Systems

This section introduces us to *iterated function systems (IFS)*, systems of equation that map points in the plane to other locations. In the following section, we combine IFS and the chaos game to generate fractals, the attractors of the IFS. Let this serve as a warning that there's math with vectors and matrices ahead. If these are unfamiliar beasts, have no worries. We can use IFS to make fantastic images without grasping all of the details behind the process. Still, it's best to understand at a deeper level if we have the option.

To build an IFS, we need to understand a few concepts. First, that matrices map vectors in the xy-plane to new vectors in the xy-plane. Second, that a vector in the xy-plane is another way to refer to a point, (x, y). Third, what a contraction mapping is. And finally, that a set of contraction mappings—with offset vectors and associated probabilities—forms an IFS, enabling us to use the chaos game to create fractal images. Let's begin.

Earlier, we used a matrix to rotate a set of points by some angle, θ, about the origin:

$$\begin{bmatrix} x' \\ y' \end{bmatrix} = \begin{bmatrix} \cos\theta & -\sin\theta \\ \sin\theta & \cos\theta \end{bmatrix} \begin{bmatrix} x \\ y \end{bmatrix}$$

We write this as $x' = Mx$, where the bold lowercase letters are vectors and the bold uppercase letter is a matrix.

To be more specific, multiplying a two-dimensional vector

$$x = \begin{bmatrix} x \\ y \end{bmatrix}$$

by a matrix

$$M = \begin{bmatrix} a & b \\ c & d \end{bmatrix}$$

maps the vector to a new two-dimensional vector, x':

$$x' = \begin{bmatrix} x' \\ y' \end{bmatrix} = Mx = \begin{bmatrix} a & b \\ c & d \end{bmatrix} \begin{bmatrix} x \\ y \end{bmatrix} = \begin{bmatrix} ax + by \\ cx + dy \end{bmatrix}$$

This also shows us how to multiply a two-dimensional vector by a matrix: multiply each element of the rows of the matrix by the corresponding element of the vector and sum. We say the matrix M is a *mapping* from x to x'.

In Chapter 3, we learned about the Euclidean distance between two points. Consider two vectors, x_0 and x_1, and their mappings under matrix M: x'_0 and x'_1. If $d(x_0, x_1)$ is the Euclidean distance between x_0 and x_1, and we have

$$d(x'_0, x'_1) = \alpha d(x_0, x_1)$$

for $0 < \alpha < 1$, then M is a *contraction mapping* that moves points together. We'll use contraction mappings for our IFS, with each mapping represented by a matrix.

A mapping may also include a constant offset vector added to the vector found by applying the matrix. We write this as

$$\begin{bmatrix} x' \\ y' \end{bmatrix} = \begin{bmatrix} a & b \\ c & d \end{bmatrix} \begin{bmatrix} x \\ y \end{bmatrix} + \begin{bmatrix} e \\ f \end{bmatrix}$$

where adding vectors is element-wise, like NumPy. However, it's more convenient to replace the 2×2 matrix with a 3×3 matrix incorporating the offset vector:

$$\begin{bmatrix} x' \\ y' \\ 1 \end{bmatrix} = \begin{bmatrix} a & b & e \\ c & d & f \\ 0 & 0 & 1 \end{bmatrix} \begin{bmatrix} x \\ y \\ 1 \end{bmatrix} = \begin{bmatrix} ax + by + e \\ cx + dy + f \\ 1 \end{bmatrix}$$

A set of contraction mapping matrices, $\{M_0, M_1, M_2, \ldots\}$, forms an IFS where each M is a 3×3 matrix combining a mapping with an offset vector. IFS have *attractors*, sets of points that subsequent mappings with the IFS send back to other points in the attractor. In this way, mapping a point on the attractor using one of the matrices in the IFS results in another point on the attractor. The points forming the attractor are the fractal we want to image, and the chaos game is the tool we use to find them.

Fractals Plotted with Points

We're almost where we need to be. We have an IFS, a collection of matrices, M. We also know that the chaos game helps us find the points on the attractor of the IFS. To adapt the chaos game to a set of matrices, we assign a probability to each map. Then, when playing the chaos game, we select the next map to apply based on that probability. The assigned probabilities alter the weighting of the attractor points.

Let's get practical. The code we'll work with is in *ifs.py*. Read through the file to give yourself an overview. The majority of the file is the IFS class. Running *ifs.py* without arguments teaches us how to configure the command line:

```
> python3 ifs.py

ifs <points> <output> <fractal> <color> [<kind> | <kind> <seed>]

    <points>   - number of points to calculate
    <output>   - output image
    <fractal>  - name from the list below or 'random'
    <color>    - <hex> (no '#')|maps
    <kind>     - randomness source
    <seed>     - seed value


circle dragon fern koch shell sierpinski tree thistle
maple spiral mandel tree2 tree3 fern2 dragon2
```

The words at the bottom of the output are assigned to different IFS, meaning different sets of mappings and probabilities. We specify the IFS we want on the command line by name. The maps are hardcoded in a dictionary embedded in the IFS class. For example, here's the definition of the sierpinski IFS:

```
"sierpinski": {
    "nmaps":3,
    "probs":[0.3333,0.3333,0.3333],
    "maps":[
            [[.5, 0, 0], [0, .5, 0], [0,0,1]],
            [[.5, 0, .5], [0, .5, 0], [0,0,1]],
            [[.5, 0, .25], [0, .5, .5], [0,0,1]]]},
```

The IFS consists of three maps, or three 3×3 matrices:

$$
\begin{bmatrix} 0.5 & 0.0 & 0.0 \\ 0.0 & 0.5 & 0.0 \\ 0.0 & 0.0 & 1.0 \end{bmatrix},
\begin{bmatrix} 0.5 & 0.0 & 0.5 \\ 0.0 & 0.5 & 0.0 \\ 0.0 & 0.0 & 1.0 \end{bmatrix},
\begin{bmatrix} 0.5 & 0.0 & 0.25 \\ 0.0 & 0.5 & 0.5 \\ 0.0 & 0.0 & 1.0 \end{bmatrix}
$$

Each is selected with probability $1/3$ and is equally likely to be chosen when playing the chaos game.

The other IFS are defined similarly, though with more or fewer maps, as necessary. The maple IFS, and those following, are by Paul Bourke (*https://paulbourke.net/fractals/ifs*), used with permission. Bourke's site is a treasure trove of fascinating computer graphics and geometry pages, including many on fractals and IFS. I highly recommend taking a look.

The remaining command line arguments are as described, with two options worth special mention. The first is using maps for the color. Specifying maps has the same effect as *sierpinski.py*; points are plotted in a color associated with the selected IFS map. This option reveals the maps in the output image. The second is using random for the IFS, which generates an IFS at random and then iterates to find the fractal it represents. We'll experiment with this option shortly. For now, let's use the fern IFS to see what sort of output we get. Give this command line a go:

```
> python3 ifs.py 1_000_000 fern.png fern maps
```

It plots 1 million points of the attractor for the fern IFS using a different color for each of the four maps. The result is Figure 7-10.

Figure 7-10: One million points of the fern attractor

The fourth map is the narrow stem of the fern. Explore the other fractals supported by *ifs.py* or browse the *misc* directory, which contains images of each using 1 million points.

The *ifs.py* code uses Matplotlib to produce the output plot. Matplotlib's graphics are interactive. Try generating 10 million points of the shell IFS and then click the magnifying glass icon and draw a box around the center to zoom in. With 10 million points, you should be able to zoom in two or three times. The spiral continues forever.

That IFS fractals resemble objects in nature can't be a mere coincidence; there must be a biological basis for self-similar patterns, even if they're not rigorous but only approximate mathematical fractals.

While it's fun to look at the pretty fractal images *ifs.py* creates, it's even more fun to understand the "how" behind them. Let's examine the code and then experiment with purely random IFS fractals.

The IFS Class

The dictionary of predefined maps constitutes the bulk of the IFS class. In terms of actual code, there are a handful of methods to contemplate: ChooseMap, GeneratePoints, StoreFractal, and RandomMaps. I'll save RandomMaps for the next section; StoreFractal is a straightforward application of Matplotlib to plot the generated points using the associated colors. That leaves GeneratePoints and ChooseMap. Let's begin with GeneratePoints, as shown in Listing 7-2.

```
def GeneratePoints(self):
    self.xy = np.zeros((self.npoints,3))
    xy = np.array([self.rng.random(), self.rng.random(), 1.0])

    for i in range(100):
        m = self.maps[self.ChooseMap(),:,:]
        xy = m @ xy

    for i in range(self.npoints):
        k = self.ChooseMap()
        m = self.maps[k,:,:]
        xy = m @ xy
        self.xy[i,:] = [xy[0],xy[1],k]
```

Listing 7-2: Finding the points on the fractal attractor

The GeneratePoints method fills in a NumPy array of npoints rows and three columns: *x*, *y*, and the index of the selected map. The code iterates a randomly initialized vector, xy. The third element, a constant 1, allows us to use 3×3 matrices for the maps.

A randomly selected point isn't likely to be on the attractor; therefore, before we store points in self.xy, we iterate 100 times to ensure that new points are on the attractor. Each iteration involves selecting a map according to the assigned probabilities followed by a matrix multiplication with that map, $x \rightarrow Mx$, which becomes xy = m @ xy in code. NumPy uses @ for matrix multiplication.

While the GeneratePoints method does the lion's share of the work, it relies on ChooseMap (Listing 7-3).

```
def ChooseMap(self):
    r = self.rng.random()
    a = 0.0
    k = 0
    for i in range(self.nmaps):
        if (r > a):
            k = i
```

```
        else:
            return k
        a += self.probs[i]
    return k
```

Listing 7-3: Choosing a map

The IFS constructor creates member variables nmaps, probs, and maps according to the selected IFS. The ChooseMap method returns an index into the list of maps by picking a random number in [0, 1) (r) and then adding per-map probabilities until the accumulated sum (a) exceeds the selected value, at which point it returns the current index, k. The effect selects maps in proportion to their assigned probabilities.

Using the IFS class is straightforward:

```
app = IFS(npoints, name, ctype, rng, show=True)
app.GeneratePoints()
app.StoreFractal(outfile)
```

The constructor accepts the number of points, fractal name, color, an initialized RE object, and a flag to show or hide the fractal when we call StoreFractal. The GeneratePoints and StoreFractal methods complete the process. Encapsulating IFS in a single class with a small main driver lets us use *ifs.py* as a program or module; we'll use it as the latter in the next section.

Random IFS

Running *ifs.py* with random as the fractal name generates a set of random maps with random probabilities. Each map is a 3×3 matrix

$$\begin{bmatrix} a & b & e \\ c & d & f \\ 0 & 0 & 1 \end{bmatrix}$$

with an associated probability, p. The six elements of the map must satisfy a set of constraints to be a contraction mapping. Specifically, the following must be true:

$$a^2 + d^2 < 1$$

$$b^2 + e^2 < 1$$

$$a^2 + b^2 + d^2 + e^2 - (ae - db)^2 < 1$$

The RandomMaps method creates a collection of maps and probabilities, as in Listing 7-4.

```
def RandomMaps(self):
    def mapping():
        while (True):
            a,b,c,d,e,f = -1 + 2*self.rng.random(6)
```

```
            if (a*a+d*d) >= 1:
                continue
            if (b*b+e*e) >= 1:
                continue
            if a*a+b*b+d*d+e*e - (a*e-d*b)**2 >= 1:
                continue
            break
        return [[a,b,c],[d,e,f],[0,0,1]]

    nmaps = 2 + int(4*self.rng.random()) # [2,5]
    probs = self.rng.random(nmaps)
    probs = probs / probs.sum()

    maps = []
    for k in range(nmaps):
        maps.append(mapping())

    return nmaps, probs, np.array(maps)
```

Listing 7-4: Creating a random IFS

The code first chooses a random number of maps, $[2, 5]$, and probabilities (probs). It then loops nmaps times, calling the embedded function, mapping, which returns a valid map matrix. The mapping method repeatedly selects random elements in $[-1, 1)$ until all constraints are satisfied.

Let's create 100 random fractals to get a feel for what the attractor of a randomly generated IFS looks like. The file *ifs_maps.py* contains the code we need. It uses a master seed value passed on the command line to generate a collection of random fractals. Here's the command line I used:

```
> python3 ifs_maps.py 100 fractals 271828 >ifs_maps_271828.txt
```

If you use the same command line, you'll get the same collection of fractals in the *fractals* directory. The file *ifs_maps_271828.txt* contains the randomly generated maps and the seed value. The code in *ifs_maps.py* is only a few dozen lines, but it demonstrates how to use the IFS class inside another program.

Figure 7-11 presents some fractals that were produced by the previous command line. (These are better in color.)

0	16	18
25	36	41
83	88	93

Figure 7-11: Randomly generated IFS fractals

Even without color, we get a taste of what random IFS can do. The numbers identify the fractal. The maps used are in *ifs_maps_271828.txt*.

IFS Maps

Now that we understand the process of iterating a function system represented as a collection of matrices and probabilities, we'll generate our own IFS maps. This section is optional, as some linear algebra is involved.

Consider Figure 7-12.

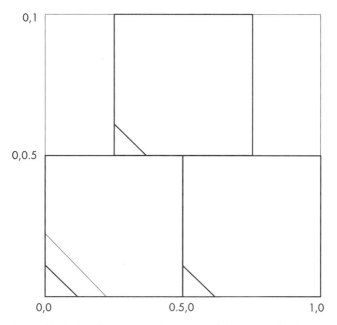

Figure 7-12: The three contraction maps of the Sierpiński triangle

This shows the effect of the three maps that make up the Sierpiński triangle. The maps take the unit square (outermost square) and map it to one of the three smaller squares. The diagonal line shows the orientation to clarify flips or rotations. Here the mappings shrink the unit square and offset it from the origin, but they don't flip or rotate. The recursive self-similar appearance of the attractor comes from repeating this mapping in each of the smaller squares, repeating in each of those, and so on forever.

Let's turn the illustration in Figure 7-12 into three $3{\times}3$ matrices, or maps. We'll track where the three points of the unit square, $(0, 0)$, $(1, 0)$, and $(0, 1)$, end up in each map.

We'll begin with the top map that maps the unit square into the range $[0.25, 0.75]$ in x and $[0.5, 1]$ in y. The map takes any point in the unit square to the proper place in this smaller square. The point at $(1, 0)$ must map to $(0.75, 0.5)$; similarly, $(0, 0) \rightarrow (0.25, 0.5)$ and $(1, 0) \rightarrow (0.25, 1)$. How do we find the matrix M that does this?

Mathematically, we need M to act like so

$$\begin{bmatrix} 0.75 \\ 0.5 \\ 1 \end{bmatrix} = \begin{bmatrix} a & b & e \\ c & d & f \\ 0 & 0 & 1 \end{bmatrix} \begin{bmatrix} 1 \\ 0 \\ 1 \end{bmatrix}$$

for properly selected a, b, c, d, e, and f. The same matrix must also map $(0, 0)$ and $(0, 1)$ to $(0.25, 0.5)$ and $(0.25, 1)$, respectively.

Performing matrix multiplication gives us two equations to map $(1, 0)$ to $(0.75, 0.5)$:

$$0.75 = a(1) + b(0) + e(1)$$
$$0.5 = c(1) + d(0) + f(1)$$

The third equation contains only constants, so we'll ignore it going forward.

We have two other points to consider that use the same matrix, so we have four more equations:

$$0.25 = a(0) + b(0) + e(1)$$
$$0.5 = c(0) + d(0) + f(1)$$
$$0.25 = a(0) + b(1) + e(1)$$
$$1.0 = c(0) + d(1) + f(1)$$

Let's find a, b, and e. The process repeats to find c, d, and f. If we group equations involving a, b, and e, we get:

$$0.75 = a(1) + b(0) + e(1)$$
$$0.25 = a(0) + b(0) + e(1)$$
$$0.25 = a(0) + b(1) + e(1)$$

This gives us three equations and three unknowns, so we can find a unique solution. First, let's rewrite the equations in matrix form:

$$\begin{bmatrix} 0.75 \\ 0.25 \\ 0.25 \end{bmatrix} = \begin{bmatrix} 1 & 0 & 1 \\ 0 & 0 & 1 \\ 0 & 1 & 1 \end{bmatrix} \begin{bmatrix} a \\ b \\ e \end{bmatrix}$$

This is a matrix equation of the form $b = Ax$, and we want to find x. One approach is to use *Cramer's rule*, which solves the equation, element by element, as the ratio of determinants. In particular, we find each element of x as

$$x_i = \frac{|A_i|}{|A|}$$

for $|A|$ the determinant of the matrix and $|A_i|$ the determinant of a matrix formed by replacing the ith column of A with b. Therefore, we get a, b, and e like so:

$$a = \frac{\begin{vmatrix} 0.75 & 0 & 1 \\ 0.25 & 0 & 1 \\ 0.25 & 1 & 1 \end{vmatrix}}{\begin{vmatrix} 1 & 0 & 1 \\ 0 & 0 & 1 \\ 0 & 1 & 1 \end{vmatrix}}, \quad b = \frac{\begin{vmatrix} 1 & 0.75 & 1 \\ 0 & 0.25 & 1 \\ 0 & 0.25 & 1 \end{vmatrix}}{\begin{vmatrix} 1 & 0 & 1 \\ 0 & 0 & 1 \\ 0 & 1 & 1 \end{vmatrix}}, \quad e = \frac{\begin{vmatrix} 1 & 0 & 0.75 \\ 0 & 0 & 0.25 \\ 0 & 1 & 0.25 \end{vmatrix}}{\begin{vmatrix} 1 & 0 & 1 \\ 0 & 0 & 1 \\ 0 & 1 & 1 \end{vmatrix}}$$

Working through the determinants, perhaps by using NumPy's np.linalg.det function, tells us that $a = 0.5$, $b = 0.0$, and $e = 0.25$. The corresponding equations for c, d, and f give 0.0, 0.5, and 0.5, respectively.

We now have the mapping matrix we need:

$$\begin{bmatrix} 0.5 & 0.0 & 0.25 \\ 0.0 & 0.5 & 0.5 \\ 0 & 0 & 1 \end{bmatrix}$$

This is indeed one of the sierpinski matrices in *ifs.py*.

We can generalize this process to write immediate solutions for a desired map. Assume three points on the unit square, (a_0, a_1), (b_0, b_1), and (c_0, c_1). If their desired positions after mapping are (A_0, A_1), (B_0, B_1), and (C_0, C_1), then we can find the map by calculating

$$a = \frac{\begin{vmatrix} A_0 & a_1 & 1 \\ B_0 & b_1 & 1 \\ C_0 & c_1 & 1 \end{vmatrix}}{\begin{vmatrix} a_0 & a_1 & 1 \\ b_0 & b_1 & 1 \\ c_0 & c_1 & 1 \end{vmatrix}}, \quad b = \frac{\begin{vmatrix} a_0 & A_0 & 1 \\ b_0 & B_0 & 1 \\ c_0 & C_0 & 1 \end{vmatrix}}{\begin{vmatrix} a_0 & a_1 & 1 \\ b_0 & b_1 & 1 \\ c_0 & c_1 & 1 \end{vmatrix}}, \quad e = \frac{\begin{vmatrix} a_0 & a_1 & A_0 \\ b_0 & b_1 & B_0 \\ c_0 & c_1 & C_0 \end{vmatrix}}{\begin{vmatrix} a_0 & a_1 & 1 \\ b_0 & b_1 & 1 \\ c_0 & c_1 & 1 \end{vmatrix}}$$

and:

$$c = \frac{\begin{vmatrix} A_1 & a_1 & 1 \\ B_1 & b_1 & 1 \\ C_1 & c_1 & 1 \end{vmatrix}}{\begin{vmatrix} a_0 & a_1 & 1 \\ b_0 & b_1 & 1 \\ c_0 & c_1 & 1 \end{vmatrix}}, \quad d = \frac{\begin{vmatrix} a_0 & A_1 & 1 \\ b_0 & B_1 & 1 \\ c_0 & C_1 & 1 \end{vmatrix}}{\begin{vmatrix} a_0 & a_1 & 1 \\ b_0 & b_1 & 1 \\ c_0 & c_1 & 1 \end{vmatrix}}, \quad f = \frac{\begin{vmatrix} a_0 & a_1 & A_1 \\ b_0 & b_1 & B_1 \\ c_0 & c_1 & C_1 \end{vmatrix}}{\begin{vmatrix} a_0 & a_1 & 1 \\ b_0 & b_1 & 1 \\ c_0 & c_1 & 1 \end{vmatrix}}$$

This process gives the maps, but says nothing about the associated probability for selecting the map when playing the chaos game. Intuition and experimentation help with assigning probabilities.

There are a plethora of IFS-related programs out there, many of which let you design IFS mappings interactively, so you need not work through the map calculations by hand. For example, Larry Riddle's IFS Construction Kit is a Windows program for creating and animating IFS fractals that includes a graphical designer (*https://larryriddle.agnesscott.org/ifskit*). I was able to run the program under Linux using wine after downloading the missing *.ocx* and *.dll* files Riddle refers to on the installation page.

Exercises

There's no end to creating generative art. Here are some exercises related to this chapter's experiments:

- The code in *moire.py* draws from a single point to the edges. Try drawing from a corner; then, try multiple points with different step sizes.

- Modify *walker.py* to select a random color table index or use the step number modulo 256 to cycle through the color table.

- Why limit *walker.py* to a single walker?

- The list of available functions in *warp.py* is rather Spartan. Add new ones that assume inputs in $[-1, 1]$ for both x and y and return new x and y values plus a color table index $[0, 255]$.

- What happens if you change the probabilities for each map in *ifs.py*?

- Create new IFS maps for *ifs.py* using the manual process described in the text. Do they produce the expected result?

- Create new IFS maps using IFS Construction Kit or a similar program and add them to the library in *ifs.py*.

Summary

This chapter introduced us to randomness in generative art, though we only scratched the surface. We explored images generated by the moiré effect, random walkers, and the application of randomly selected warping functions and color tables.

Fractals came next. We learned that IFS have attractors, or self-similar fractals. The chaos game, using a random sequence of mappings, falls onto the attractor, thereby giving us a means for generating an infinite number of fractal images. Randomly selected sets of maps produce fascinating and mesmerizing collections of fractal images. There seems to be a deep connection between the formation of structures in nature and the attractors of IFS.

This chapter focused on what is pleasant to look at. In the next chapter, we'll delve into what is pleasant to listen to.

8

MUSIC

In this chapter, we'll continue exploring randomness in art with sound and music. We'll begin generating sound via random samples, random walks through frequency space, and random walks up and down a musical scale. These projects will prepare us for the chapter's most ambitious experiment: evolving pleasant melodies from scratch. While we can't really quantify such a melody, that won't stop us from trying.

Creating Random Sounds

At first blush, generating a random sound seems straightforward. For instance, if we have some way of creating a sound file, like a WAV (*.wav* file extension), it follows that we should just need random sound samples at a specified playback rate—right? Let's implement this, as it will introduce us to the audio tools we need for this section.

WAV files are easy to read and write via SciPy's wavfile module. To write a WAV file, we need two things: a specified sampling rate and the samples themselves in some range that programs like mplayer or Audacity will understand.

We measure the sampling rate, the speed with which the samples are played back, in samples per second. The higher the sampling rate, the better the audio quality. A sampling rate of 22,050 Hz (cycles per second) is sufficient for our purposes. This is half the rate of a compact disc.

The samples are quantized voltages, a continuous range partitioned into a specified number of discrete steps, with each discrete value specifying a particular analog voltage level. The discrete samples correspond to the output audio waveform. Samples are usually 16-bit signed integers, but we'll work with 32-bit floating-point samples in the range [−1, 1]. Most audio programs will have little difficulty with floating-point samples.

To make random sounds, we need to generate random samples, set up the WAV output, and write the samples to disk for playback. Let's give it a go and hear what happens. The code we want is in *random_sounds.py*.

Let's run it first:

```
> python3 random_sounds.py 3 tmp.wav
```

Play the three-second output file, *tmp.wav*. I recommend turning down the volume first. Did you hear what you expected to? Consider Listing 8-1.

```
from scipy.io.wavfile import write as wavwrite

def WriteOutputWav(samples, name):
    s = (samples - samples.min()) / (samples.max() - samples.min())
    s = (-1.0 + 2.0*s).astype("float32")
    wavwrite(name, rate, s)

rate = 22050
duration = float(sys.argv[1])
oname = sys.argv[2]

nsamples = int(duration * rate)
samples = -1.0 + 2.0*np.random.random(nsamples)

WriteOutputWav(samples, oname)
```

Listing 8-1: Generating random samples

I excluded the usual message about the proper form for the command line to focus on the relevant code.

First, we import wavwrite from SciPy. I renamed write as wavwrite to clarify what the function does. Ignore the WriteOutputWav function for a moment.

The main part of the file fixes the sampling rate and reads the duration in seconds from the command line, along with the output WAV filename (oname), before calculating nsamples.

If the samples are played at a given rate, and we want a total duration in seconds, then the product, rounded to an integer, provides us with the number of samples we must generate. The `samples` are randomly selected in [-1, 1) using NumPy's pseudorandom generator. There's no point in using RE here.

All that remains is to use `WriteOutputWav` to create the output WAV file. We'll use this function for all the experiments in this section. The first line rescales the samples to be in the range [0, 1], which lets us be a bit freer with how we generate samples. The second line changes from [0, 1] to [-1, 1], the valid range for floating-point samples. The last line uses `wavwrite` to dump the WAV file.

The output of *random_sounds.py* is so grating due to how humans perceive sound. We like sound that is represented as nice collections of sine waves summed together; in other words, tones with a fundamental frequency and overtones (harmonics). A random collection of unrelated samples can be represented only by summing a large number of sine waves.

Figure 8-1 shows a sine wave at 440 Hz on the left and random noise on the right.

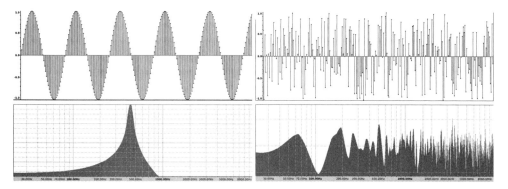

Figure 8-1: Top left: a sine wave; top right: random noise; bottom left: the frequency spectrum of the sine wave; bottom right: random noise

The top of Figure 8-1 shows the actual sound samples over time. The bottom shows the frequency spectrum, the strength of the various sine waves that go into the signal, so the *x*-axis is no longer time but frequency.

The sine wave is, fundamentally, a single frequency at 440 Hz. The energy at other frequencies is likely due to an imperfect approximation of the pure sine wave. The vertical scale is logarithmic, meaning there is very little energy outside of 440 Hz. The random noise spectrum, however, is roughly uniform over the entire frequency range up to 8,000 Hz, reflecting the number of sine waves that must be summed to approximate the random signal. The *x*-axis is similarly logarithmic.

The following two sections explore other approaches to random sound generation, both utilizing the idea of a random walk—not in space, but in frequency. We'll produce sound using the sum of sine waves. The first section pays no attention to the mix of frequencies, while the second uses frequencies from the notes of a C major scale.

Sine Waves

If we add two sine waves with different frequencies, they merge to become a new wave. Where the two sine waves are positive, they reinforce each other, and the resulting wave is more positive. When one is positive and the other negative, they cancel each other. For example, consider Figure 8-2.

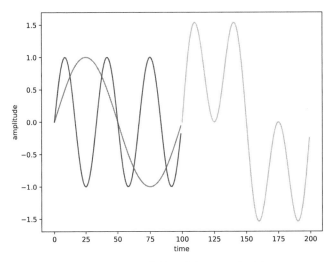

Figure 8-2: Two sine waves (left) and their sum (right)

On the left are two sine waves with frequencies in the ratio of 3:1. On the right is the sum of the two sine waves on the left. Sum enough waves and any desired output waveform is possible.

The code in *sine_walker.py* creates a collection of random walkers that each generate 0.5-second sine waves before altering the frequency used for the next 0.5 seconds. For each 0.5-second block of time, the final wave is the sum of all the walkers. Let's run the code and then walk through it:

```
> python3 sine_walker.py 5 3 walk.wav
```

This should produce a 5-second output file consisting of three independent sine wave random walks. Give *walk.wav* a listen; it reminds me of sound effects in 1950s science fiction movies.

The *sine_walker.py* file parses the command line and then configures the values we need for the random walks:

```
nsamples = int(duration * rate)
samples = np.zeros(nsamples, dtype="float32")
dur = 0.5
step_samp = int(dur * rate)
fstep = 5
freq = np.zeros(nwalkers, dtype="uint32")
freq[:] = (440 + 800*(rng.random(nwalkers)-0.5)).astype("uint32")
```

First, we use nsamples to define samples, which holds all output. The next two lines define dur, the step duration, and step_samp, the number of samples in a step. Each sine wave, for a specific frequency, creates this many samples. Next, fstep sets the step size in Hertz, and freq is a vector of initial frequencies in [40, 840) Hertz. The double-definition handles the case where there's only one walker.

We then loop until we have generated all samples. Each step is 0.5 seconds long, and each walker generates a sine wave with step_samp samples using its current frequency and a randomly chosen amplitude. We sum the walkers and assign the summed wave to the next 0.5 second's worth of samples; see Listing 8-2.

```
    k = 0
❶ while (k < nsamples):
  ❷ for i in range(nwalkers):
            r = rng.random()
            if (r < 0.33333):
                freq[i] += fstep
            elif (r < 0.66666):
                freq[i] -= fstep
            freq[i] = min(max(100,freq[i]),4000)
      ❸ amp = rng.random()
            if (i == 0):
                t = amp*np.sin(2*np.pi*np.arange(rate*dur)*freq[i]/rate)
            else:
                t += amp*np.sin(2*np.pi*np.arange(rate*dur)*freq[i]/rate)
        n = 1
  ❹ while (np.abs(t[-n]) > 1e-4):
            n += 1
        t = t[:-n]
        if ((k+len(t)) < nsamples):
          ❺ samples[k:(k+len(t))] = t
        k += len(t)

    lo = np.quantile(samples, 0.1)
    hi = np.quantile(samples, 0.9)
    samples[np.where(samples <= lo)] = lo
    samples[np.where(samples >= hi)] = hi
    WriteOutputWav(samples, oname)
```

Listing 8-2: Generating sine wave walks

A while loop runs over all the output samples ❶. The next for loop ❷ is over all walkers for the current step. A random value decides whether to increment, decrement, or leave each walker's frequency unchanged. Then a quick check with min and max keeps the frequency in the range [100, 4000].

A sine wave can be written as $y = A \sin \omega x$ for amplitude A and frequency ω (omega). We select a random amplitude ❸ and use it to create a step's worth of samples that are added to any existing samples, t, thereby summing across all the walkers for the current step.

Each step's waveform begins at zero amplitude because the sine function starts at zero. Therefore, we want the end of the previous step to also be at zero amplitude. The second while loop ❹ attempts to scan from the end of the step waveform to find a sample reasonably close to zero.

Finally, we place the samples for the step in the output samples vector ❺ if they fit. When samples is full, it's clipped to keep samples above the 10th percentile and below the 90th, and then written to disk via WriteOutputWav.

Generate a 15-second or longer sample with one walker. Do you hear the walk? It might help to temporarily set amp=1 to make each step equally loud. Use an app on your smartphone to show the frequency spectrum in real time.

What happens if you add more walkers? Examine the waveform for 50 walkers using a program like Audacity. It should start to resemble noise with little structure.

Making strange sounds with arbitrary combinations of sine waves is fun, but let's see if we can be more musical in our approach.

C Major Scale

The sine walker stepped in frequency by a fixed interval of 5 Hz. The code in *note_walker.py* is nearly identical to that in *sine_walker.py*, but instead of altering the frequency of the sine waves by a constant number of Hertz, the walk takes place over the frequencies of the notes in a C major scale:

```
frequencies = np.array([
146.83 ,  164.81 ,  174.61 ,  196.   ,
220.   ,  246.94 ,  261.63 ,  293.66 ,  329.63 ,  349.23 ,
392.   ,  440.   ,  493.88 ,  523.25 ,  587.33 ,  659.26 ,
698.46 ,  783.99 ,  880.   ,  987.77 , 1046.5  ])
```

In this list, middle C is 261.63 Hz, and A above middle C is 440 Hz. The *note_walker.py* file uses the same command line as *sine_walker.py*. Read through the code and give it a go. Is the result the same as *sine_walker.py*? What instrument does the output remind you of?

NOTE   *Electronic oscillators of various frequencies pieced together with other circuits to modulate the resulting final waveform were the backbone of early analog music synthesizers. We can emulate an analog synthesizer in software; see* https://github.com/yuma-m/synthesizer. *It's Python based, and the page provides complete instructions for installing dependencies. The examples on the GitHub page use sine waves as the base waveform, just as we used here.*

Let's make the leap from randomly varying sine waves to evolving a melody from scratch.

Generating Melodies

We'll use swarms in the service of generating melodies, with the goal of creating a "pleasant" sounding melody from a clean slate.

First, we'll set up our environment. Then we'll learn how to use the program *melody_maker.py* to generate melodies. Finally, we'll walk through the essential parts of the code. The objective function is quite a bit more complicated than what we've worked with previously.

Swarm Search

We'll use MIDI files in this section instead of directly generating WAV files. *MIDI (Musical Instrument Digital Interface)* is the standard format for digital music. It can be complicated, but our use is as simple as it gets: a single melody line. Therefore, for us, MIDI becomes a NumPy vector of pairs of numbers, the first a note number (60 is middle C), and then a duration where the ratio between the durations marks whole, half, quarter, eighth notes, and so on.

The melodies we evolve are expressed, ultimately, as MIDI files. Therefore, we need additional software beyond the usual toolkit to play MIDI files, work with MIDI files in code, and turn a MIDI file into an image of the musical score. Let's install `wildmidi`, `midiutil`, and `musescore3`.

Install `wildmidi` with:

```
> sudo apt-get install wildmidi
```

The `wildmidi` add-on plays MIDI files (*.mid*) from the command line. For macOS and Windows, see the main website at *https://github.com/Mindwerks/wildmidi/releases*.

We need `midiutil` to work with MIDI files in Python:

```
> sudo pip3 install midiutil
```

The `midiutil` library reads and writes MIDI files, though we'll only ever write them.

Finally, to generate sheet music of our evolved melodies, we will need `musescore3`:

```
> sudo apt-get install musescore3
```

Versions for macOS and Windows are available from the main site (*https://musescore.org/en/download*). If `musescore3` is not found, install the latest version (`musescore4`) and update *melody_maker.py* accordingly.

Once everything's installed, we're good to go. If you didn't install `musescore3`, the code will still run, but you won't get to see the final result visually.

The melody_maker.py code

The code we need to evolve melodies is in *melody_maker.py*:

```
> python3 melody_maker.py

melody_maker <length> <outfile> <npart> <max_iter> <alg> <mode> [<kind> | <kind> <seed>]

    <length>    - number of notes in the melody
    <outdir>    - output directory
    <npart>     - swarm size
    <max_iter>  - maximum number of iterations
    <alg>       - algorithm: PSO,DE,RO,GWO,JAYA,GA,BARE
    <mode>      - mode
    <kind>      - randomness source
    <seed>      - random seed
```

Many command line arguments are familiar or self-evident, like the number of notes in the melody. The mode argument refers to the musical mode or scale we want to use. Traditionally, there are seven modes, all of which are supported, plus the blues and the pentatonic (rock) scales. The other modes use their classical Greek names, or major or minor for standard major and minor keys. The mode names are in Table 8-1 along with a sequence of intervals and words often associated with the mode. The intervals of Table 8-1 refer to the steps between the notes of the scale with H a half step (semitone) and W a whole step (tone).

Table 8-1: The Modes, Intervals, and Characteristics

Mode	Intervals	Characteristics
Ionian (major)	W W H W W W H	Bright, positive, strong, simple
Aeolian (minor)	W H W W H W W	Sad
Dorian	W H W W W H W	Light, cool, jazzy
Lydian	W W W H W W H	Bright, airy, sharp
Mixolydian	W W H W W H W	Celtic
Phrygian	H W W W H W W	Dark, depressing
Locrian	H W W H W W W	Darker still, "evil"

For example, if the first note of a candidate melody is middle C (MIDI note 60), and the desired mode is major, then the notes of the scale are:

$$\begin{array}{ccccccccccccc}
\textbf{C} & \text{\tiny W} & \textbf{D} & \text{\tiny W} & \textbf{E} & \text{\tiny H} & \textbf{F} & \text{\tiny W} & \textbf{G} & \text{\tiny W} & \textbf{A} & \text{\tiny W} & \textbf{B} & \text{\tiny H} & \textbf{C} \\
60 & \to & 62 & \to & 64 & \to & 65 & \to & 67 & \to & 69 & \to & 71 & \to & 72
\end{array}$$

The objective function will score a candidate melody, in part, by how closely it follows the desired scale.

If you're not familiar with music, scales, or music theory of any kind, have no fear. All we need to know is that there are different scales, or different sets of intervals between notes, that, when played, affect the sound of the melody. For example, melodies in a major scale (Ionian mode) sound bright, while those in a minor key (Aeolian mode) tend to sound sad. These rules are not hard and fast—just guidelines. We'll generate many melodies in different modes.

Let's run *melody_maker.py*:

```
> python3 melody_maker.py 20 tmp 20 10000 bare mixolydian

Melody maker:

npart = 20
niter = 10000
alg = BARE
Optimization time = 114.775 seconds
63,0.90 60,0.60 63,0.60 60,0.60 63,0.60 60,0.60 61,1.20
65,0.60 68,0.60 65,0.60 68,1.20 72,1.20 68,0.60 65,1.20
70,0.60 67,1.20 63,1.20 67,0.60 70,1.20 73,0.60

31 best updates, final objective value 1.6637
```

I didn't specify a seed, so your run will be entirely different. The output tells us about the search, then spits out a long string of integers and floats. This is the evolved melody as written to the output MIDI file, which is in the *tmp* directory along with several other files, including a NumPy vector, a Python pickle file, and the score (score.png). The melody is in pairs, so the first note is (63,0.9), a dotted eighth note E-flat above middle C.

The output MIDI file is in *tmp* as well. Play it with wildmidi

```
> wildmidi tmp/melody_BARE.mid
```

where BARE is replaced by whatever swarm algorithm we select. I chose mixolydian for the mode, so, in theory, the melody should sound somewhat "celtic." Does it? I really don't know.

The evolved melody is in *score.png*; see Figure 8-3.

Figure 8-3: An evolved melody

I glossed over the objective function value for the returned melody. We'll explore that in more detail when we glance at the code. However, as with all our optimization experiments, lower is better.

Try experimenting with melodies, modes, algorithms, swarm sizes, and iterations of different lengths. More iterations generally lead to better performance, which should mean a better sounding melody, or, at least, a melody more faithful to the desired mode.

You might wish to run the examples in *melody_examples*. The file works as a shell script

```
> sh melody_examples
```

and produces directories *ex0* through *ex8* inside *example_melodies* using different swarm algorithms and modes. I fixed the seeds, so you'll hear what I heard, which hints at the range of possible outcomes.

The following sections experiment with *melody_maker.py*. The first queries the melodies as they evolve, the second focuses on the algorithms to understand what sort of melodies they favor, and the last builds a library of melodies in four modes.

Evolving a Melody

Let's evolve a melody in a major scale using bare-bones PSO. This experiment aims to listen to the melody as it evolves. It should go from erratic and far from the desired mode to a tune that a beginning piano student might play (or so I've been told).

The file *evolve.py* runs *melody_maker.py* to evolve a major scale melody of 20 notes using 20 particles and bare-bones PSO. The generator and seed are fixed; all that changes between runs is the number of iterations, which vary from a low of 1 to a high of 50,000, as in Figure 8-4.

The fixed seed means that the best melody found for 10 iterations passed through the best found in 1 iteration. Each higher number of iterations tells us where any earlier iteration would have ended if it were left to run. In other words, the same initial configuration is allowed to evolve for a varying number of iterations.

Figure 8-4: Progressive melodies. From top: 1, 1,000, 10,000, and 50,000 iterations.

The directory *evolve_results* contains the MIDI files and score images for each number of iterations: 1, 10, 100, 1,000, 5,000, 10,000, and 50,000. I recommend using `wildmidi` to play the files. How does the melody found after 1 iteration of the swarm compare to that after 50,000? Figure 8-4 shows the score for select melodies by iteration. While the early melodies are a mess, there is little change—other than the key—between the melody after 10,000 iterations and after 50,000.

The objective function value, which we have yet to understand, decreases as the number of iterations increases. This is, of course, all it can do, but the rate at which it decreases levels off after 1,000 iterations. The slight change in the score between 10,000 iterations and 50,000 implies that we may not want to run the search for too long, as we risk eliminating potentially exciting melodies in the process.

What, precisely, happens when a melody evolves? Swarm algorithms are initialized randomly over an appropriate range of MIDI note numbers and durations. The selected musical mode effectively alters the objective function used by the swarm as it searches. While the swarm search includes randomness, it is the initial configuration of the swarm that most strongly influences the final result—at least, that's what I think is happening. The combination of initial swarm configuration, algorithm approach, and randomness leads to convergence on a melody that more or less fits the objective function.

Exploring the Algorithms

The *algorithms.py* file runs *melody_maker.py* 10 times for each of the 7 swarm algorithms we've used throughout the book. Each run generates a 36-note melody in the Lydian mode utilizing a swarm of 20 particles and 10,000 iterations. You can run this file to produce output in the *algorithms* directory. Alternatively, since the seed values are fixed, you can listen to the seven MP3 files that concatenate the output by algorithm.

Even though the algorithms are tasked with the same overall goal—learning a melody in the Lydian mode—hopefully the resulting melodies reveal distinctions between the algorithms. Let's see whether it matters which swarm algorithm we use, and whether some produce "nicer" results than others.

I listened to all the MP3 files (made by using `wildmidi`'s `-o` option followed by `lame`) and ranked the resulting melodies by how nice I thought they sounded. Here's my ranking from best to worst, including ties where I couldn't choose one algorithm over another:

1. Bare-bones PSO
2. Genetic algorithm, PSO
3. Differential evolution, Jaya
4. GWO
5. Random optimization

Your ranking might be different, but I suspect you'll agree that bare-bones PSO works best in this case and random optimization is the worst. Both GWO and random optimization produce rushed output; the melodies play faster, so the resulting MP3 files are about 30 seconds shorter than the other algorithms.

We'll perform one more experiment before exploring the code in which we use the "best" algorithm, bare-bones PSO, to create a library of songs in different modes.

Building a Library of Melodies

The file *songs.py* is similar to *algorithms.py*, but uses bare-bones PSO repeatedly to generate 36-note melodies in four modes: major, minor, Dorian, and blues. In this case, there are 32 particles in the swarm and 30,000 iterations. The code takes some time to run, so I created MP3 files of the output: *major.mp3*, *minor.mp3*, *dorian.mp3*, and *blues.mp3*. Give them a listen while keeping the descriptions of Table 8-1 in mind. If you agree with them, it means the objective function has at least captured something of the modes, if not much of what makes a good melody.

Implementation

The time has come to explore *melody_maker.py*. At a high level, it's no different from any of the other swarm optimization experiments: we parse the command line, initialize swarm framework objects, and call Optimize to perform the search. The result is then converted into a MIDI file object and written to disk.

Let's understand the structure of the swarm—the mapping between particle position and melody—and then walk through the objective function class, as that's the heart of the process.

If we want n notes in a melody, each particle becomes a $2n$-element vector, a collection of n pairs, (MIDI note number, duration). MIDI note numbers are restricted (see MusicBounds) to $[57, 81]$ with 57 interpreted as a rest. Again, middle C on the piano is note 60, and each increment or decrement corresponds to a semitone. Durations are integers that we multiply by 0.3 when creating the MIDI file to control the tempo. The ratio between the notes matters so that duration 4 is twice as long as duration 2, and so on. Particles, then, *are* the melody under consideration, and the search seeks to find, given the random initial collection of melodies and the particulars of the selected algorithm, a best melody as decided by the objective function.

Everything depends on the MusicObjective class. It's rather elaborate, with more than 150 lines of code. I'll start at the end, the Evaluate method, and then fill in the pieces—standard top-down design. Recall, the swarm uses the score to decide the quality of the melody. The lower it is, the better. Here's the code:

```
def Evaluate(self, p):
    self.fcount += 1
    s = self.Distance(p[::2], self.mode)
    d = self.Durations(p)
    i = self.Intervals(p[::2], self.mode)
    l = self.Leaps(p[::2], self.mode)
    return 4*s+3*d+2*i+l
```

The score is a multipart function of the output from the `Distance`, `Durations`, `Intervals`, and `Leaps` methods. The respective values are summed but not equally weighted so that the output of `Distance` is four times as important as the output from the `Leaps` method. Each part of the final score is in the range [0, 1], with lower being better.

The `Distance` method measures the Hamming distance between the notes of the current melody (particle) and the notes expected for a melody in the given mode. The `ModeNotes` method returns two binary vectors where a 1 indicates that the corresponding note is in the mode. The first vector is for the current melody, and the second includes the notes in that mode, assuming the first note of the melody to be the root. In other words, `ModeNotes` returns two binary numbers expressed as vectors of 0s and 1s. The Hamming distance between two binary numbers is the number of differing bits. For example, the Hamming distance between 10110111 and 10100101 is 2 because two corresponding bit positions differ. The `Distance` method scales the Hamming distance by the number of notes to return a value in [0, 1]; it's deemed the most important part of the objective function because a good melody in a specified mode should consist mostly of notes in that mode.

The `Durations` method is an ad hoc measure using a scaled root squared error distance between the count of the different note durations in the melody and the preferred favoring of quarter and half notes. The idea is to minimize dotted notes.

The `Intervals` method is another ad hoc metric that looks at the spacing from one note in the melody to the next. We humans generally prefer major or minor thirds and fifths, meaning the interval from note i to note $i + 1$ should be 3, 4, or 7 semitones.

Finally, `Leaps` tries to minimize leaps, or intervals between notes that exceed 5 semitones. This competes with what `Intervals` is measuring, but `Leaps` is weighted half as much.

Now that we have a high-level understanding of what the objective function measures, let's look at the corresponding code and its essential parts.

Distance

The `Distance` method uses the Hamming distance between the notes of the melody and the notes that should be in the melody if it conforms to the desired mode. Here's the code:

```
def Distance(self, notes, mode):
    A,B = self.ModeNotes(notes, mode)
    lo = int(notes.min() - self.lo)
    hi = int(notes.max() - self.lo)
    a = A[lo:(hi+2)]
    b = B[lo:(hi+2)]
    score = (np.logical_xor(a,b)*1).sum()
    score /= len(a)
    return score
```

The ModeNotes method, not shown, returns two lists where each element is 1 if the corresponding note is in the melody (A) or belongs in the melody given the mode (B). The versions in a and b cover the given melody's range.

The score variable holds the Hamming distance between a and b. The Hamming distance is the number of mismatched bit positions. Scaling by the length of the melody transforms the count into a fraction of the melody, [0, 1], which is returned.

Durations

The Durations method calculates a score reflecting how closely the distribution of note durations matches the ad hoc predefined "best" mix that favors quarter and half notes. In code:

```
def Durations(self, p):
    d = p[1::2].astype("int32")
    dp = np.bincount(d, minlength=8)
    b = dp / dp.sum()
    a = np.array([0,0,100,0,60,0,20,0])
    a = a / a.sum()
    return np.sqrt(((a-b)**2).sum())
```

First, we set d to the durations of the current melody so that bincount can create the corresponding distribution, b, which is scaled to a probability. The desired mix of note durations is in a and likewise scaled to be a probability.

The sum of the squared distance between the two distributions is returned as the duration score.

Intervals

The interval between two notes is measured in semitones. The Intervals method counts the number of major thirds (4 semitones), minor thirds (3 semitones), and fifths (7 semitones) in the melody and transforms those numbers into a score. In code this becomes:

```
def Intervals(self, notes, mode):
    _,B = self.ModeNotes(notes, mode)
    minor = major = fifth = 0
    for i in range(len(notes)-1):
        x = int(notes[i]-self.lo)
        y = int(notes[i+1]-self.lo)
        if (B[x] == 1) and (B[y] == 1):
            if (abs(x-y) == 3):
                minor += 1
            if (abs(x-y) == 4):
                major += 1
            if (abs(x-y) == 7):
                fifth += 1
    w = (3*minor + 3*major + fifth) / 7
    return 1.0 - w/len(notes)
```

The code examines each pair of notes in the melody. We use the difference in semitones to count the number of thirds and fifths. We then assign w the weighted mean of these counts where, by fiat, I'm favoring thirds over fifths by 3 to 1.

The higher w is, the more the melody conforms to the desired interval arrangement; therefore, subtract the scaled w score from 1 to minimize.

Leaps

The final part of the objective function score is Leaps:

```
def Leaps(self, notes, mode):
    _,B = self.ModeNotes(notes, mode)
    leaps = 0
    for i in range(len(notes)-1):
        x = int(notes[i]-self.lo)
        y = int(notes[i+1]-self.lo)
        if (B[x] == 1) and (B[y] == 1):
            if (abs(x-y) > 5):
                leaps += 1
    return leaps / len(notes)
```

A leap is any difference between a pair of notes that exceeds 5 semitones, either up or down. Smaller distances imply a smoother melody. We return the fraction of the melody that are leaps.

Why use these components in the objective function and not others? No reason other than a perusal of thoughts on what makes a good melody mentions some of these. Music is subjective, and it isn't possible to create an objective objective function. "Exercises" on page 254 asks you to think of other terms that might fit well in MusicObjective.

GENERATIVE AI

The discussions of generative art and music in Chapters 7 and 8 make no mention of artificial intelligence, and thus are incomplete. However, throwing AI into the mix would turn these chapters into a book. Instead, I'll point you toward AI-based examples of generative art and music. Most of these use *generative adversarial networks, deep style transfer, variational autoencoders,* or related techniques that depend on deep neural networks to either sample from some learned representation space or merge features from embedded representations to build new output from multiple inputs.

If you want to explore what AI can do in this area, *https://aiartists.org* is a good place to start and has links to artists and tools to make AI-based art and music. A fun, advanced approach to evolutionary algorithms and music is found in Al Biles' GenJam at *https://genjam.org*. I recommend the video examples, especially the TEDx talk demonstrating and explaining the system.

(continued)

We've witnessed an explosion of powerful AI-based text, image, and video generation systems, including Stable Diffusion, DALL-E 2, and ChatGPT. New systems and updates appear weekly, but these should get you started:

DALL-E 2 *https://openai.com/dall-e-2*
Stable Diffusion *https://beta.dreamstudio.ai/*
ChatGPT *https://openai.com/blog/chatgpt*

Exercises

Like generative art, there's no end to what we can do with generative music. Here are some exercises related to this chapter's experiments.

- Alter the clipping range in *sine_walker.py*. How does this affect the overall sound? What does the waveform look like in Audacity?

- You can change the key in *note_walker.py* by altering the frequency table. For example, to change from C major to D minor, flatten the B notes: 246.94 → 233.08, 493.88 → 466.16, and 987.77 → 932.33.

- The last argument to `addProgramChange` in *melody_maker*'s `StoreMelody` function specifies the MIDI instrument number. The default is 0 for an acoustic piano. Change this number using the *MIDI_instruments.txt* list. For example, try 30 for a distorted electric guitar or 13 for a xylophone. The oboe, barely breathing, is 68. Or go for broke with 114, steel drums.

- Alter *melody_maker*'s objective function weighting in the `Evaluate` method. Does it sound as nice if all components are weighted the same? What happens if you reverse the weighting?

- What other terms can you add to *melody_maker*'s objective function?

Summary

This chapter introduced us to randomness in generative music beginning with audio and a random walk of sine waves to produce otherworldly sound effects. Restricting the frequencies to those of a musical scale transformed the odd sounds into something akin to a pipe organ.

We closed the chapter by evolving melodies from scratch using swarm intelligence and evolutionary algorithms. We were moderately successful in that the evolved melodies did, for the most part, conform to the desired musical mode. Along the way, we saw an example of how to create simple MIDI files in code.

We'll change gears in the next chapter to explore randomness in an entirely different domain: recovering a signal from a small collection of measurements.

9

AUDIO SIGNALS

In this chapter, we continue exploring the relationship between continuous and discrete signals. The *Nyquist–Shannon sampling theorem* relates continuous signals and discrete signals. To properly discretize a continuous signal, we must sample it at a rate at least twice that of the highest frequency present in the signal. The theorem is why compact discs sample audio signals at 44.1 kHz, or 44,100 times per second. At that rate, any frequency up to 22,050 Hz will be captured. Note that 22 kHz is the theoretical upper limit on the highest frequency a human can hear, though most adults have a much lower upper limit; mine is about 13.5 kHz.

This chapter explores *compressed sensing* (or compressive sensing), a technique for beating Nyquist and Shannon at their own game. With compressed sensing, it becomes possible to acquire less data when digitizing a signal than the Nyquist-Shannon theorem says is required. This is an exciting real-world inverse problem involving randomness.

We'll begin by walking through the main points of compressed sensing; we'll cover some of the math, but I encourage you to explore the rest on your own. Then we'll explore compressed sensing in one dimension, audio, to see how it lets us break the Nyquist limit. Finally, as unraveled images are, to compressed sensing, no different from signals in time, we'll apply compressed sensing to reconstruct images from what seems like too little data.

There's some matrix-vector math in the first section, but it doesn't go much beyond what we encountered in Chapter 7 with iterated function systems.

Compressed Sensing

Digitizing a signal usually means reading the output of an analog-to-digital converter at a specified but constant time interval. The number of readings per second is the sampling rate, which the Nyquist-Shannon theorem is concerned with. If we acquire the samples according to the Nyquist-Shannon theorem, we can accurately reconstruct the signal from the samples.

When we sample at a fixed time interval, we are *uniform sampling*. However, there are times when uniform sampling isn't desired, possibly because it's too expensive or there's too much risk associated with it (for example, in X-ray tomography). In such situations, it would be nice to acquire less data but still reconstruct the entire signal. For example, if the signal we want is denoted as x, we'll measure some subset of the signal, y, and from y reconstruct x. Mathematically, we can cast this process as a matrix equation

$$y = Cx \tag{9.1}$$

where we know vector y because we measured it and matrix C because it dictates the parts of x we sampled. We want x, the vector we would've measured following standard sampling theory. Keep Equation 9.1 in the back of your mind for the time being.

Let's return to algebra class, which asks us to solve systems of equations, usually two equations and two unknowns:

$$ax + by = e$$
$$cx + dy = f$$

Here, a through f are constants. Because there are two equations and two unknowns, we can find x and y values that satisfy both, assuming one equation isn't a multiple of the other. In matrix form, we write the system of equations as:

$$\begin{bmatrix} e \\ f \end{bmatrix} = \begin{bmatrix} a & b \\ c & d \end{bmatrix} \begin{bmatrix} x \\ y \end{bmatrix} \quad \rightarrow \quad b = Ax$$

The rules of matrix-vector multiplication tell us to multiply each row of the matrix, A, by the corresponding elements of the vector, x, then sum. This transforms the matrix equation into the system. Regardless of the number of elements in the vectors, this rule applies.

The system of equations works because there are as many unknowns as there are equations, meaning the number of elements in the vectors, here b and x, matches the number of rows in the matrix, A. For such an equation, the solution (if there is one), or the x vector that makes the equation true, is $A^{-1}b = x$ for A^{-1}, the inverse matrix of A. For example, this system

$$3x + 2y = 19$$
$$x + y = 8$$

becomes

$$\begin{bmatrix} 19 \\ 8 \end{bmatrix} = \begin{bmatrix} 3 & 2 \\ 1 & 1 \end{bmatrix} \begin{bmatrix} x \\ y \end{bmatrix} \quad \rightarrow \quad b = Ax$$

with solution

$$x = A^{-1}b \quad \rightarrow \quad \begin{bmatrix} x \\ y \end{bmatrix} = \begin{bmatrix} 1 & -2 \\ -1 & 3 \end{bmatrix} \begin{bmatrix} 19 \\ 8 \end{bmatrix} = \begin{bmatrix} 3 \\ 5 \end{bmatrix}$$

where the inverse of a matrix A is A^{-1} such that $AA^{-1} = A^{-1}A = I$.

Here, I is the *identity matrix*—the matrix of all zeros with ones along the diagonal. In the world of matrices, I is akin to the number 1. Use NumPy's `linalg.inv` function to find A^{-1}.

A is a *square matrix*, meaning it has as many rows as columns. If A is square, and the number of rows matches the number of elements in b and x, then we can use A^{-1} to find x.

Now comes the fun part. Return to Equation 9.1. By design, there are *fewer* elements in y, the values we measure, than in x, the full signal. If there are N elements in y and M elements in x, then C is an $N \times M$ matrix with N rows and M columns. There are more unknowns in Equation 9.1 than there are equations. Such a system is called *underdetermined*. Underdetermined systems have an infinite number of solutions; there are an infinite number of vectors, x, that, when multiplied by C, give y.

We want to get x by measuring y, but y alone doesn't have enough information to tell us *which* of the infinite set of x vectors we want. Compressed sensing comes to the rescue—at least in some cases.

According to compressed sensing theory, if x is *sparse*, meaning most of its elements are essentially zero, then we can recover x from y by solving the *inverse problem*, which searches for the x that minimizes some measure of the difference $Ax - b$ while strongly encouraging x to be sparse. As we'll see, algorithms capable of this kind of optimization exist.

Great! We're in business. We measure y containing some subset of the elements that would be in x, and we get x from y by solving a minimization problem. But it isn't that simple; the optimization trick only works if x is sparse. Most signals are not sparse; an audio signal isn't likely to be. Recall working with the waveforms in Chapter 8. Are we doomed? Not necessarily.

While an audio signal isn't sparse, there are Fourier-like transformations that map from a signal changing in time to one changing in frequency, and it is often the case that the frequency domain signal *is* sparse. Therefore, if we can write $x = \Psi s$ for some transformation matrix Ψ (psi) and sparse vector s, Equation 9.1 becomes

$$y = Cx = C\Psi s = \Theta s \tag{9.2}$$

where $\Theta = C\Psi$.

While x isn't sparse and is therefore unrecoverable, s is, meaning the optimization trick might have a chance of working. The measurements in y combined with the external knowledge that s is sparse will let us find s. Once we have s, we get x.

But, what are all these matrices floating around? The *measurement matrix*, C, can mathematically be any matrix of values such that the values in C satisfy some notion of *incoherence* in relation to the elements of Ψ. For us, the elements of C are binary, zero or one, and serve to select specific elements of x that are actually measured. This requirement supplies the necessary mathematical incoherence between C and Ψ. The most important point is that C is somehow *random*. In practice, we don't explicitly define C, but our measurement process uses it implicitly. The random bit is essential to the entire operation, however.

The Ψ matrix is a transformation matrix that transforms the sparse vector, s, into a new representation, x, which we ultimately want to measure. For us, Ψ is a Fourier-like *discrete cosine transformation (DCT)*. Signals are often sparse in this domain, thereby making it very useful for compressed sensing.

Finally, Θ represents the combination of the measurement process working on Ψ.

We are now in a position to try solving $y = \Theta s$ for some s that both is sparse and leads to y, the set of measurements we have. There are multiple algorithms available, but we'll use the Lasso algorithm, courtesy of scikit-learn. Likewise, we need the DCT and its inverse, which SciPy dutifully supplies:

```
from sklearn.linear_model import Lasso
from scipy.fftpack import dct, idct
```

Lasso minimizes the following

$$\frac{1}{2n} \|y - \Theta s\|_2^2 + \alpha \|s\|_1 \tag{9.3}$$

where n is the number of samples, or the number of elements in y.

The double vertical bar notation refers to a *norm*, which is a metric measuring distance of some kind. The first term uses the square of the ℓ^2 norm, while the second term multiplies the ℓ^1 norm by α. The ℓ^p norm of a vector, x, is defined as:

$$\|x\|_p = \left(\sum_{i=0}^{n-1} |x_i|^p \right)^{1/p}$$

The ℓ^2 norm is the Euclidean distance. The ℓ^1 norm, sometimes called the Manhattan or taxicab distance, is the sum of the absolute values of the elements of x. Lasso uses this term, scaled by α, to find an s vector that minimizes the Euclidean distance between the measurements, y, and Θs while simultaneously minimizing the sum of the absolute values of the elements of s. This latter constraint forces many elements of s toward zero, thereby ensuring sparsity.

To understand why the ℓ^1 norm term is present in the Lasso objective function, consider a simple case where we have a two-element vector and a single-element output. This is akin to finding a solution to $x + y = c$ that is as sparse as possible, where $x = 0$ or $y = 0$. Geometrically, minimizing the ℓ^1 norm leads to a situation as on the left of Figure 9-1, while minimizing the ℓ^2 norm is shown on the right. Minimizing the ℓ^2 norm is standard least-squares regression.

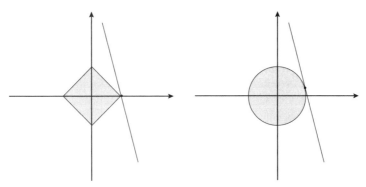

Figure 9-1: Minimizing the ℓ^1 norm (left) and the ℓ^2 norm (right)

The line in Figure 9-1 represents the infinite set of solutions to $ax + by = c$ for some c. The diamond on the left and the circle on the right correspond to constant ℓ^1 and ℓ^2 norms, respectively. Minimizing the ℓ^1 norm intersects the line at a point where y is zero, while minimizing ℓ^2 intersects the line at a point where neither x nor y is zero. This trend continues as the dimensionality increases. In each case, minimizing the ℓ^1 norm implies sparsity in the solution while minimizing the ℓ^2 norm distributes the "energy" throughout each dimension, the opposite of enforcing sparsity.

Let's sum up. We want to acquire x by measuring a subset of it, y. To solve for random-ish measurement matrix A, we want $y = Ax$. This expression is underdetermined, meaning there are an infinite number of x that work as solutions, so we need extra information to find the one we (likely) want. We get this information from expressing x in some other form (basis) where it becomes sparse. If sparse, the probability of finding a meaningful and parsimonious solution presents itself. A commonly used basis for this comes from the Fourier family of transformations, like the DCT, $x = \mathbf{\Psi}s$, where $\mathbf{\Psi}$ encapsulates the DCT and s is a sparse vector we want to find. If we find s, we find x.

Combining the measurement matrix and the DCT gives us a new equation, $y = \mathbf{\Theta}s$, where we know y and $\mathbf{\Theta}$. It's still underdetermined, but we know s is sparse. To find s, we use an optimization algorithm that knows how to minimize the ℓ^1 norm of s in the process. This enforces sparsity and gives us some confidence that we might find a suitable s.

Let's give the recipe a go and see what happens.

Signal Generation

We'll walk through *cs_signal.py*, which illustrates the compressed sensing process and why we need to use random measurements. First, let's run it, then I'll explain the various plots it generates. Here's the command line:

```
> python3 cs_signal.py 0.2 minstd 65536
```

Several plots should appear in succession; close each to move to the next. Output files are created as well.

The code first generates a one-second signal, the sum of three sine waves forming a C major chord. Standard Nyquist sampling gives this signal for a sample rate of 4,096 Hz, that is, *x*. This is a demonstration, so we start with *x* and then throw much of it away to create a *y* that we might have plausibly measured in the first place. The command line includes an argument of 0.2, the fraction of *x* to retain, meaning that *y* has 20 percent of the samples; we throw the remaining 80 percent away. The remainder of the command line specifies the randomness source (minstd) and a seed value (65536).

The signal comes from:

```
rate = 4096
dur = 1.0
f0,f1,f2 = 261.63, 329.63, 392.0
samples  = np.sin(2*np.pi*np.arange(rate*dur)*f0/rate)
samples += np.sin(2*np.pi*np.arange(rate*dur)*f1/rate)
samples += np.sin(2*np.pi*np.arange(rate*dur)*f2/rate)
```

We used similar code in Chapter 8. The three frequencies (f0, f1, f2) are the C major chord. The samples vector is the final signal, *x*. It's a vector of 4,096 elements because the sampling rate is 4,096 Hz and the duration is one second.

Let's build *y* from *x*. This process makes implicit use of the measurement matrix. We'll keep 20 percent of the samples in *x*, first by selecting samples at a uniform interval, and then randomly. The uniform samples correspond to measuring the signal at some rate below the Nyquist limit:

```
nsamp = int(frac*len(samples))
u = np.arange(0, len(samples), int(len(samples)/nsamp))
bu = samples[u]
r = np.argsort(rng.random(len(samples)))[:nsamp]
br = samples[r]
```

There are nsamp samples in *y*. The first *y* vector is bu, uniformly sampled, and the second is br, randomly sampled. Figure 9-2 shows the original signal with the uniform and random samples marked (*cs_signal_samples.png*).

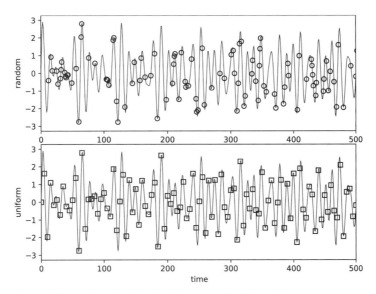

Figure 9-2: Random (top) and uniform samples (bottom)

We have the measurements. Now we need Θ, the combination of Ψ and the measurement matrix. Once we have that, we're ready to use Lasso. We have two y vectors, so we need two Θ matrices:

```
D = dct(np.eye(len(samples)))
U = D[u,:]
R = D[r,:]
```

The first, U, keeps only the uniformly selected measurements. The second, R, uses the randomly selected measurements. Here, D is the discrete Fourier transform matrix, Ψ, and U and R are Θ_u and Θ_r, respectively.

We optimize twice, first $y_u = \Theta_u s_u$ and then $y_r = \Theta_r s_r$, where the subscripts refer now to the uniformly and randomly sampled measurements:

```
lu = Lasso(alpha=0.01, max_iter=6000)
lu.fit(U, bu)
su = lu.coef_
lr = Lasso(alpha=0.01, max_iter=6000)
lr.fit(R, br)
sr = lr.coef_
```

Lasso follows the scikit-learn convention of creating an instance of a class and then calling fit to do the optimization. For Lasso, the solution vector is buried in the coef_ member variable, which we extract to get su and sr, the uniform and random s vectors, respectively. Figure 9-3 shows the two s vectors (*cs_signal_sparse.png*).

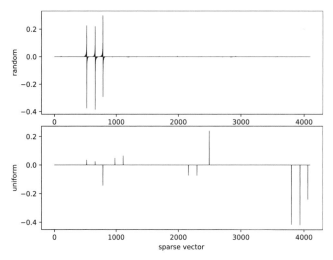

Figure 9-3: Random (top) and uniform (bottom) solution vectors, s

The top plot shows s_u and the bottom shows s_r. The spikes correspond to DCT components. Both s vectors are sparse, with most of the 4,096 elements near zero, but the bottom vector has more than 10 nonzero elements while the top has only 3 (the shape is sometimes both positive and negative). Recall, x is the sum of three sine waves, so the three sine waves and three spikes in the s_r vector seems promising.

Lasso has solved $y = \Theta s$ for us. Now we need $x = \Psi s$, which we find by calling the inverse DCT:

```
ru = idct(su.reshape((len(samples),1)), axis=0)
rr = idct(sr.reshape((len(samples),1)), axis=0)
```

Figure 9-4 shows us x_u (ru) and x_r (rr); see *cs_signal_recon.png*.

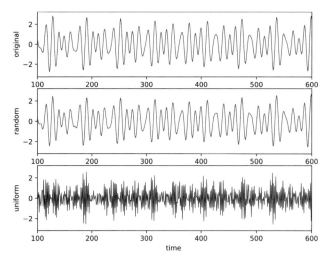

Figure 9-4: From top: the original, reconstructed randomly sampled, and reconstructed uniformly sampled signals

The topmost plot shows the original signal. The middle plot shows the signal reconstructed from 20 percent of the original signal using randomly selected measurements. Finally, the bottom plot shows the signal reconstructed from the uniformly selected measurements, likewise 20 percent of the original number. Which do you think more faithfully captured the original signal?

The final step is to output the signal as a WAV file:

```
WriteOutputWav(samples, "original.wav")
WriteOutputWav(rr, "recon_random.wav")
WriteOutputWav(ru, "recon_uniform.wav")
```

See Listing 8-1 to review how WriteOutputWav works. Play the output files. I think you'll agree that random sampling produced the better result.

Uniform sampling failed for deep mathematical reasons related to coherence between the measurement matrix and the DCT transform basis. However, we can intuitively understand the failure due to uniform sampling at less than the Nyquist rate, which means *aliasing* where higher frequency signals look like lower frequency signals, and there's no way to disentangle the two. On the other hand, with random sampling, the likelihood of aliasing decreases, making Lasso more likely to find a suitable *s* vector.

Rerun *cs_signal.py*, but alter the fraction from 20 percent to smaller and higher values. Is there a place where everything falls apart? See if you can re-create the signal from only 10 percent, 5 percent, or even 1 percent of the original, and then try the opposite direction. Sampling even slightly above 50 percent appears to have a dramatic effect on the quality of the uniform sample. Why might that be? Consider the Nyquist-Shannon sampling theorem requirements.

Unraveled Images

The file *cs_image.py* applies compressed sensing to images. It's similar to *cs_signal.py*, but it unravels the image before selecting the measured components (pixels). The image is *x*, with the selected mask pixels forming *y*. The code expects these command line arguments:

```
> python3 cs_image.py

cs_image <image> <output> <fraction> <alpha> [ <kind> | <kind> <seed> ]

   <image>    - source image (RGB or grayscale)
   <output>   - output directory (overwrittten)
   <fraction> - fraction of image to sample
   <alpha>    - L1 lambda coefficient
   <kind>     - randomness source
   <seed>     - seed value
```

The input image may be grayscale or RGB. If RGB, each channel is processed individually using the same random mask. The output directory contains the original image, the reconstructed image, and a parameter file.

The code tries to import from scikit-image. It will run if scikit-image isn't installed, but you can install it with:

```
> pip3 install scikit-image
```

If scikit-image is present, the code imports structural_similarity, which measures the mean structural similarity between two images—here the original image and the reconstructed image. Higher similarity is better, with 1.0 indicating an exact match.

The code loads the input image, converts it to RGB, and tests to see if it's really grayscale:

```
simg = np.array(Image.open(sname).convert("RGB"))
grayscale = False
if (np.array_equal(simg[:,:,0],simg[:,:,1])):
    grayscale = True
```

A grayscale image converted to RGB ends up with every channel the same, hence the call to array_equal. The test is not entirely foolproof, but it's good enough for us.

The next step generates the random mask, the subset of actual image pixels that construct *y*:

```
row, col, _ = simg.shape
mask = np.zeros(row*col, dtype="uint8")
M = int(fraction*row*col)
k = np.argsort(rng.random(row*col))[:M]
mask[k] = 1
```

The mask vector is 1 for selected pixels.

The remainder of the code calls CS for each image channel, if RGB, or the first channel, if grayscale, before dumping the original image, reconstructed image, and parameters to the output directory. All the action is in CS:

```
def CS(simg, mask, fraction, alpha, rng):
    row, col = simg.shape
    f = simg.ravel()
    N = len(f)
    k = np.where(mask != 0)[0]
    y = f[k]
    D = dct(np.eye(N))
    A = D[k, :]
    seed = int(10000000*rng.random())
    lasso = Lasso(alpha=alpha, max_iter=6000, tol=1e-4, random_state=seed)
    lasso.fit(A, y.reshape((len(k),)))
    r = idct(lasso.coef_.reshape((N, 1)), axis=0)
```

```
r = (r - r.min()) / (r.max() - r.min())
oimg = (255*r).astype("uint8").reshape((row,col))
return oimg
```

The CS function is a compact version of the essential code in *cs_signal.py*. It forms the unraveled image (f) and then selects the masked regions to form y.

To make the code reproducible from a given seed value, we define the local variable, seed, and pass it to the Lasso constructor before calling fit.

When fit exits, the inverse DCT uses the sparse vector (s) to recover the image. The image isn't scaled to [0, 255], so we first scale it to [0, 1] and then multiply by 255 and reshape (oimg).

Let's find out whether *cs_image.py* works. This command line

```
> python3 cs_image.py images/peppers.png peppers 0.1 0.001 mt19937 66
```

attempts to reconstruct the peppers image. It will take several minutes to run before producing Figure 9-5.

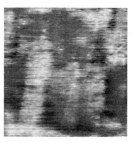

Figure 9-5: The original image (left), mask (middle), and reconstructed image (right)

The original image is on the left, the 10 percent mask in the middle, and the reconstructed image on the right. This is best viewed in color; look at the files in the *peppers* directory. I inverted the mask image to show the selected pixels in black.

The reconstructed image isn't particularly impressive until you remember that 90 percent of the original image information was discarded or, in practice, never measured in the first place.

I claimed that Lasso finds sparse *s* vectors. The signal example was sparse, but what about images? The test images are 128×128 = 16, 384 pixels, meaning *s* has that many elements. A quick test with the *barbara.png* image, keeping 20 percent of the pixels, returned an *s* that's 70 percent zeros. Dropping down to 10 percent jumps to 81 percent zeros, while moving up to 80 percent drops to only 15 percent zeros. Fewer measurements imply a sparser *s*, which seems reasonable. Recall that *s* is the representation of the image in the discrete cosine transform space. If we can find only a few presumably low frequency components when attempting to best fit the few measurements in *y*, we might expect most of *s* to be zero after imposing ℓ^1 regularization.

The *cs_image_test* script runs *cs_image.py* repeatedly on the same test image while varying the measured fraction of pixels from 1 percent up to 80 percent. Figure 9-6 shows the resulting reconstructed images.

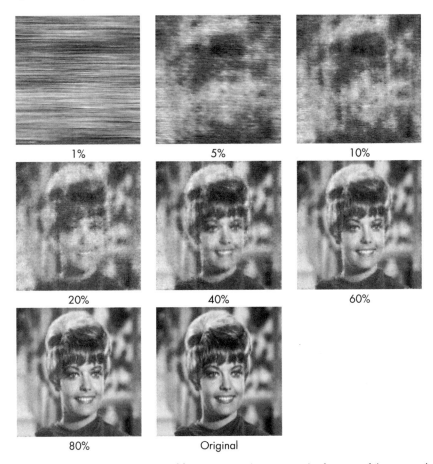

Figure 9-6: Reconstructions of the zelda.png image by varying the fraction of the original pixels

At 10 percent, we can start to recognize the image, but it isn't clear that it's a person's face until 20 percent. Note that I altered the intensity of the original *zelda.png* image to use the entire range $[0, 255]$; this makes it as bright as the reconstructions.

Figure 9-7 shows a plot of the mean structural similarity index (SSIM) between the reconstructions and the original image.

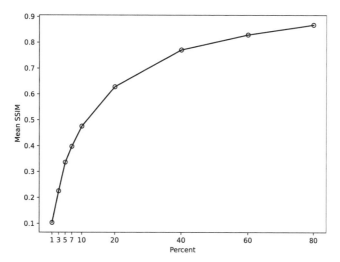

Figure 9-7: The mean structural similarity index as a function of measurements

As we might anticipate, the index increases rapidly as the number of pixels measured increases. The results are encouraging, because there's little perceptual difference between the original image and the one made from 20 percent fewer measurements.

Compressed Sensing Applications

Compressed sensing is used in many places, including medical imaging, where its use has improved acquisition times in magnetic resonance imaging and various forms of tomography. Applying compressed sensing to tomography implies collecting fewer projections, leading to a substantial reduction in the amount of X-ray energy used (ionizing radiation).

Magnetic resonance imaging is a natural target for compressed sensing. The image acquisition process literally measures in *k-space*, or Fourier space, equivalent to measuring *s* directly. The desired image is recovered by a two-dimensional inverse Fourier transform, just as we recovered *x* from *s* via the inverse discrete cosine transform. Many k-space sampling strategies have been developed to speed image acquisition while still producing clinically valuable images. How magnetic resonance image acquisition works makes the simple random sampling in this chapter impractical, but alternative approaches for sampling k-space in a mathematically incoherent manner exist and lead to reduced acquisition times. For example, GE's *HyperSense*, an advanced compressed sensing method, can reduce scan times by up to 50 percent. Faster scan times mean less scanner time for the patient.

The future of compressed sensing is, however, a bit unclear. Deep neural networks are also quite good at solving inverse linear problems—in fact, likely better than traditional compressed sensing. Using deep neural networks in place of traditional CS, or in combination with it, is an active research area.

Exercises

This brief chapter included two experiments, first with one-dimensional signals, then with images expressed as one-dimensional vectors. Here are some possible avenues for further exploration:

- The code in *cs_signal.py* worked with the entire one-second sound sample. How might you modify this basic approach to compress an arbitrary WAV file? Hint: try keeping only a random subset of each few hundred milliseconds of sound and reconstructing each.

- Assuming you build an arbitrary WAV filesystem, can you get away with using the same measurement matrix (the same random sampling) for each subset, or is it better to alter that in some way—maybe by using a fixed pseudorandom seed and selecting measurements in blocks as needed?

- All of our image experiments used α = 0.001. Try varying α from near 0 up to, or even beyond, 1. If α = 0, the ℓ^1 regularization term in Lasso vanishes, and the optimization becomes standard least-squares using only the ℓ^2 norm. Does compressed sensing work well when α is very small? Note that the scikit-learn documentation for Lasso warns not to use α = 0, so, for that case, replace `Lasso` with `LinearRegression`.

- The *cs_image.py* file includes checks to see if the supplied randomness source is `quasi` and, if so, to interpret the seed value as the quasi-random generator base. What happens if you use `quasi` for different prime bases like 2, 3, or 13? Can you explain the results you see?

- We process RGB images color channel by color channel. As an alternative, we can unravel the full RGB image into a vector three times larger and then perform the optimization (remember to re-form the RGB image on output). Alter *cs_image.py* to do this. Does it matter? Does it help or hurt?

- Are all random measurement matrices created equal?

Summary

Compressed sensing breaks the Nyquist-Shannon sampling theorem limit and allows signals to be reconstructed from fewer samples than initially thought possible. In this chapter, we experimented with a basic form of compressed sensing and applied it to audio signals and images.

First, we discussed the core concepts in compressed sensing, including sparsity and ℓ^1 regularization. We then expressed the compressed sensing problem as an inverse linear problem of the form $y = Cx$ for measured vector y and desired output vector x. In practice, sparsity constraints means using an alternate form of $x = \Psi s$ for sparse vector s and basis Ψ. For us, Ψ came from the discrete cosine transform in which signals are known to be sparse.

The compressed sensing problem then became one of finding solution vector s such that the ℓ^2 distance between y and $C\Psi s = \Theta s$ was as small as possible, subject to the constraint that $\|s\|_1$ was also as small as possible. We discovered that Lasso regression accomplishes this goal quite nicely.

Theory in hand, we performed two sets of experiments. The first sought to reconstruct a one-second audio signal, a C major chord, using uniform sampling below the Nyquist limit and random sampling. Uniform sampling couldn't recover the signal until the sampling rate exceeded half the playback rate (in which more than 50 percent of samples were kept). On the other hand, random sampling with compressed sensing returned good results even after discarding up to 90 percent of the original data.

In the second experiment, we worked with both grayscale and RGB images. As with signals, we successfully used compressed sensing and the discrete cosine transform to recover images from as little as 10 percent of the original pixels, often with considerable noise. The DCT isn't necessarily the best basis for images, but better ones, like wavelets, are beyond the scope of this book.

We closed the chapter by pointing out that compressed sensing has been a boon to medical imaging to improve patient comfort and reduce exposure to ionizing radiation. Finally, we noted that recent advances in deep neural networks will likely substantially impact the future of compressed sensing.

We'll take a break from experimenting in the next chapter to explore how we use randomness in experiments themselves. Modern science critically depends on properly designed experiments, and randomness is a powerful player in that process.

10

EXPERIMENTAL DESIGN

The *scientific method* is the backbone of science. It involves the creation of theories from hypotheses that are tested through experiments and supported by evidence gleaned from those experiments. In this chapter, we'll explore the design of experiments, or *experimental design*, a foundational part of the scientific method. Randomness is critical to successful experimental design for two main reasons. First, many measurements, regardless of the field, involve uncertainty or other factors outside of the researcher's control, called *random noise*. We use randomness in experimental design to combat noise, like fighting fire with fire. Second, randomness makes the results conform to the expectations of statistics.

This chapter uses simulation (Chapter 3) to explore three common approaches to randomization in experimental design. Our running example mimics medical research, but the concepts involved apply everywhere.

Randomization in Experiments

Say a researcher wants to know the general public's opinion regarding a ballot proposal to enact a noise ordinance restricting loud parties after 8 PM. He decides to use a phone survey, picking 100 names at random from the phone book and calling each on a Wednesday afternoon. He gets 64 answering machines, 36 pickups, and 17 willing to talk. Of those 17, 15 support the ordinance and 2 oppose it. Are these results a fair representation of the general public's stance on this issue?

I suspect your answer is no because there are many possible sources of bias in the results. The researcher used the phone book, which generally lists only landlines; relied on those called to be willing to offer their opinion; and called on a weekday in the afternoon.

His sample is strongly biased toward retired people who tend to be older and (stereotypically) less inclined to have parties, so his results do not necessarily reflect the population as a whole. His sample excludes the group most likely to be affected by the ordinance—young people who use cell phones and were probably at school or work during the time he collected his data.

Avoiding such *sample bias* is an excellent reason to use randomness when conducting experiments. However, bias in surveys and polling is tough to correct and often leads to contrary results. More subtle, and perhaps more dangerous, is sample bias in selecting cohorts for medical studies.

The code in *bad_sample.py* generates a population where individuals are a combination of four randomly selected characteristics: age, income, smoker or not, and the average number of alcoholic drinks per week. The characteristics are linked to the person's age, so an older person is likelier to have a higher income, not smoke, and drink less alcohol.

The Population function generates a population of individuals as a NumPy array:

```
def Population(npop):
    pop = []
    for i in range(npop):
        age = 20 + int(55*rng.random())
        income = int(age*200 + age*1000*rng.random())
        income = int(income/1000)
        smoker = 0
        if (rng.random() < (0.75 - age/100)):
            smoker = 1
        drink = 1.0 - age/100
        drink = int(14*drink*rng.random())
        pop.append([age, income, smoker, drink])
    return np.array(pop)
```

The code loops append four-element lists of characteristics, each derived from the selected age. Notice, age is in years, income is in thousands, smoker is binary, and drink is average number of alcoholic drinks per week.

The code selects a random subset of the population to simulate a random sample as a cohort for a medical study. For example, here's one run of *bad_sample.py*:

```
> python3 bad_sample.py 1000 10 1 mt19937 4004
age    : 47.22  42.20  (t= 0.9782, p=0.32823)
income: 33.20  24.80  (t= 1.4284, p=0.15350)
smoker:  0.27   0.10  (t= 1.2126, p=0.22555)
drink :  3.14   3.20  (t=-0.0714, p=0.94309)
```

The population size is 1,000, from which we randomly select 10. The 1 argument samples once. As usual, the randomness source and seed follow.

We sort the output by age, income, smoker, and drink. The first column shows the mean of these values for the entire population. The second is the mean for the 10-person random sample. The values in parentheses are t-test results where t is the *t* statistic and p the p-value. A significant difference between the population and sample produces a low p-value. The sign of the *t* statistic is such that a positive *t* signifies that the population mean exceeds the sample mean.

In this case, the 10-person random sample was similar to the population regarding drinking and overall age. However, income was different, which might impact the results of a study that claims this sample is a suitable stand-in for the general population.

Let's run the same command again:

```
> python3 bad_sample.py 1000 10 1 mt19937 6502
age    : 47.35  56.10  (t=-1.7258, p=0.08468)
income: 32.43  31.80  (t= 0.1118, p=0.91097)
smoker:  0.29   0.10  (t= 1.3255, p=0.18530)
drink :  3.20   2.60  (t= 0.7774, p=0.43713)
```

This random sample is significantly older than the population and is consequently much less likely to smoke. Try a few runs of *bad_sample.py*. While some samples are similar to the population, others deviate significantly.

Run *bad_sample.py*, but change the number of samples from 1 to 40:

```
> python3 bad_sample.py 1000 10 40 mt19937 8080
10 6.36529487 0.49581353
```

Each person in the population is a 4-tuple (age, income, smoker, drink). As such, they become a point in a four-dimensional space. The entire population becomes a single point in space if we consider the means of the individual age, income, smoker, and drink characters. The *bad_sample.py* file reports the mean ($\pm$ standard error) Euclidean distance between the mean point of the population and the mean point of the sample, for each of the 40 samples taken. In this case, with a sample size of 10, the mean distance was 6.37 ± 0.50.

Let's increase the sample size to 100:

```
> python3 bad_sample.py 1000 100 40 mt19937 8080
100 2.12064926 0.23245698
```

The mean distance between the population and the samples decreases. If we have a larger sample, it should be a better representation of the population as a whole. This effect is the motivation behind working with as much data as possible, both for scientific research and machine learning models.

The file *bad_sample_test.py* repeats the process for a population of 10,000 and sample sizes from 10 to 5,000. For each sample size, we collect 40 samples. The results are amenable to plotting, as Figure 10-1 illustrates.

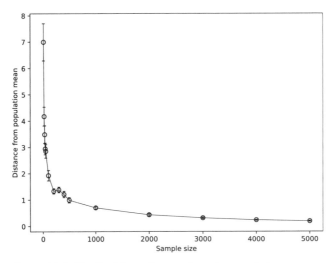

Figure 10-1: The Euclidean distance from the population mean and sample mean as a function of the sample size

The *x*-axis in Figure 10-1 is the number of samples in each set. We plot the average deviation of the mean of those samples from the population mean with error bars representing the standard error. I included end caps on the error bars to make the range easier to spot. Each point is the mean over 40 population-sample pairs.

Larger sample sizes lead to a smaller distance between the population and sample means, indicating that larger samples are more likely to be better representations of the population than smaller samples. Note that the error bars shrink as the sample size increases. The noise in the samples, or the deviation from the population mean, drops to nearly zero.

The error bars for a sample size of 10 are large, and the mean deviation is about 7, roughly seven times the mean deviation for samples of size 500. Notice also how quickly the curve decreases as the sample size increases. A large sample size might not be necessary for many experiments, but the strength of the effect being investigated influences sample size choice. More is always better.

Experiments often report mean results along with the standard deviation or standard error of the mean. Then they perform a significance test, like a t-test; if the p-value is below the (arbitrary) threshold of 0.05, they declare victory and label the result as statistically significant. However, this is only part of the story. If the sample sizes are substantial, as might be the case in some cohort studies based on widely collected health data, then it's entirely possible that the result is $p < 0.05$, but, because of the large cohort, the *effect size* as measured by Cohen's d is small. In other words, the effect is real but perhaps relatively meaningless in practice because the effect size is small. The moral of the story is to report effect sizes whenever possible.

The noise in Figure 10-1 for small sample sizes justifies my rant in the previous paragraph. If the effect is small, and we expect a slight deviation from a population mean, then a small sample size is unlikely to capture the effect because it's swamped by the inherent noise in the sample itself. We'll return to this notion later in the chapter when using experimental design techniques to compensate for selection bias.

Finally, the very thing that makes small sample sizes in Figure 10-1 a poor choice for research purposes is what drives evolution via genetic drift. In Chapter 3, we saw that the random mix of characters in a subpopulation separated by chance from a larger population led to a drift that ultimately produced a new species. In that sense, evolution is a poor researcher but a clever tinkerer able to take sample bias and turn it into something useful in the long run.

Now that we know sample bias can lead to poor research results, let's try to find ways to compensate for it.

Randomization in experimental design comes in three broad categories. The code we'll use throughout the remainder of the chapter supports all three approaches.

Simple

If the experiment is testing one outcome—such as the effectiveness of a particular treatment—we might build the treatment and control groups by recruiting members one at a time, flipping a coin to decide group assignment. We can simulate such *simple randomization* with a snippet of Python code:

```
>>> from RE import *
>>> RE(mode="int", low=0, high=2).random(10)
array([1, 1, 1, 0, 0, 0, 0, 1, 0, 1])
>>> RE(mode="int", low=0, high=2).random(10)
array([0, 1, 1, 0, 1, 1, 0, 0, 1, 1])
```

We configure RE to return coin flips, 0 or 1. Both calls return 10 flips each. Therefore, we have the assignments for 20 participants: 11 for the treatment group and 9 for the control group.

Simple randomization seems like a good approach, but it has an obvious drawback, similar to what we observed when working with *bad_sample.py*.

Let's try one more assignment for an experiment with 10 participants:

```
>>> RE(mode="int", low=0, high=2).random(10)
array([0, 1, 1, 1, 0, 0, 1, 1, 1, 1])
```

If we follow simple randomization, we'll have seven in the treatment group and only three in the control group—this seems unwise. Note that this isn't a contrived example; the output is from a single run of that line of code.

While simple randomization isn't the best approach when the study size is small, it often works well for larger study sizes:

```
>>> s = RE(mode="int", low=0, high=2).random(10000)
>>> np.bincount(s)
array([4852, 5148])
```

In this case, the difference between treatment and control group size is a smaller fraction of the total study size, so we might expect each group to be similar.

Simple randomization's weakness, especially for small study sizes, is that the number of subjects in the treatment and control groups might be highly imbalanced. We can remedy this with block randomization.

Block

Block randomization ensures that the number of subjects in each group is the same. For a binary experiment with only two groups—treatment and control—we first select a block size, usually between 4 and 6. We'll use a constant block size of 4 for our example. Next, create all possible blocks of four subjects where the assignments are balanced. For us, this means all combinations of four-digit binary numbers where the number of 1s and 0s is the same:

1100, 1010, 1001, 0110, 0101, 0011

Each block has two 1s and two 0s.

We select as many blocks as needed to cover the subjects to generate the final group assignments. For completely balanced groups, the number of subjects must be a multiple of the block size, in this case, a multiple of four. So, for 32 subjects total, we need a sequence of eight blocks because $32 / 4 = 8$. We randomly select the eight blocks from the set, but no matter which blocks we select, the total number of 1s and 0s will be the same.

For example, here's a randomly selected sequence of eight blocks:

```
>>> from RE import *
>>> b = ["1100","1010","1001","0110","0101","0011"]
>>> r = RE(mode="int", low=0, high=6)
>>> "".join([b[i] for i in r.random(8)])
'11000011010101010100100110110'
```

There are 16 subjects assigned to both the treatment and the control groups. Run the code again and the sequence will be different, but the number assigned to each group is still 16. Apply the sequence to the 32 subjects selected for the study, perhaps by matching the assigned subject ID number in order with the sequence. See Table 10-1.

Table 10-1: Matching Subject ID and Assigned Group

	0	1	2	3	4	5	6	7	...
Subject	0	1	2	3	4	5	6	7	...
Group	1	1	0	0	0	0	1	1	...

This continues for all 32 subjects.

Block randomization is an improvement over simple randomization in that it balances the number of subjects in the treatment and control groups. However, neither approach pays attention to other characteristics that may impact or mask treatment effects. If the outcome of the treatment is dramatic and broadly applicable, both simple and block randomization will show it. But if the treatment effect is weaker or relevant only to a specific demographic, then neither approach on its own is desirable. Enter stratified randomization.

Stratified

In *stratified randomization*, we assign subjects to treatment and control groups such that each group contains a consistent mix of subjects with other characteristics, called *covariates*, that we (the experimenters) believe are likely to influence the results. For example, if we use a block design for a research project that tests running endurance after taking a supplement for three months, and our treatment group happens to have an abundance of nonsmokers under age 25, there will likely be a strong difference in endurance between the treatment and control groups at the end of the experiment.

Stratified randomization manages the group composition as much as possible by matching the covariates of the controls to the treatment group.

There are different approaches to stratified randomization. In our simulation, we'll pick a treatment group subject at random from a large population of potential subjects. Based on the covariates for that subject, we'll search for a matching subject to put in the control group. This way, we have balanced numbers in the treatment and control groups, and we'll know that every person in the treatment group has a matched control to make the overall characteristics of both groups the same. In other words, we'll adjust for variation in the covariates.

Our experience with *bad_sample.py* demonstrated that small study sizes are susceptible to dramatic variation in factors like age, income, smoking, and drinking. These are the covariates we'll use in our simulation.

Defining the Simulation

The remainder of the chapter works with the code in *design.py*. As always, I recommend reading the code before continuing. We'll do a detailed walkthrough in the next section; in this section, we'll define what we hope to accomplish with the simulation.

Here's the scenario: we want to evaluate the impact a novel supplement has on overall health after one year of use. To do this, we first select a treatment group that takes the supplement for a year. Then, we compare their overall health score to the score of a control group that did not take the supplement.

NOTE   *Modern clinical trials often use* placebos, *or "fake" treatments, with the control group. Here the placebo would be a sugar pill that looks like the supplement. Neither the participants nor the researchers would be aware of who received what, thereby making the study double-blind. Alternatively, the control group might not be told anything about the supplement group. They would be queried or tested at the beginning of the study and again at the end, with no placebo given. While we'll use the latter approach, either will work in this case.*

The simulation is rigged; there will always be a positive health benefit for the treatment group. The strength of the effect is under our control. Naturally, there is no supplement; the point of the simulation is to understand how each of the three experimental design choices affects the experiment's outcome. Is it easy to detect the treatment effect? How convinced might we be if we see an improvement that the measured effect is real? We'll seek to answer these questions.

As we did with *bad_sample.py*, we'll generate populations of individuals using the same covariates: age, income, smoking, and drinking. Unlike *bad_sample.py*, we won't use age to influence the other covariates, and we'll immediately place covariates into bins. In other words, we won't say this person is 48 or 11; instead, we'll select from three age bins, meaning age will be 0, 1, or 2. We'll do the same for income and drinks per week. Smoking is binary, 0 or 1.

Each run of the code accumulates statistics for a user-selected number of experiments. We randomly generate a population, and select treatment and control groups according to the desired randomization scheme. We then apply the treatment (supplement), and use the health score for both groups to generate metrics. When all experiments are complete, we use the collected information to create output that provides insight into how well the randomization scheme worked.

As with all simulations, we must be careful to convince ourselves that the thing being simulated is a fair approximation of reality to the level we require. Remember Box's adage: all models are wrong, but some are useful.

Implementing the Simulation

Let's understand the essential parts of *design.py*, beginning with the main loop near the bottom of the file. Then, as needed, we'll discuss functions that loop calls. After that comes the code to analyze the results. Algorithmically, the main loop repeats the following for the desired number of experiments (nsimulations):

1. Create a population (a list of Person objects).
2. Select control and treatment cohorts according to the user-supplied randomization scheme (typ).
3. Apply the treatment to the treatment group by calling the Treat method, passing the desired treatment effect size (beta).
4. Accumulate stats on the control and treatment cohorts. This includes the per-subject health score and the average value of each covariate.
5. Append the collected data to the results list, with each element being a dictionary that holds the results for that simulated experiment.

The main loop is in Listing 10-1.

```
results = []
for nsim in range(nsimulations):
    pop = []
    for i in range(npop):
        pop.append(Person())

    control, treatment = [Simple, Block, Stratified][typ](pop, nsubj)

    for subject in treatment:
        subject.Treat(beta)

    ch, c_age, c_income, c_smoker, c_drink = Summarize(control)
    th, t_age, t_income, t_smoker, t_drink = Summarize(treatment)

    results.append({
        "c_age": c_age,
        "c_income": c_income,
        "c_smoker": c_smoker,
        "c_drink": c_drink,
        "t_age": t_age,
        "t_income": t_income,
        "t_smoker": t_smoker,
        "t_drink": t_drink,
        "ttest": ttest_ind(th,ch),
        "d": Cohen_d(th,ch),
    })
```

Listing 10-1: The main loop

There are five code paragraphs. The first creates pop, a list of Person objects from which we select control and treatment subjects.

The second code paragraph, a single line, selects the proper randomization function passing the population and the total number of subjects in the experiment (nsubj). For Block and Stratified randomization, the number of subjects in each group is always half the desired total. Both control and treatment are lists of Person objects selected from pop.

The next code paragraph calls the Treat method on each subject in the treatment group. The argument, beta, [0, 1), is the user-supplied treatment effect size, where a higher beta implies a stronger positive treatment effect. In other words, the supplement's effectiveness improves as beta increases.

The fourth code paragraph calculates the per-subject health scores and covariate averages for the control and treatment groups.

The last code paragraph appends the results of this experiment to the results list. The call to ttest_ind performs a t-test between the treatment and control health scores while Cohen_d calculates Cohen's *d* to measure the effect size.

Let's explore the supporting cast called by the main loop, followed by each randomization function. We want to simulate the latter correctly.

Functions and Classes

Populations are lists of Person instances (Listing 10-2).

```
class Person():
    def __init__(self):
        self.age = int(3*rng.random())
        self.income = int(3*rng.random())
        self.smoker = 0
        if (rng.random() < 0.2):
            self.smoker = 1
        self.drink = int(3*rng.random())
        self.adj = 2*(rng.random() - 0.5)

    def Health(self):
        return 3*(2-self.age) + 2*self.income - 2*self.smoker - self.drink + self.adj

    def Treat(self, beta=0.03):
        self.adj += 3*binomial(300, beta, rng) / 300   # [0,1]
```

Listing 10-2: The Person class

A person is a collection of five characteristics. Three of them are integers in [0, 2]: age, income, and drink. Notice the call to the globally defined rng, an instance of the RE class according to the supplied randomness source and seed. The smoker characteristic is binary, so a person has a 20 percent chance of being a smoker. The fifth characteristic, adj, is a random float in [−1, 1]. We'll learn why we use this binning scheme in the next section.

The Health method returns a float indicating the person's overall health:

$$\text{Health} = 3 \times (2 - \text{age}) + 2 \times \text{income} - 2 \times \text{smoker} - \text{drink} + \text{adj}$$

Better health (a higher float) is associated with younger age, higher income bracket, not being a smoker, and minimized drinking, plus the random adjustment.

The Treat method applies the treatment by altering adj based on beta, the desired treatment effect. It does this by increasing adj by three times a random draw from a binomial distribution using a fixed number of trials (300).

We haven't worked with binomial distributions before. If there's an event with a probability p of happening on any given trial—think of flipping a loaded coin with probability p of landing heads up—and there are n trials, then the number of successful events in n trials follows a binomial distribution for an expected number of outcomes, $[0, n]$.

For example, if the probability of an event is $p = 0.7$, or 70 percent, and there are $n = 10$ trials, then the expected number of events for repetitions follows the distribution in Figure 10-2.

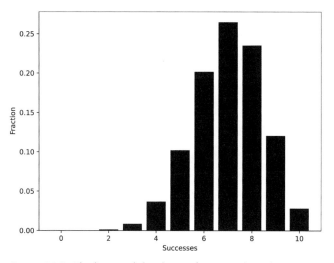

Figure 10-2: The binomial distribution for 10 trials and 70 percent probability of success per trial

The most frequent number of successes in 10 trials at 70 percent probability of success is 7, as we might expect. However, about 2.5 percent of the time, every trial was successful.

As the probability of success increases, so does the number of possible successful events over a given number of trials. In Treat, we scale the number of successful trials for probability beta by 300; thus, the entire expression becomes a number in $[0, 1]$, which we multiply by 3 and add to adj. When the number of trials is large, the binomial distribution looks like a narrow normal distribution, where the treatment effect is narrowly centered on the supplied beta value.

The `binomial` function simulates a draw from a binomial distribution using the uniform distribution supplied by rng. This algorithm is inefficient because it uses many calls to rng to return one sample from the desired binomial distribution. Still, it's good enough for us (Listing 10-3).

```
def binomial(n,a,rng):
    k = 0
    p = a if (a <= 0.5) else 1.0-a
    for i in range(n):
        if (rng.random() <= p):
            k += 1
    return k if (a <= 0.5) else n-k
```

Listing 10-3: Sampling from a binomial distribution

The number of successful trials is in k. The conditional return and assignment to p makes use of symmetry in the binomial distribution to return the proper sample for any desired probability of success for a given probability (a).

The main loop calls two more functions, `Cohen_d` and `Summarize`, as in Listing 10-4.

```
def Cohen_d(a,b):
    s1 = np.std(a, ddof=1)**2
    s2 = np.std(b, ddof=1)**2
    return (a.mean()-b.mean())/np.sqrt(0.5*(s1+s2))

def Summarize(subjects):
    h = []
    age = income = smoker = drink = 0.0
    for subject in subjects:
        h.append(subject.Health())
        age += subject.age
        income += subject.income
        smoker += subject.smoker
        drink += subject.drink
    age /= len(subjects)
    income /= len(subjects)
    smoker /= len(subjects)
    drink /= len(subjects)
    return np.array(h), age, income, smoker, drink
```

Listing 10-4: Additional helper functions

The `Cohen_d` function measures the effect size by calculating the difference in the means of two datasets along with the square root of their averaged variances.

The `Summarize` function queries a collection of subjects from either the control or treatment group to return a vector of subject health scores along with the mean age, income, smoke, and drink values over the subjects.

Let's get to the heart of the matter and discuss the different randomization schemes themselves.

Schemes

Listing 10-5 uses simple randomization to flip a coin, assigning selected subjects to either the treatment or control group.

```
def Simple(pop, nsubj):
    order = np.argsort(rng.random(len(pop)))
    c = []; t = []
    for k in range(nsubj):
        if (rng.random() < 0.5):
            c.append(pop[order[k]])
        else:
            t.append(pop[order[k]])
    return c,t
```

Listing 10-5: Simple random assignment

The Simple function first generates a random ordering of the population. This isn't strictly necessary, as we randomly generated pop, but we shouldn't assume that if we don't need to.

The for loop picks subjects sequentially and, based on the coin flip, assigns them to the treatment or the control group. Note that the coin flip means the number of subjects in each group might not be the same. When done, we return both lists.

Block randomization ensures balanced treatment and control group sizes, as shown in Listing 10-6.

```
def Block(pop, nsubj):
    ns = 4*(nsubj//4)
    blocks = ["1100","1010","1001","0110","0101","0011"]
    nblocks = ns//4
    seq = ""
    for i in range(nblocks):
        n = int(len(blocks)*rng.random())
        seq += blocks[n]
    order = np.argsort(rng.random(len(pop)))
    c = []; t = []
    for i in range(ns):
        if (seq[i] == "1"):
            t.append(pop[order[i]])
        else:
            c.append(pop[order[i]])
    return c,t
```

Listing 10-6: Block randomization

The first line of `Block` uses integer division to set `ns` to the nearest multiple of 4 less than or equal to `nsubj`. Then come the block definitions as strings of binary numbers. These blocks are the six ways we can split four items evenly among two groups, treatment (1) and control (0).

We create the `seq` string with the number of blocks (`nblocks`) by randomly concatenating block definitions until the string is `ns` characters long.

The final `for` loop mimics `Simple` in randomizing the order of `pop` before assigning subjects to treatment and control groups based on the current character of `seq`.

Our approach to stratified randomization assigns a random subject from the population to the treatment group, searches the population for an unassigned person with the same characteristics, and assigns them to the control group. This way, both groups are balanced in number and in overall characteristics. In code, this becomes Listing 10-7.

```
def Stratified(pop, nsubj):
    def match(n,m,pop,selected):
        if (selected[m]):
            return False
        if (pop[n].age != pop[m].age):
            return False
        if (pop[n].income != pop[m].income):
            return False
        if (pop[n].smoker != pop[m].smoker):
            return False
        if (pop[n].drink != pop[m].drink):
            return False
        return True

    selected = np.zeros(len(pop), dtype="uint8")
    c = []; t = []
    while (len(t) < nsubj//2):
        n = int(len(pop)*rng.random())
        while (selected[n] == 1):
            n = int(len(pop)*rng.random())
        selected[n] = 1
        t.append(pop[n])
        m = int(len(pop)*rng.random())
        while (not match(n,m, pop, selected)):
            m = int(len(pop)*rng.random())
        selected[m] = 1
        c.append(pop[m])
    return c,t
```

Listing 10-7: Stratified randomization

Listing 10-7 creates a flag vector, `selected`, to mark members of the population already assigned to a group. The `while` loop then runs until the treatment group contains half the requested number of subjects. The body of the while loop sets n to the index of an unselected member of the population, who becomes a member of the treatment group. We update selected once n is found, and append the person to the treatment group (t).

Next, we find another unselected member of the population who has characteristics matching those of the person we just added to the treatment group; this is why the Person class uses bins for characteristics. There are only $3 \times 3 \times 2 \times 3 = 54$ possible combinations of characters. With a large population, we're virtually assured of finding a match, which happens when the match embedded function returns True. After we append the match to the control group, the outer while loop continues until it assigns all subjects.

The remaining code in *design.py* analyzes the results of multiple experiments. While I won't walk through it, I'll describe the analysis in the next section.

Exploring the Simulation

We're ready to run some experiments. First, let's orient ourselves with a quick recap:

- We are simulating experiments using a known positive treatment effect of a size we supply.

- We are supplying the randomization scheme: simple, block, or stratified.

- We are evaluating the results of many experiments to understand the randomization scheme's influence.

We aren't simulating a single experiment, but dozens to hundreds. We'll evaluate the combined results from these experiments to (hopefully) arrive at a picture of the subtle influence randomization schemes have on experiment outcomes. The *design.py* file is a stand-in for an entire universe of treatment studies, all performed using the given randomization scheme.

Let's dive in. The *design.py* file expects several command line arguments:

```
> python3 design.py
design <npop> <nsubj> <beta> <nsim> <type> <plot> [<kind> | <kind> <seed>]

    <npop>   -  population size (e.g. 1000)
    <nsubj>  -  number of subjects in the experiment (e.g. 40)
    <beta>   -  supplement effect strength [0..1]
    <nsim>   -  number of simulations to run (e.g. 100)
    <typ>    -  selection type: 0=simple, 1=block, 2=stratified
    <plot>   -  1=show plot, 0=no plot
    <kind>   -  randomness source
    <seed>   -  seed value
```

We must supply the population size (npop), number of subjects in each experiment (nsubj), treatment effect strength (beta), number of experiments to simulate (nsim), and type of randomization to use (type). As always, the randomness source and optional seed value are last.

Simple

We can run our first simulation like so:

```
> python3 design.py 10000 32 0.3 40 0 1 minstd 6809

mean p-value (lowest) : 0.01138
mean p-value (highest): 0.82919

mean Cohen's (lowest) : 1.00568
mean Cohen's (highest): 0.07732

delta age    : (high, low, t, p) = (0.16556, 0.41945, -3.27589, 0.01691)
delta income: (high, low, t, p) = (0.21191, 0.30570, -0.64742, 0.54132)
delta smoker: (high, low, t, p) = (0.12377, 0.11255,  0.21945, 0.83357)
delta drink : (high, low, t, p) = (0.38284, 0.22193,  0.83403, 0.43620)
```

A plot should appear along with the text output; we'll get to it momentarily. We asked for 40 simulations of an experiment where we selected 32 subjects from a pool of 10,000 using simple randomization (0). The treatment effect size was 0.3, meaning the Treat method on each Person object added a random value centered on 0.3 to the overall health of the individual.

These stats are drawn from the results of the simulations. The first two lines show the mean p-value for the lowest and highest 10 percent of experiments. Recall, the p-value for an experiment is from the health scores for the treatment and control groups. In this case, each experiment had approximately 16 subjects per group and there were 40 simulated experiments, so the highest and lowest 10 percent p-values correspond to 4 experiments. Because we used simple randomization, the treatment and control group size for a particular simulated experiment isn't always 16.

The next two lines report the mean Cohen's d for the same set of experiments. Cohen's d is a scaled version of the difference in the means between the treatment and control groups. Therefore, as we'll see when using stratified randomization, we might expect the mean effect size for at least the lowest 10 percent group to be about the same as the requested effect size. Here, we asked for a smaller effect size of 0.3, but the mean was a substantial 1.0. For the highest 10 percent group, the effect size was insignificant, which matches the p-value of 0.83. For this group of experiments, we found no meaningful difference between the treatment and control groups.

The final four lines of output present the means of the per-experiment covariate *differences* for each group, low p-value and high p-value. We use a t-test here to see if the covariate differences between treatment and control significantly differed between the groups. For this run, the age difference was significant.

The t statistic for age is negative, meaning the low p-value experiments consisted of treatment and control groups with large age differences. In other words, for the low p-value experiments, the age covariate for the treatment and control groups was more likely to be different than for the high p-value experiments. Age is the most prominent factor in the health score, so a significant imbalance between treatment and control groups strongly affects measured results.

So, what should we make of these results? We ran many simulations, and, even for the best performing of them—those that showed a significant difference after treatment—we shouldn't believe the results, as the covariates of those experiments were quite different. If we ran a single experiment, we'd likely see no effect and judge the treatment as a failure. For those cases where we did see an effect, it was likely artificially inflated because the treatment group was quite different from the control group in an important covariate.

Remember that these results are for 40 experiments, meaning 32 subjects and simple random sampling into treatment and control groups tend toward unreliable results that either exaggerate the effectiveness of the treatment or mask it entirely.

The left side of Figure 10-3 shows the *design.py* plot, a histogram displaying the t-test p-values between the health of the treatment and control groups for the 40 simulated experiments. The results to the left of the vertical dashed line at 0.05 indicate results that would, typically, be heralded as statistically significant.

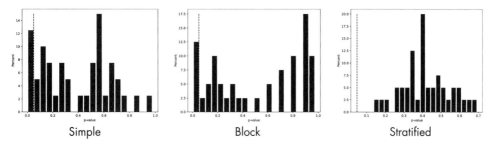

Figure 10-3: Example plot output from design.py by randomization type

From the leftmost plot in Figure 10-3, we see that about 12.5 percent of the experiments yielded a statistically significant difference. These are the experiments we'd likely read about in a scientific journal. The remaining 87.5 percent of the experiments didn't result in a statistically significant treatment effect. Proper scientific adherence states that these experiments should be published as negative results, but such papers seldom appear—a known issue in the scientific literature.

While we don't know a priori that the treatment effect is present and positive in the real world, in our case we do—and that all 40 simulated experiments should have found a positive result. So, why wasn't it found? The size of the cohort is a contributing factor that we'll experiment with shortly, but as shown in the simulation, so is the imbalance of covariates due to simple randomization, both in masking and exaggerating the effect.

Block

Switching to block randomization requires changing a 0 to a 1 on the command line:

```
> python3 design.py 10000 32 0.3 40 1 1 minstd 6809

mean p-value (lowest) : 0.02021
mean p-value (highest): 0.94917

mean Cohen's (lowest) : 0.89363
mean Cohen's (highest): 0.00087

delta age    : (high, low, t, p) = (0.15625, 0.45312, -2.12870, 0.07735)
delta income: (high, low, t, p) = (0.21875, 0.42188, -1.53562, 0.17554)
delta smoker: (high, low, t, p) = (0.10938, 0.14062, -0.35729, 0.73310)
delta drink : (high, low, t, p) = (0.37500, 0.17188,  1.80858, 0.12051)
```

Not much has changed when compared to the simple randomization results. Selecting balanced treatment and control group sizes for each experiment has eliminated any possible effect due to bad splits. Still, that effect is unlikely to show up at the level of many experiments, which we're considering here. The middle plot in Figure 10-3 is much like the simple randomization plot; in both, 12.5 percent of experiments were statistically significant at the $p < 0.05$ level. However, age is again highly different between the lowest and highest p-value experiment groups, though not at the level it was for the simple randomization case. We might improve this by increasing the number of subjects, but let's switch to stratified randomization first and see what happens.

Stratified

To switch to stratified randomization, change the 1 on the command line to a 2:

```
> python3 design.py 10000 32 0.3 40 2 1 minstd 6809

mean p-value (lowest) : 0.19833
mean p-value (highest): 0.64602

mean Cohen's (lowest) : 0.46910
mean Cohen's (highest): 0.16422
```

Note that the output lacks any "delta" lines because treatment and control groups have the same covariate makeup in stratified randomization, so all the deltas are zero; there is nothing to report. Recall, the "delta" lines refer to the means of the individual experiment differences between treatment and control groups. Those differences are zero because we selected a matched control group subject for each treatment group subject.

The plot is on the right in Figure 10-3. It doesn't look like the simple or block randomization plots, as the covariate effect has been compensated for by the stratified (matching) selection process. What the plot shows is due to other factors.

However, no experiment produced a p-value below the 0.05 threshold. After properly accounting for the covariates we believe influence health, the treatment effect, while present, is too weak to detect with only 32 subjects in the study.

Cohen's d for the lowest p-value group is 0.47—not too far from the 0.3 we should get if we detect the treatment effect consistently. To improve matters further, we'll adjust the study size by increasing the number of subjects.

Increasing the Number of Subjects

Let's move from 32 subjects to 128 subjects:

```
> python3 design.py 10000 128 0.3 40 2 1 minstd 6809

mean p-value (lowest) : 0.06149
mean p-value (highest): 0.20789

mean Cohen's (lowest) : 0.33364
mean Cohen's (highest): 0.22381
```

The lowest 10 percent of experiments has a mean p-value of 0.06 and a mean d of 0.334, so we're getting closer, but no experiments passed the magic threshold of 0.05.

Let's double the study size again:

```
> python3 design.py 10000 256 0.3 40 2 1 minstd 6809

mean p-value (lowest) : 0.00781
mean p-value (highest): 0.05495

mean Cohen's (lowest) : 0.33582
mean Cohen's (highest): 0.24168
```

Now even the highest p-value group is near 0.05, and both d values are in the 0.3 ballpark. The plot (not shown) illustrates that almost all experiments produce $p < 0.05$ results.

With stratified randomization, we've accounted for covariates, so we're seeing the actual treatment effect in these means. Still, we needed a sufficiently large number of subjects to tease the treatment effect out of the data. Let's find out how many subjects we need for a given desired effect size.

The code in *cohen_d_test.py* uses *design.py* to estimate the number of subjects necessary to achieve a mean p-value of 0.05 or less in the highest 10 percent of experiments. A run of the code produces:

```
0.9:   50 subjects, p=0.03350000, d=0.7079
0.8:   60 subjects, p=0.03402000, d=0.6172
0.7:   80 subjects, p=0.02784000, d=0.5373
0.6:   90 subjects, p=0.03872000, d=0.4702
0.5:  110 subjects, p=0.04107000, d=0.4141
0.4:  170 subjects, p=0.04417000, d=0.3213
0.3:  280 subjects, p=0.04682000, d=0.2431
0.2:  580 subjects, p=0.04017000, d=0.1724
0.1: 2370 subjects, p=0.04922000, d=0.0811
```

The requested effect size is on the left, followed by the estimated number of subjects necessary, the p-value, and *d* for the highest 10 percent of simulated experiments. My code run took about five hours, mostly working the *d* = 0.1 case. Generally, the printed *d* values are close to the requested values, which is a good sign.

The code is a loop over the desired effect sizes with an inner while loop that increments the number of subjects until the mean p-value is 0.05 or less:

```
def RunTest(beta, nsubj):
    cmd = "python3 design.py 100000 %d %0.1f 20 2 minstd 6809 >/tmp/xyzzy"
    os.system(cmd % (nsubj,beta))
    lines = [i[:-1] for i in open("/tmp/xyzzy")]
    pv = float(lines[2].split()[-1])
    d  = float(lines[5].split()[-1])
    return pv,d


base = 10
for beta in [0.9,0.8,0.7,0.6,0.5,0.4,0.3,0.2,0.1]:
    pvalue = 10.0
    k = 1
    while (pvalue > 0.05):
        pvalue,d = RunTest(beta, k*base)
        k += 1
    print("%0.1f: %3d subjects, p=%0.8f, d=%0.4f" % (beta, k*base, pvalue, d), flush=True)
```

The number of subjects starts at 10 (base) and is multiplied by a value k that increments by 1 inside the while loop until the p-value is at or below 0.05. The RunTest function constructs the *design.py* command line, executes it, and parses the output file to get *p* and *d*.

Calculating the Simulation's Statistical Power

Let's use TTestIndPower from the statsmodels package to see whether the simulation captures the correct number of subjects necessary for a desired *d* value.

The `TTestIndPower` class performs a *power* analysis for two independent samples, the treatment and control groups. See *power_analysis.py*, which produces output for the same set of effect sizes as *cohen_d_test.py*. Comparing the two in terms of the number of subjects results in Table 10-2.

Table 10-2: Comparing the Calculated and Estimated Number of Subjects

d	Calculated	Estimated
0.9	26	50
0.8	33	60
0.7	43	80
0.6	59	90
0.5	85	110
0.4	132	170
0.3	234	280
0.2	526	580
0.1	2,102	2,370

The calculated results are for a threshold of 0.05 (alpha) and a power of 0.9, meaning we want a 90 percent chance of achieving at least 0.05 as the p-value for the t-test between the treatment and control groups. I selected a power of 0.9 because the empirical code seeks to make the mean of the top 10 percent of the highest p-values 0.05 or less, implying that almost all simulated experiments will find a result at a p-value of 0.05 or lower. The agreement between the calculated and estimated numbers of subjects is rough but reasonable, which validates the simulation.

Exercises

This chapter has only two exercises, both of which are challenging.

Working with Two Treatments

Apply block randomization to a study with three conditions: control, treatment 1, and treatment 2. The block size should be a multiple of the number of conditions; fix the block size at 6, meaning each block will use each condition twice. For example, [0,0,1,1,2,2] is a valid block of six subjects where 0 is control, 1 receives treatment 1, and 2 receives treatment 2.

We need a way to select at random from the set containing all possible permutations of the set of three conditions used twice. In other words, we'd like a NumPy array where each row is a valid block. Use code like this:

```
import numpy as np
from itertools import permutations
b = [0,0,1,1,2,2]
l = list(set(permutations(b)))
blocks = np.array(l)
```

The blocks array is a 90×6 matrix with 90 unique blocks to select from when building a balanced cohort. The permutations function returns an iterable that set uses to return all unique permutations of the basic block, b. To make a NumPy array, convert the set into a list before passing it to array.

Your task: alter a copy of *design.py* to select subjects for simple and block randomization for the control, treatment 1, and treatment 2 groups. For simple randomization, roll a three-sided die. Ignore stratified randomization for now.

Combining Block and Stratified Randomization

To combine block and stratified randomization, create blocks where all subjects are matched according to the desired covariates, and each block includes all conditions in a balanced manner. If there are three conditions—as in the previous exercise—and the blocks have six subjects each, then the subjects in each block must be matched to some level in terms of covariates.

Alter the copy of *design.py* from Possibility 1, but account for the 56 possible combinations of the four covariates. The subjects placed into a block must all have the same covariate values.

If the block is [0,1,2,1,2,0] and the covariate set is [2,1,0,1], for example, then search the population for six subjects in age group 2, income group 1, smoker 0, and drink 1, assigning them to the conditions as dictated by the block. Continue for *n* subjects where *n* is a multiple of six, the block size. Add this capability to your altered *design.py* from Exercise 1.

Summary

This chapter focused on randomization in experimental design, particularly during subject selection. After briefly discussing the need for randomization, we explored the most common types: simple, block, and stratified.

Simple randomization uses coin flips or dice rolls to place subjects in treatment or control groups. While this process can introduce bias due to unequal group sizes, block randomization removes this possibility by ensuring that treatment and control group sizes are balanced.

We learned about covariates, factors outside the study scope that may affect the results by enhancing or masking any possible treatment effect. Stratified randomization uses covariates to construct treatment and control groups with matching characteristics.

We then created a simulation to test randomization schemes on the outcome of an experiment where we knew the treatment effect we should see. We learned that small cohorts using simple randomization are prone to bias; while block randomization removes bias due to imbalanced group sizes, stratified randomization corrects for bias in the treatment and control groups. This allows us to measure the effect we expect from the simulation, provided the number of subjects is of some minimal size.

Lastly, we matched the number of subjects necessary in a stratified setting to achieve the desired effect size at a $p < 0.05$ level for most of our simulated experiments to the number of subjects indicated by a t-test-based power analysis. The empirical number was in reasonably good agreement with the model, so we claimed this as validation of our approach.

Let's continue our journey with randomized algorithms, an essential class of algorithms in computer science.

11

COMPUTER SCIENCE ALGORITHMS

While everything we've looked at so far can be called a "randomized algorithm," in computer science the phrase refers to two broad categories of algorithms—the subject of this chapter.

A *randomized algorithm* employs randomness as part of its operation. The algorithm succeeds in accomplishing its goal, either by producing the correct answer quickly, but sometimes not, or by running rapidly with some probability of returning a false or nonoptimal result.

We'll begin by defining the two broad categories of randomized algorithms with examples. Next, we'll learn about estimating the number of animals in a population. Following that, we'll learn how to demonstrate that a number is a prime to any desired level of certainty while avoiding the brute force approach of searching for all possible divisors. We'll end with randomized Quicksort, the textbook example of a randomized algorithm.

Las Vegas and Monte Carlo

Las Vegas and Monte Carlo are locations famously associated with gambling, that is, with games of chance dependent on randomness and probability. However, when computer scientists refer to Las Vegas and Monte Carlo, they are (usually) referring to the two main types of randomized algorithms.

A *Las Vegas algorithm* always returns a correct result for its input in a random amount of time; that is, how long the algorithm takes to execute isn't deterministic, but the output is correct.

On the other hand, a *Monte Carlo algorithm* offers no assurance that its output is correct, but the runtime is deterministic. There is a nonzero probability that the output isn't correct, but for a practical Monte Carlo algorithm, this probability is small. Most algorithms we've encountered, including swarm intelligence and evolutionary algorithms, are Monte Carlo algorithms. Las Vegas algorithms can be transformed into Monte Carlo algorithms by allowing them to exit before locating the correct output.

The first example we'll investigate is a sorting algorithm that is, at our discretion, a Las Vegas or Monte Carlo algorithm. The second is a Monte Carlo algorithm for verifying matrix multiplication.

Permutation Sort

A *permutation* is a possible arrangement of a set of items. If there are n items in the set, there are $n! = n(n-1)(n-2)\ldots 1$ possible permutations. For example, if the set consists of three things, say $A = \{1, 2, 3\}$, then there are six possible permutations:

$$\{1, 2, 3\}, \{1, 3, 2\}, \{2, 1, 3\}, \{2, 3, 1\}, \{3, 1, 2\}, \{3, 2, 1\}$$

Notice that one permutation sorts the items from smallest to largest. Therefore, if given a vector of unsorted numbers, we might use a sort algorithm that generates permutations until finding the one that sorts the items. While we can implement this deterministically, we can also use random permutations with the hope that we might stumble across the correct ordering before testing too many candidate permutations. The *permutation sort* algorithm (also known as *bogosort* or *stupid sort*) implements this idea.

Run the file *permutation_sort.py* with no arguments:

```
permutation_sort <items> <limit> [<kind> | <kind> <seed>]

    <items> - number of items in the list
    <limit> - number of passes maximum (0=Las Vegas else Monte Carlo)
    <kind>  - randomness source
    <seed>  - seed value
```

The code generates a random vector of integers in $[0, 99]$ and sorts it by trying random permutations up to limit. To score each permutation, the code returns the fraction of pairs of elements that are out of order, where $a > b$ for a at index i and b at index $i + 1$. If the array is sorted, the score is zero.

If limit is zero, the algorithm runs until it finds the correct permutation, which depends on the number of possible permutations. As the number of permutations increases ($n!$), the runtime rapidly increases if we insist on trying until we succeed. In this way, a limit of 0 turns *permutation_sort.py* into a Las Vegas algorithm.

For example, to run *permutation_sort.py* as a Las Vegas algorithm, use:

```
> python3 permutation_sort.py 6 0 minstd 42
sorted: 0  25  44  57  65  96  (268 iterations)
```

The code found the proper order of elements after testing 268 of the possible 6! = 720 permutations. Changing the randomness source from `minstd` to `pcg64` sorts in 59 iterations while `mt19937` uses 20. We set the limit to 0 to make the code run until success, but the number of permutations tested was often far less than the maximum.

If we switch to Monte Carlo:

```
> python3 permutation_sort.py 6 100 minstd 42
partially: 0  57  25  44  65  96  (score = 0.16667, 100 iterations)
```

we get a partially sorted array with a score > 0. Because of the fixed randomness source and seed, we know we need 268 iterations to sort the array:

```
> python3 permutation_sort.py 6 268 minstd 42
sorted: 0  25  44  57  65  96  (268 iterations)
```

Listing 11-1 shows the main loop in *permutation_sort.py*.

```
v = np.array([int(rng.random()*100) for i in range(N)], dtype="uint8")
k = 0
score = Score(v)
while (score != 0) and (k < limit):
    k += 1
    i = np.argsort(rng.random(len(v)))
    s = Score(v[i])
    if (s < score):
        score = s
        v = v[i]
```

Listing 11-1: The main loop in permutation_sort.py

We create the vector (v), along with an initial score. The main loop, `while`, runs until the score is zero or the `limit` is exceeded. If Las Vegas, we set `limit` to a huge number to play the odds that we'll find the true ordering long before that many trials.

The body of the `while` loop creates a random ordering of v and calculates the score. Whenever it finds a lower score, the code reorders v to return at least a partially ordered vector should the limit be reached; however, this is not strictly necessary.

The remainder of the file displays the results or calculates the score (Listing 11-2).

```
def Score(arg):
    n = 0
    for i in range(len(arg)-1):
        if (arg[i] > arg[i+1]):
```

```
                n += 1
          return n / len(arg)
```

Listing 11-2: Scoring a permutation

Let's plot the mean number of permutations tested as a function of the number of items to sort, *permutation_sort_plot.py*, which plots the mean and standard error for 10 calls to *permutation_sort.py* for *n* in [2, 10]. The result is Figure 11-1.

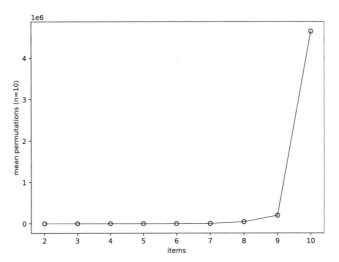

Figure 11-1: The mean number of permutations (in millions) tested as a function of the number of items

Figure 11-1 illustrates *combinatorial explosion*, which is the rapid growth in a problem's runtime or resource use as a function of the size of its input. Permutation sort works decently when sorting lists of up to nine items; any more and the complexity explodes.

We also see this effect in *permutation_sort_plot.py*'s output:

```
 2:  0.127855 +/- 0.002026
 3:  0.128128 +/- 0.001737
 4:  0.129859 +/- 0.002469
 5:  0.131369 +/- 0.002483
 6:  0.136637 +/- 0.003704
 7:  0.172775 +/- 0.008236
 8:  0.534369 +/- 0.081601
 9:  1.987567 +/- 0.488691
10: 44.133984 +/- 10.929158
```

The output shows, as a function of *n*, the mean ($\pm$ SE) time in seconds to sort a vector of that size. After seven elements, runtimes quickly increase.

Combinatorial explosion is the bane of many otherwise useful algorithms, often reaching a point where many lifetimes of the universe are insufficient to find the correct output.

Permutation sort is tied closely to the factorial, which is why we're getting these results:

2! = 2	5! = 120	8! = 40,320
3! = 6	6! = 720	9! = 362,880
4! = 24	7! = 5,040	10! = 3,628,800

The factorial grows at a tremendous rate. If we want to sort 20 items, we have

$$20! = 2,432,902,008,176,640,000$$

permutations to check. At 1 millisecond per permutation, we'd need over 77 million years of computing time to check them all. This doesn't mean permutation sort couldn't, by pure chance, sort 20 items in less than a second, but the probability is exceedingly low. This is the paradox of randomized algorithms.

Matrix Multiplication

I have three matrices, A, B, and C. We'll use bold, uppercase letters to represent matrices. Does $AB = C$? How can we know?

Introducing Matrix Multiplication

First, let's ensure we're on the same page regarding matrix multiplication. A *matrix* is a 2D array of numbers, so the matrices here might be:

$$A = \begin{bmatrix} 1 & 2 \\ 3 & 4 \end{bmatrix}, \quad B = \begin{bmatrix} 1 & 0 \\ 2 & 3 \end{bmatrix}, \quad C = \begin{bmatrix} 5 & 6 \\ 11 & 12 \end{bmatrix}$$

These are 2×2 matrices with two rows and two columns. If the number of rows equals the number of columns, we're working with *square matrices*. However, the number of rows and columns need not match; for example, we might have a matrix of 3×5 or $1,000\times13$. The latter case is typical in machine learning, where rows represent observations and columns represent features associated with those observations. An $n\times1$ matrix is a *column vector*, while a $1\times n$ matrix is a *row vector*.

Asking whether $AB = C$ implies that we know how to find AB. In NumPy, in order to multiply two 2D arrays, we multiply each corresponding element (Listing 11-3).

```
>>> A = np.array([[1,2],[3,4]])
>>> B = np.array([[1,0],[2,3]])
>>> A*B
array([[ 1,  0],
       [ 6, 12]])
```

Listing 11-3: Multiplying element-wise in NumPy

Unfortunately, multiplying matrices is more involved. We begin by checking that the number of columns of the first matrix equals the number of rows of the second. If not, then we can't multiply the matrices. Therefore, to multiply an $n \times m$ matrix by a $u \times v$ matrix requires $m = u$. If this is true, we can multiply to produce an $n \times v$ result. The square matrices in this section automatically satisfy this requirement.

The matrix multiplication process requires multiplying each column of the second matrix by the rows of the first matrix, where the elements of the column multiply the corresponding elements of the row. We then sum these products to produce a single output value. For example, in symbols, multiplying two 2×2 matrices returns a new 2×2 matrix:

$$\begin{bmatrix} a_{00} & a_{01} \\ a_{10} & a_{11} \end{bmatrix} \begin{bmatrix} b_{00} & b_{01} \\ b_{10} & b_{11} \end{bmatrix} = \begin{bmatrix} a_{00}b_{00} + a_{01}b_{10} & a_{00}b_{01} + a_{01}b_{11} \\ a_{10}b_{00} + a_{11}b_{10} & a_{10}b_{01} + a_{11}b_{11} \end{bmatrix}$$

We're indexing matrices from 0, as we would NumPy arrays. However, many math books index from 1 so that the first element of the first row of matrix A is denoted a_{11}, not a_{00}.

Mathematically, if A is an $n \times m$ matrix and B is an $m \times p$ matrix, then the elements of $C = AB$, an $n \times p$ matrix, are:

$$c_{ij} = \sum_{k=0}^{m-1} a_{ik}b_{kj}, \quad i = 0, 1, 2, \ldots, n-1; \quad j = 0, 1, 2, \ldots, p-1 \qquad (11.1)$$

Remember that matrix multiplication does not commute; in general, $AB \neq BA$. The sum over k in Equation 11.1 illustrates why: the single index accesses by row for A and by column for B so that swapping the order of A and B means different elements of the matrices are mixed.

The sum in Equation 11.1 uses index variable k with two more implied sums over all values of i and j to fill in the output matrix, C. These observations point toward an implementation: matrix multiplication becomes a triple loop indexing elements of 2D arrays.

Listing 11-4 translates the loops of Equation 11.1 into code.

```
def mmult(A,B):
    n,m = A.shape
    p = B.shape[1]
    C = np.zeros((n,p), dtype=A.dtype)
    for i in range(n):
        for j in range(p):
            for k in range(m):
                C[i,j] += A[i,k]*B[k,j]
    return C
```

Listing 11-4: Naive matrix multiplication

We'll use this implementation even though NumPy supports matrix multiplication natively in several ways, for example, via the @ operator. To understand why, we'll learn how computer scientists measure algorithm performance.

Introducing Big O Notation

Computer scientists characterize the resource use of an algorithm by comparing the algorithm's performance as input size increases to a similar function that captures how the algorithm's resource use changes as the input grows. Here, resource refers to either memory or time. For example, an $\mathcal{O}(n)$ algorithm linearly increases the memory used as the size of the input, n, increases. A linear function can be written as $y = mx + b$ for input x, but in big O notation, we ignore multiplicative and constant factors so that $y = x$ is functionally the same as x gets very large.

The matrix multiplication code in Listing 11-4 contains a triply nested loop. If the input matrices are square ($n \times n$), then I = J = K = n. Each loop runs n times, making the innermost loop run n times for every increment of the next outer loop, which must run n times to increment the outermost loop. Therefore, the total number of operations required to multiply two $n \times n$ matrices scales as n^3. The time needed to create the output matrix, C, and evaluate the first two lines of the function doesn't alter the essential character of the function—namely that it takes n^3 passes through the three loops.

A computer scientist would therefore write that Listing 11-4 is an $\mathcal{O}(n^3)$ algorithm and an "n cubed" implementation. In general, we want algorithms that scale as $\mathcal{O}(n)$ or better. As n increases, the work required by the algorithm scales linearly or, better still, sublinearly like $\mathcal{O}(\log n)$ or $\mathcal{O}(n \log n)$. In other words, a plot of the work as a function of n is a straight line. Ideally, we want $\mathcal{O}(1)$ algorithms that run in constant time regardless of the size of their input, but that isn't always possible. An algorithm that runs in $\mathcal{O}(n^2)$ is often tolerable, but $\mathcal{O}(n^3)$ is suitable only for small values of n.

Note that $\mathcal{O}(n)$, $\mathcal{O}(n^2)$, and $\mathcal{O}(n^3)$ are all powers of n. Such algorithms are known as *polynomial time* algorithms. We never want algorithms that run in *superpolynomial time*, with a runtime (or resource use) such that no polynomial tracks it. For example, an algorithm running in $\mathcal{O}(2^n)$ time is an *exponential time* algorithm, and its resource use grows dramatically with the size of the input at a rate no polynomial can match. Worse still is the permutation sort we experimented with previously; it's an $\mathcal{O}(n!)$ algorithm that runs in *factorial time*. To see how factorial time is worse than exponential time, make a plot comparing 2^n and $n!$ for $n = [1, 8]$.

Matrix multiplication as in Listing 11-4 is $\mathcal{O}(n^3)$. Our goal is to quickly check whether $AB = C$ when given three matrices. We first multiply A and B and then check whether each element of the result matches the corresponding element in C. The multiplication is $\mathcal{O}(n^3)$ and the check runs in $\mathcal{O}(n^2)$ time because we have to examine each element. As the cube grows much faster than the square, the overall naive algorithm runs in essentially $\mathcal{O}(n^3)$ time. Let's see if we can do better.

Introducing Freivalds' Algorithm

In 1977, Latvian computer scientist Rūsiņš Freivalds invented a randomized algorithm that correctly answers the question of whether $AB = C$ with high probability, yet runs in $\mathcal{O}(n^2)$ time.

For the following, we'll assume that A, B, and C are $n \times n$ matrices. The algorithm works for non-square matrices, but this restriction makes things easier to follow.

The algorithm itself is straightforward:

1. Pick a random n-element vector, $r = \{0, 1\}^n$, that is, a random vector of 0s and 1s.

2. Calculate $D = A(Br) - Cr$. (Note: the parentheses matter.)

3. If all elements of D are zero, claim "yes," $AB = C$; otherwise, claim "no."

At first glance, Freivalds' algorithm doesn't look like it will help. However, recall how matrix multiplication works. The expression Br is multiplying an $n \times n$ matrix by an $n \times 1$ vector, which returns an $n \times 1$ vector. The next multiplication by A returns another $n \times 1$ vector. Likewise, Cr is also an $n \times 1$ vector. At no point is a full $n \times n$ matrix multiplication happening. As n grows, the savings in the number of calculations grows all the faster. Freivalds' algorithm runs in $\mathcal{O}(n^2)$ time—a considerable improvement over the $\mathcal{O}(n^3)$ runtime of the naive algorithm.

Multiplying B by r is the equivalent of selecting a random subset of B's columns and adding their value across the rows. For example:

$$\begin{bmatrix} 1 & 3 & 4 & 0 \\ 2 & 4 & 3 & 1 \\ 0 & 0 & 1 & 2 \\ 2 & 2 & 1 & 0 \end{bmatrix} \begin{bmatrix} 1 \\ 0 \\ 1 \\ 0 \end{bmatrix} = \begin{bmatrix} 1+4 \\ 2+3 \\ 0+1 \\ 2+1 \end{bmatrix} = \begin{bmatrix} 5 \\ 5 \\ 1 \\ 3 \end{bmatrix}$$

The algorithm is betting that examining random elements of the three matrices will, if they are equal, result in D being a vector of all zeros more often than D being all zeros by chance. An analysis of the probabilities involved, which we won't cover, demonstrates that the probability of D being all zeros given $AB \neq C$ is less than or equal to $1/2$.

If the probability of one calculation involving a randomly selected r returning the wrong answer is at most $1/2$, then two random vectors (if we run the algorithm twice) have a probability of returning the wrong answer of at most $(1/2)(1/2) = 1/4$. Here the wrong answer is an output of "yes" when in fact $AB \neq C$.

Each application of the algorithm is independent of any previous application. For independent events, like the flip of a fair coin, probabilities multiply, so k runs of Freivalds' algorithm implies that the probability of a false "yes" result is $1/2^k$ or less. This means we can be as confident of the result as we like by running the algorithm multiple times.

The algorithm will always return "yes" when $AB = C$, meaning it is *one sided*—an error in the output happens only if $AB \neq C$. In a *two-sided* error, the algorithm could be wrong in either case, with some probability.

Testing Freivalds' Algorithm

Let's give the algorithm a try using *freivalds.py*, which generates 1,000 random triplets of $n \times n$ matrices, with n given on the command line. In all cases, $AB \neq C$, so we report failures as a fraction of 1,000.

Run *freivalds.py* like so:

```
freivalds <N> <mode> <reps> [<kind> | <kind> <seed>]

  <N>     - matrix size (always square)
  <mode>  - 0=Freivalds', 1=naive
  <reps>  - reps of Freivalds' (ignored for others)
  <kind>  - randomness source
  <seed>  - seed
```

The first argument is the dimensionality of the matrices. The second decides whether to use the naive algorithm that calculates $AB - C$ or Freivalds'. The third is the number of times to repeat the test with random r vectors. We'll use this option shortly to track the error rate. As usual, the other arguments enable any randomness source and a seed to repeat the same sequence of random matrices.

For example:

```
> python3 freivalds.py 3 0 1 mt19937 19937
0.08598161 0.132
```

tells us that testing 1,000 3×3 matrices using Freivalds' algorithm once each took some 0.09 seconds and failed 13.2 percent of the 1,000 tests.

To use the naive algorithm, change the 0 to 1 on the command line:

```
> python3 freivalds.py 3 1 1 mt19937 19937
0.05456829 0.000
```

As expected, there are no failures because the complete calculation always catches when $AB \neq C$. While the naive algorithm seems to run faster than Freivalds', this is an illusion; as n increases, the two diverge.

Failing 13 percent of the time when checking 3×3 matrices isn't too inspiring. Let's repeat the test, but check twice instead of once:

```
> python3 freivalds.py 3 0 2 mt19937 19937
0.14664984 0.016
```

Now we fail only 1.6 percent of the tests at the expense of nearly doubling the running time. Let's try four tests instead of two:

```
> python3 freivalds.py 3 0 4 mt19937 19937
0.26030445 0.000
```

With four tests, Freivalds' algorithm is 1,000 out of 1,000.

Freivalds' algorithm is probabilistic. The likelihood of error decreases quickly as matrix size increases. To see this effect, alter the matrix size while fixing the repetitions at 1. By the time $n = 11$, the error is generally below 0.1 percent.

It makes sense that the error rate goes down with matrix size. The probability that a random selection of values sum by accident to two equal values ($A(\boldsymbol{Br})$ and $\boldsymbol{Cr}$) should decrease as the number of values summed increases.

Let's explore running time as a function of n. Run *freivalds_plots.py* to produce the graphs in Figure 11-2.

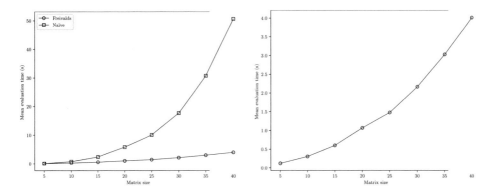

Figure 11-2: Comparing Freivalds' running time to the naive algorithm as a function of matrix size (left) and plotting Freivalds' running time alone to show the $\mathcal{O}(n^2)$ complexity—note the y-axis range (right)

On the left of Figure 11-2, we see the growth in running time between Freivalds' and the naive algorithm as the size of the matrices increases. The naive algorithm, $\mathcal{O}(n^3)$, grows substantially faster than Freivalds' $\mathcal{O}(n^2)$, shown by itself on the right.

The combination of performance gain and decreasing likelihood of error as n increases makes Freivalds' algorithm particularly nice. Yes, it's probabilistic, but in the places where it's most desirable (large n), it's also most likely to be correct.

Before examining *freivalds.py*, I have a confession. We can do matrix multiplication better than $\mathcal{O}(n^3)$, especially for matrices with $n > 100$. Volker Strassen's 1969 matrix multiplication algorithm has a runtime of about $\mathcal{O}(n^{\log_2 7}) \approx \mathcal{O}(n^{2.807})$, which is slightly better than the naive algorithm. NumPy, based on the BLAS library, makes use of Strassen's algorithm, which is why we didn't use NumPy in this section. However, $\mathcal{O}(n^2)$ is better than $\mathcal{O}(n^{2.807})$, so Freivalds' algorithm is still useful, even with Strassen matrix multiplication.

GALACTIC ALGORITHMS

There are matrix multiplication algorithms with even better asymptotic behavior than Strassen's algorithm. The current best have complexity $\mathcal{O}(n^{2.373})$ or so. However, these algorithms are, in practice, completely useless. The seeming contradiction has to do with Big O notation, which shows the overall behavior

---

but ignores multiplicative factors and constants. This means that an algorithm running in $10n^3$ time is the same as one running in $10{,}000n^3 + 10{,}000$ time. Both scale as $\mathcal{O}(n^3)$, but in practice, the first is more likely to be helpful.

Matrix multiplication algorithms that beat Strassen's algorithm in overall complexity, like the *Coppersmith–Winograd* algorithm, have constants so large that the algorithm becomes practical only once n is some number far larger than anything computers can currently handle, if ever.

Such algorithms have been christened *galactic algorithms* by Kenneth W. Regan. We cannot effectively use galactic algorithms in practice even if they are "the best" in terms of asymptotic behavior. While these algorithms are of theoretical importance, they won't show up in our toolkits any time soon.

Exploring the Code

Listing 11-5 contains the code implementing Freivalds' algorithm. The mmult function is in Listing 11-4. The array_equal function asks whether the absolute maximum of the difference between $A(Br)$ and Cr is below eps, and returns True if so.

```
def array_equal(a,b, eps=1e-7):
    return np.abs(a-b).max() <= eps


k = 0
m = 1000
s = time.time()
for i in range(m):
    A = 100*rng.random(N*N).reshape((N,N))
    B = 100*rng.random(N*N).reshape((N,N))
    C = A@B + 0.1*rng.random(N*N).reshape((N,N))
    if (mode == 0):
        t = True
        for j in range(reps):
            r = (2*rng.random(N)).astype("uint8").reshape((N,1))
            t &= array_equal(mmult(A,mmult(B,r)), mmult(C,r))
    else:
        t = array_equal(mmult(A,B), C)
    k += 1 if t else 0
print("%0.8f %0.3f" % (time.time()-s, k/m))
```

Listing 11-5: Freivalds' algorithm

The outer for loop executes 1,000 trials using a randomly selected set of matrices each time. C is such that it never equals AB, so every call to array_equal should return False.

The body of the outer for loop either multiplies A and B directly (mode==1), or uses Freivalds' algorithm by generating a random binary vector, r. Note that r is reshaped to be an $n \times 1$ column vector, as required for matrix multiplication.

The inner for loop applies Freivalds' repeatedly (reps) each time, AND-ing the result with t. The AND operation means that after reps tests with different r vectors each time, the only way t is still true is if all tests give a wrong result. Each test should see array_equal return False because $AB \neq C$ by design. Once t becomes False, it remains False for all remaining tests, so even one correct output from array_equal causes t to have the expected value.

If t is still True after the inner loop, then the trial failed and we increment k. After all trials, we print the total runtime and the fraction of the 1,000 trials that failed.

Freivalds' algorithm is a Monte Carlo algorithm because it might, with a probability we can minimize, produce a false output and claim $AB = C$ when it isn't true.

Let's turn to a different type of question for the next section: to get an estimate of the number of things in a collection, is it necessary to count them individually?

Counting Animals

Ecologists often want to know how many animals of a particular species live in an area, though counting each one is often impossible. Enter *mark and recapture*, a strategy for estimating population size from a small sample.

In mark and recapture, the ecologist first goes into the field and captures n specimens, which they then mark and release. A short while later, they revisit the field and capture animals again until they get at least one that is marked. If they capture K animals to get k that are marked, they now have everything necessary to estimate the full population size, N. They do this by using ratios.

Initially, the ecologist marked n of the N animals, meaning the fraction of the total population marked is n/N. The recapture phase netted k marked animals out of K. Assuming no births, deaths, or migrations, the two ratios should be approximately equal, so solving for N gives:

$$N_L \approx \frac{nK}{k}$$

This equation results in the *Lincoln-Petersen population estimate*, hence N_L.

A slightly less biased estimate of the population (or so it's claimed) comes from the *Chapman population estimate*:

$$N_C \approx \frac{(n+1)(K+1)}{k+1} - 1$$

Finally, we have a Bayesian approach to mark and recapture:

$$N_B \approx \frac{(n-1)(K-1)}{k-2}$$

This approach requires at least three marked animals in the recapture group to avoid dividing by zero.

Let's compare these three different estimates for the same population size with *mark_recapture.py*:

```
> python3 mark_recapture.py
mark_recapture <pop> <mark> <reps> [<kind> | <kind> <seed>]
  <pop>  - population size (e.g. 1000)
  <mark> - number to mark (< pop)
  <reps> - number of repetitions
  <kind> - randomness source
  <seed> - seed
```

The code simulates marking and recapturing by randomly marking a specified number of animals before recapturing a fraction of the population to count how many are marked. Let's run the code a few times to get a feel for the output. We'll fix the true population size at 1,000 and initially mark 100, or 10 percent. Setting the repetitions to 1 takes a single sampling, which is similar to what an ecologist might do in practice. We get:

```
> python3 mark_recapture.py 1000 100 1 mt19937 11
Lincoln-Petersen population estimate = 1250
Chapman population estimate          = 1132
Bayesian population estimate         = 1633

> python3 mark_recapture.py 1000 100 1 mt19937 12
Lincoln-Petersen population estimate = 666
Chapman population estimate          = 636
Bayesian population estimate         = 753

> python3 mark_recapture.py 1000 100 1 mt19937 13
Lincoln-Petersen population estimate = 833
Chapman population estimate          = 783
Bayesian population estimate         = 980
```

The estimates vary widely from run to run, as we might expect from a randomized algorithm. While the Lincoln-Petersen and Chapman estimates are generally low, the Bayesian estimates are closer to or even exceed the population size.

Using a single repetition is akin to attempting to generalize from a single collected data point, so let's increase the repetitions:

```
> python3 mark_recapture.py 1000 100 25 mt19937 11
Lincoln-Petersen population estimate = 1028.4713 +/-  78.4623
Chapman population estimate          =  940.0367 +/-  61.6982
Bayesian population estimate         = 1345.2015 +/- 166.3078

> python3 mark_recapture.py 1000 100 25 mt19937 12
Lincoln-Petersen population estimate = 1052.0985 +/-  61.3198
Chapman population estimate          =  963.4317 +/-  49.9620
Bayesian population estimate         = 1345.9986 +/- 108.4192
```

```
> python3 mark_recapture.py 1000 100 25 mt19937 13
Lincoln-Petersen population estimate = 1112.8340 +/-  80.5759
Chapman population estimate          = 1009.5146 +/-  63.3742
Bayesian population estimate         = 1485.2546 +/- 169.0492
```

The output now reflects the mean and standard errors for 25 repetitions, providing a better idea of how the estimates behave. The Lincoln-Petersen and Chapman results are closer to the actual population size, while the Bayesian estimate is consistently too high. The standard errors are illustrative as well, with the Bayesian standard error being larger than the others, indicating more trial-to-trial variation.

Try experimenting with different combinations of population size and number of animals initially marked.

Figure 11-3 presents three somewhat crowded graphs.

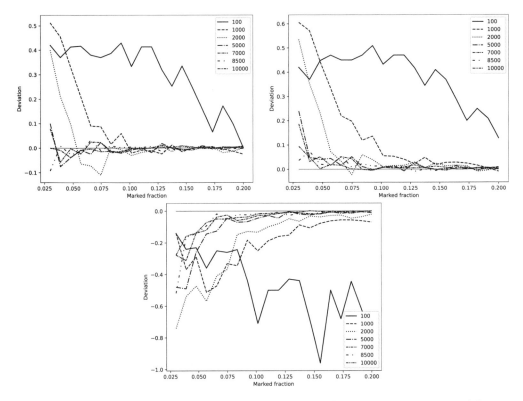

Figure 11-3: The three mark and recapture estimators as a function of the true population size and the fraction of that size initially marked. The plots show the signed deviation from the true population size: Lincoln-Peterson (top left), Chapman (top right), and Bayesian (bottom).

In the top-left graph in Figure 11-3, each of the seven plotted lines represents a different true population size from 100 to 10,000. The x-axis indicates the fraction of the true population marked by the ecologist on their first trip to the field. The value plotted is the median signed difference between the Lincoln-Petersen estimate for that combination of population size and fraction initially marked and the true population size. If the curve is above zero, the estimate is too low; below zero, it's too high. In other words, the graph shows $N_{true} - N_{est}$, so underestimating is a positive difference and overestimating is negative. The remaining two graphs show the same information for the Chapman (top-right) and Bayesian (bottom) estimators.

For populations above 1,000, the Lincoln-Petersen estimator is generally useful when initially marking more than 10 percent of the population, which may not be feasible in practice. However, for small populations, the estimator requires some 20 percent of the population to be marked to achieve reliability. One might use a simulation to generate a correction function for the Lincoln-Petersen estimator based on the suspected population size and the number of animals initially marked.

The Chapman estimator consistently underestimates the true population size to the point where one questions its utility compared to the Lincoln-Petersen estimate. However, the underestimate is relatively consistent for populations above 1,000, so again, a fudge factor might be derived from a simulation.

The Bayesian estimator's performance is quite different. It consistently overestimates the actual population, converging to the true population value only when the population becomes large and the percent initially marked is also significant (at least 15 percent). In practice, these conditions are unlikely to be met.

Figure 11-3 is the output of *mark_recapture_range.py*, which can be understood by examining the relevant parts of *mark_recapture.py* in Listing 11-6.

```
lincoln = []
chapman = []
bayes = []

for j in range(nreps):
    pop = np.zeros(npop, dtype="uint8")
    idx = np.argsort(rng.random(npop))[:nmark]
    pop[idx] = 1

    K = nmark
    while (True):
        idx = np.argsort(rng.random(npop))[:K]
        k = pop[idx].sum()
        if (k > 2):
            break
        K += 5

    lincoln.append(nmark*K/k)
    chapman.append((nmark+1)*(K+1)/(k+1) - 1)
    bayes.append((nmark-1)*(K-1)/(k-2))
```

Listing 11-6: Simulating mark and recapture estimates

The outer for loop over nreps handles the trials. For each trial, we create a population (pop) vector where npop is the population size from the command line. The vector is initially zero as we haven't marked any animals yet.

The next two lines represent the ecologist's first field trip. The argsort trick, coupled with keeping only the first nmark elements of the sort order, sets idx to the indices of pop that the ecologist has initially captured and marked (pop[idx]=1).

The second code paragraph represents the recapture phase in which the ecologist returns to the field and captures as many animals as were initially marked (K). We represent the captured animals by the K indices in idx as assigned in the inner while loop.

Marks are binary, so the sum of the selected elements of pop is the number of marked animals, k. If k is 3 or greater, break out of the while loop. Otherwise, increase K by five and try again. The final code paragraph calculates the three estimates of the true population size for this trial.

When the outer for loop exits, we have three vectors of estimates for the given population size and number initially marked. The remainder of *mark_recapture.py* displays the results. Given the simulation results, my money's on the Lincoln-Petersen estimator.

Let's move on from counting to the mathematically important task of primality testing.

Testing Primality

Prime numbers—integers evenly divisible by only themselves and one—are greatly beloved by number theorists. Primes have fascinated humanity since antiquity, and significant computing power is currently devoted to locating *Mersenne primes* of the form $2^p - 1$, where p is a prime number.

The largest known primes are Mersenne primes. As of this writing, the largest known Mersenne prime, discovered in 2018, is:

$$M_{82,589,933} = 2^{82,589,933} - 1$$

$M_{82,589,933}$ is a 24,862,048-digit number. Mersenne primes are sometimes denoted by their number and not their exponent. Therefore, $M_{82,589,933}$, the 51st Mersenne prime, might be given as M_{51}.

NOTE *To contribute in locating Mersenne primes, visit* https://www.mersenne.org *and sign up for the Great Internet Mersenne Prime Search.*

How do we know if n is a prime number? The definition gives us a natural starting point for a primality testing algorithm: if the only numbers that evenly divide n (resulting in no remainder) are 1 and n, then n is a prime.

Let's turn this definition into an algorithm. The brute force approach is to test every number that could be a factor of n. In practice, this means testing every integer up to $\sqrt{n}$ because any factor of n larger than $\sqrt{n}$ will necessarily be multiplied by some number less than $\sqrt{n}$, and will be caught before reaching $\sqrt{n}$.

When contemplating numbers comprising nearly 25 million digits, the amount of work involved increases dramatically. And if n is prime, must we test *every* integer up to $\sqrt{n}$?

The *Miller-Rabin test* is a fast, randomized algorithm to decide whether a positive integer, n, is prime. However, to understand the Miller-Rabin test, we need to know a bit about modular arithmetic.

Modular Arithmetic

We're used to the set of integers, denoted $\mathbb{Z}$ from the German for number. Integers are unbounded and extend infinitely in both directions from zero. If we restrict the range to the set $\{0, 1, 2, 3\}$, we can define arithmetic operations over this set by wrapping around as needed. Adding works as expected if the sum is less than 4: $1 + 1 = 2$ and $2 + 1 = 3$. However, if the sum exceeds 4, we wrap around. For example, $2+3 = 1$ because $2+3 = 5$, and we subtract 4 from 5 to get 1. Another way to view these operations is to apply the modulo operator after each addition to return the remainder after dividing by 4. For example, 5 mod 4 = 1.

Pierre Fermat, a 17th-century French mathematician, realized that if n is a prime number, then

$$a^{n-1} \equiv 1 \ (\text{mod } n), \ 0 < a < n$$

where $\equiv$ means:

$$a^{n-1} \bmod n = 1 \bmod n = 1$$

Great! We have a primality test: pick an integer $0 < a < n$, raise it to the $n - 1$ power, divide by n, and see if the remainder is 1. If it is, then n is a prime number, so the algorithm works and identifies n as a prime. However, some composite numbers also pass the test for many values of a, meaning this alone isn't sufficient to prove n is a prime. If this test fails, then n is definitely not a prime.

The Miller-Rabin test combines Fermat's test with another fact—if n is prime, the following is likely also true

$$a^{2^r d} \equiv -1 \ (\text{mod } n), \ 0 \leq r < s \text{ and } 0 < a < n$$

for some r in $[0, s)$ where $n = 2^s d + 1$ and d is odd. It's likely true because there are sometimes a values satisfying the congruence even if n is composite. We'll discuss these non-witness numbers shortly.

The first condition, Fermat's test, is straightforward enough, but let's unpack this second condition. We need to express n as $2^s d + 1$ or, equivalently, as $n-1 = 2^s d$. For suitable choices of s and d, $2^s d$ is another way of writing the exponent in the Fermat condition.

All of the math in $\equiv -1 \ (\text{mod } n)$ is modulo n, meaning the numbers are in the set $\{0, 1, 2, \ldots, n-1\}$, usually denoted as $\mathbb{Z}_n$. We view a negative number as counting backward, so $-1 \equiv n - 1$.

The second condition checks to see if $x^2 \equiv -1 \ (\text{mod } n)$ for some x. The Miller-Rabin test uses a sequence of such values of x, looking for one that, when squared modulo n, gives -1 (that is, $n - 1$). The sequence begins with

$r = 0$ and d as the exponent. The next check uses the square, which is the same as $r = 1$:

$$(a^{2^0 d})^2 = a^{2(2^0 d)} = a^{2^{0+1} d} = a^{2^1 d}$$

This is all modulo n. The next squaring returns $r = 2$, and so on.

If any of the sequence of such expressions is equivalent to $n - 1$, then n has a reasonably high probability of being a prime number. Otherwise, n is definitely *not* prime, and a is a *witness* to this fact.

The Miller-Rabin Test

Let's put Miller-Rabin into code, as in Listing 11-7.

```
def MillerRabin(n, rounds=5):
    if (n==2):
        return True
    if (n%2 == 0):
        return False

    s = 0
    d = n-1
    while (d%2 == 0):
        s += 1
        d //= 2

    for k in range(rounds):
        a = int(rng.random()) # [1,n-1]
        x = pow(a,d,n)
        if (x==1) or (x == n-1):
            continue
        b = False
        for j in range(s-1):
            x = pow(x,2,n)
            if (x == n-1):
                b = True
                break
        if (b):
            continue
        return False
    return True
```

Listing 11-7: Miller-Rabin in code

The function MillerRabin accepts n and rounds with a default value of 5. The first code paragraph captures trivial cases. As half of all numbers are even, testing directly for 2 and evenness saves time.

The second code paragraph locates s and d so that $n = 2^s d + 1$. It's always possible to find an s and d decomposition for any n (positive integer).

For now, we'll focus on the body of the outer for loop in the third paragraph, which implements a pass through the Miller-Rabin test for a randomly selected a and initial x value, $a^d \pmod{n}$.

The built-in Python function, pow, computes exponents and accepts an optional third argument so that pow(a,d,n) efficiently implements $a^d \pmod{n}$.

The following if checks for 1 or −1. If that's the case, the Fermat test has passed, so this pass through the outer for loop is over.

Otherwise, the inner for loop initiates the sequence of successive squarings of $x = a^d$ while looking for one equivalent to −1. If such a squaring is found, the inner loop breaks and the outer loop cycles; otherwise, n is composite and MillerRabin returns False.

When all rounds (the loop over k) are complete, and every test supports the notion that n is a prime, MillerRabin returns True.

The outer for loop applies the Miller-Rabin test repeated for randomly selected a values. Since an a value demonstrating n to be a composite number is a witness number, an a value that leads to a claim of prime when n is not prime is a *non-witness* number. It is never the case that all possible a values for a given n are non-witness numbers, so repeated applications of the outer loop body minimize the probability that a non-prime input will return True.

You'll find MillerRabin in the file *miller_rabin.py*. It expects a number to test, the number of rounds (a values to try), and the randomness source:

```
> python3 miller_rabin.py
miller_rabin <n> <rounds> [<kind> | <kind> <seed>]
   <n>      - number to test
   <rounds> - number of rounds
   <kind>   - randomness source
   <seed>   - seed
```

For example:

```
> python3 miller_rabin.py 73939133 1
73939133 is probably prime
> python3 miller_rabin.py 73939134 1
73939134 is composite
```

The output must be correct for these cases as 73,939,133 is a prime, and the closest two primes can be to each other is 2 away:

```
> python3 miller_rabin.py 8675309 1
8675309 is probably prime
> python3 miller_rabin.py 8675311 1
8675311 is probably prime
```

Both 8,675,309 and 8,675,311 are twin primes, so the test is correct.

Miller-Rabin always labels a prime a prime. Let's explore when Miller-Rabin fails.

Non-witness Numbers

As mentioned, a witness number, a, testifies to the fact that n isn't prime. Also, there are composite numbers for which the Miller-Rabin test fails if it selects a non-witness number as a.

We'll force the Miller-Rabin algorithm to fail, probabilistically, using a composite number with a known set of non-witness numbers to see if we can detect the expected number of failures.

Our target is $n = 65$. As a multiple of 5, 65 is composite. There are 64 potential witness numbers, from 1 through 64. Of these potential witness numbers, it's known that 8, 18, 47, 57, and 64 are non-witness numbers. If the Miller-Rabin test runs for one round and selects a non-witness number for a, it fails and claims that 65 is prime.

Because there are five non-witness numbers out of 64 possible, the Miller-Rabin test for a single round should fail about $5/64 = 7.8$ percent of the time. I checked this by running *miller_rabin.py* 1,000 times and counting the number of times the output indicated a prime, which it did precisely 78 times, implying a failure rate of $78/1,000 = 7.8$ percent.

At worst, the Miller-Rabin single-round failure probability for arbitrary n is $1/4$. Since each round is independent of the previous, running the test for k rounds means the worst possible failure probability is $(1/4)^k = 4^{-k}$. However, for most n values, the actual failure probability is far less than this.

Let's stick with 65. Knowing that its single-round failure rate is about 7.8 percent, running two rounds should fail $(5/64)^2 \approx 0.61$ percent of the time. Running *miller_rabin.py* 20,000 times produced 131 failures, giving a failure rate of $131/20,000 = 0.655$ percent. Three rounds puts the failure rate at about 0.05 percent. We can achieve any desired precision by setting k high enough.

Miller-Rabin Performance

Let's compare Miller-Rabin's runtime performance to the brute force approach implemented in *brute_primes.py*. The code in *prime_tests.py* runs both Miller-Rabin and the brute force algorithm for the largest $1, 2, 3, \ldots,$ 15-digit prime numbers. Recall, the brute force algorithm runs the longest when the input is a prime.

The largest single-digit prime is 7, while the largest 15-digit prime is 999,999,999,999,989. In Figure 11-4, we plot the mean of five runs of *miller_rabin.py* and *brute_primes.py* for each prime to show how the runtime changed as the inputs grew.

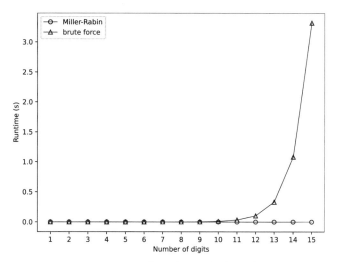

Figure 11-4: Comparing Miller-Rabin to the brute force primality test

The runtime complexity of the brute force algorithm is $\mathcal{O}(\sqrt{n})$ while that of Miller-Rabin is $\mathcal{O}(\log^3 n)$. The brute force algorithm quickly becomes unmanageable, even though it's sublinear, because $\sqrt{n} = n^{1/2}$ and $1/2 < 1$.

Miller-Rabin is a Monte Carlo algorithm because it claims n is a prime when there's a nonzero probability that it isn't. If n truly is a prime, Miller-Rabin always correctly labels it, but it also calls some composite numbers prime regardless of the number of rounds. Therefore, Miller-Rabin's false positive rate is nonzero, but its false negative rate is identically zero. In practice, however, increasing the number of rounds can make the false-positive rate as low as desired.

We have one more randomized algorithm to contemplate.

Working with Quicksort

Quicksort was developed by British computer scientist Tony Hoare in 1959 and is probably still the most widely used sorting algorithm. It's NumPy's default, for example.

If you take an undergraduate course in algorithms, you'll almost assuredly run across Quicksort, as it's easy to implement and understand, even if it's recursive. While most courses focus on characterizing its runtime complexity, we'll discuss the algorithm at a high level instead, and then run experiments on two versions: the standard nonrandom version and a randomized version.

Quicksort is a recursive, *divide-and-conquer* algorithm, meaning it calls itself on smaller and smaller versions of the problem until it encounters a base condition, at which point the implementation pieces everything back together to produce a sorted output.

The algorithm is as follows:

1. If the input array is empty or has only one element, return it.

2. Pick a *pivot* element, either the first in the array or at random.

3. Separate the array into three subsets: those elements less than the pivot, those equal to the pivot, and those greater than the pivot.

4. Return the concatenation of Quicksort called on the lower elements, the elements matching the pivot, and Quicksort called on the higher elements.

Step 1 is the base condition. If the array is empty or contains a single element, it's sorted. step 2 picks a pivot value, an array element we use in Step 3 to split the array into three parts: those less than, equal to, and greater than the pivot. Step 2 is where randomness comes into play. Non-random Quicksort always picks a specific element of the array, as it's already assumed to be in random order. Randomized Quicksort, however, selects its pivot element at random. We'll experiment with the subtle difference between nonrandom and random Quicksort.

Step 4 is the recursive part. The array is sorted if we merge the sorted lower partition with the same partition followed by the sorted higher partition. We sort the lower and higher partitions by using the sorting routine, that is, by calling Quicksort again. Each call on a portion of the array will, we assume, work with a smaller number of elements until we have single elements, the base condition of step 1.

Naive sorting methods, like bubble sort or gnome sort, run in $\mathcal{O}(n^2)$ time where n is the number of elements to sort. As we've learned, $\mathcal{O}(n^2)$ algorithms are acceptable for small n values, but quickly become unmanageable as n grows. Quicksort's average runtime complexity is $\mathcal{O}(n \log n)$, which grows at a much slower rate. This is why Quicksort is still widely used over 50 years after its introduction.

While Quicksort's *average* complexity is $\mathcal{O}(n \log n)$, if the array passed to Quicksort is already mostly or completely sorted, the complexity becomes $\mathcal{O}(n^2)$, which is no better than bubble sort. This happens if the array is in order or reverse order. Let's find out whether randomized Quicksort can help us here.

Running Quicksort in Python

The file *Quicksort.py* implements Quicksort twice. The first implementation uses a random pivot (QuicksortRandom), and the second implementation always uses the first element of the array as the pivot (Quicksort). The functions are in Listing 11-8.

```
def QuicksortRandom(arr):
    if (len(arr) < 2):
        return arr
    pivot = arr[np.random.randint(0, len(arr))]
    low  = arr[np.where(arr < pivot)]
```

```
        same = arr[np.where(arr == pivot)]
        high = arr[np.where(arr > pivot)]
        return np.hstack((QuicksortRandom(low), same, QuicksortRandom(high)))

def Quicksort (arr):
    if (len(arr) < 2):
        return arr
    pivot = arr[0]
    low  = arr[np.where(arr < pivot)]
    same = arr[np.where(arr == pivot)]
    high = arr[np.where(arr > pivot)]
    return np.hstack((Quicksort(low), same, Quicksort(high)))
```

Listing 11-8: Randomized and nonrandom Quicksort

For this example, we use NumPy instead of our RE class because it's already loaded, which minimizes the overhead when calling QuicksortRandom. The implementations differ only in how they assign pivot.

Both implementations follow the Quicksort algorithm step-by-step. First, we check for the base condition where arr is already sorted. We then split into low, same, and high based on the selected pivot. Finally, NumPy's hstack function concatenates the vectors returned by the recursive calls to Quicksort.

A high-performance implementation wouldn't call where three times, as each makes a full pass over arr, but we're interested only in relative performance differences as the input size changes.

Experimenting with Quicksort

The file *quicksort_tests.py* generates two graphs. The first, on the left in Figure 11-5, compares randomized Quicksort and nonrandom Quicksort as the input array size increases. In all cases, the input arrays are in random order. Therefore, the left side of Figure 11-5 represents the average case runtime. The points plotted are the mean over five runs. The dashed line represents $y = n \log n$.

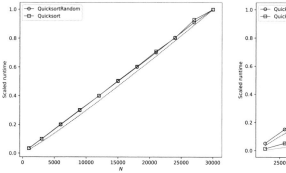 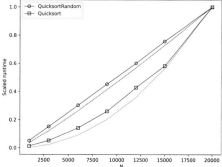

Figure 11-5: Randomized and nonrandom Quicksort on random inputs (left) and the same algorithms on pathological inputs (right)

The rightmost graph in Figure 11-5 shows the runtime for the case with already sorted input. This situation forces deterministic Quicksort into becoming an $\mathcal{O}(n^2)$ algorithm, which is why it tracks the curved plot, $y = n^2$. Randomized Quicksort, on the other hand, is unaffected by the order of the input and runs as before.

Correctly interpreting Figure 11-5 requires an explanation. Asymptotic runtime performance of algorithms ignores multiplicative factors and constants as they don't alter the overall form of the function as n increases. The randomized Quicksort function takes slightly longer to run than the nonrandom Quicksort because of the extra step of selecting a random index into the array. Therefore, plotting both runtimes together would make it somewhat difficult to see that the overall functional form is the same between QuicksortRandom and Quicksort. Moreover, plotting $y = n \log n$ follows an entirely different scale in terms of y-axis values, but again, the form of the function is the same. Therefore, to plot all three together, Figure 11-5 divides each y value by the maximum y value to map the output to $[0, 1]$ regardless of the actual range. This clarifies that randomized Quicksort and nonrandom Quicksort scale in the same way, and are following $\mathcal{O}(n \log n)$—all curves lie essentially on top of each other.

Now reconsider the right side of Figure 11-5 showing the case where the input array is already sorted. Again, the dashed line shows $y = n \log n$, and randomized Quicksort still follows that form. However, nonrandom Quicksort, which selects the first element of the array as its pivot, does not. Instead, it follows the dotted line, $y = n^2$, meaning the pathological input case alters nonrandom Quicksort, turning it into an $\mathcal{O}(n^2)$ algorithm.

Randomized Quicksort is a Las Vegas algorithm because it always returns the proper output—a sorted array. While the randomness involved doesn't make the implementation easier, it protects against a pathological case that's more frequent in practice than we might initially suspect. Therefore, I recommend always using randomized Quicksort.

To understand why nonrandom Quicksort behaves so poorly with sorted input, consider what happens during a pass when the pivot is the smallest value in the array; for example, when picking the first element as the pivot and the input array is already sorted. When this happens, the low vector is empty and, ignoring duplicates of the pivot, all the remaining values in the array end up in the high vector. This happens every time the function recurses, turning the recursion into a list of function calls n deep. Add the $\mathcal{O}(n)$ pass through the array on each recursive call (implicit in our implementation via where), and we arrive at an $\mathcal{O}(n^2)$ algorithm, which is no better than a bubble sort.

Selecting a random pivot at each level ensures that this situation won't happen in the long run, as it would amount to a string of n rolls of an n-sided fair die each landing on 1—an increasingly unlikely event as n grows.

Exercises

Consider the following exercises to further explore randomized algorithms:

- Write a Las Vegas algorithm to locate positive integers, a, b, and c, that satisfy $a^2 + b^2 = c^2$. Your code will be a Las Vegas algorithm because there are an infinite number of solutions, namely all the Pythagorean triples.

- Can you write a successful Las Vegas program to find positive integers a, b, and c such that $a^n + b^n = c^n$ for some $n > 2$? If not, how about a Monte Carlo algorithm? What might your stopping criteria be? I recommend searching for "Fermat's last theorem."

- Extend the permutation sort runtime plot for $n = 11$, 12, or even 13. How long do you have to wait?

- Make a plot of the mean number of trials of Freivalds' algorithm to get a failure case as a function of n, the size of the square matrices.

- The file *test_mmult.py* generates output suitable for *curves.py* from Chapter 4. Use that output and *curves.py* to generate fits. Is the fit exponent what you expect for the naive algorithm? What about NumPy, with the knowledge that it uses Strassen's algorithm?

- I have a bag full of marbles. I want to estimate how many are in the bag. Therefore, I pick one randomly, mark it, and put it back in the bag. I then pick another marble randomly, mark it, and put it back in the bag. I repeat this process, counting the number of marbles selected, until I pick a marble I've already marked. If the number of marbles picked and marked is k, then the number of marbles in the bag is approximately

$$N \approx \left\lfloor \frac{2k^2}{\pi} \right\rceil$$

where the combination of floor ($\lfloor$) on the left and ceiling ($\rceil$) on the right means "round to the nearest integer." I encountered this algorithm via a brief description of the process, but the description had no derivation for the formula and no references. Nonetheless, it sort of works. After experimenting some, I realized that the estimate is better if the formula is tweaked slightly to become:

$$N \approx \left\lfloor 0.8 \left(\frac{2k^2}{\pi} \right) \right\rceil$$

Implement this algorithm and explore how well it works on average. Then examine *count.py*, which runs the algorithm for many iterations, averages the results, and produces plots. For example

```
> python3 count.py 1_000_000_000 40 pcg64 6502
N = 1023827699, iterations 41414, total 1656576
```

estimates slightly more than 1 billion marbles in the bag. The correct number is exactly 1 billion. It uses 40 iterations of the algorithm for a total of 1.7 million marbles marked. Figure 11-6 is the resulting plot, *count_plot.png*, which shows each of the 40 estimates, the true value (solid line), and overall estimate (dashed).

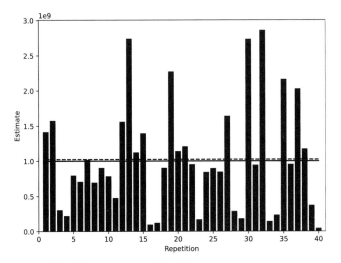

Figure 11-6: Forty estimates of marbles

Please contact me if you know a reference for this algorithm or how to derive the estimate formula.

- Can you think of a "fudge factor" for the Lincoln-Petersen population estimate for the case where the population is believed to be small?

- How does the runtime performance of nonrandom Quicksort vary as the array becomes more disordered? To figure this out, fix the array size (n) but change the degree of disorder in the array. For example, begin with a sorted array, then swap two elements, then three, and so on. Is the transition from $\mathcal{O}(n^2)$ to $\mathcal{O}(n \log n)$ linear with the number of elements swapped? Or does it seem more rapid?

Summary

In this chapter, we explored randomized algorithms, differentiating between Las Vegas and Monte Carlo. The former always produced correct output, eventually, while the latter may produce incorrect output. We considered permutation sort and Freivalds' algorithm for testing matrix multiplication. We learned that we can turn permutation sort from a Las Vegas algorithm into a Monte Carlo algorithm by imposing a limit on the number of candidate permutations considered. In general, we can transform Las Vegas algorithms into Monte Carlo algorithms, but not vice versa.

We then discussed the mark and recapture algorithm that ecologists use to estimate animal populations. We estimate the number of animals in a population by marking a known number and then recapturing animals and looking at the number marked. With sufficient numbers, the ratio of marked animals to animals recaptured should match the ratio of animals originally marked to the population size. We explored three estimators associated with this process and saw how they behave.

The Miller-Rabin algorithm quickly decides whether a positive integer is a prime. However, as a randomized algorithm, there's a certain probability that it'll falsely claim a composite number is prime. We learned how to decrease the likelihood of a false positive by repeated applications.

We concluded the chapter by comparing nonrandom and randomized Quicksort implementations. Randomized Quicksort adds little to the runtime while protecting against pathological inputs that are already (or mostly) sorted.

In our final chapter, we'll consider randomness as it relates to sampling from probability distributions.

12

SAMPLING

We've powered the majority of our experiments by extracting samples from the uniform distribution. While we've also worked with the normal distribution (Chapter 1), beta distribution (Chapter 3), and binomial distribution (Chapter 9), the tried-and-true uniform distribution is our oldest and dearest friend.

In this chapter, we'll sample from arbitrary probability distributions, be they discrete or continuous. This ability is critical to simulation and fundamental to Bayesian inference.

First, we'll discuss terminology and unpack the term *Bayesian inference*. Following that, we'll dive into sampling from arbitrary discrete probability distributions, first in one dimension, then in two.

Discrete distributions dealt with, we'll move on to sampling from continuous distributions via inverse transform sampling, rejection sampling, and Markov Chain Monte Carlo sampling using the Metropolis-Hastings algorithm.

This is perhaps our most mathematical chapter, but don't let that throw you. If you want to continue exploring the algorithms or apply them to different situations, the code is what matters.

Introduction to Sampling

Before diving into sampling, let's agree on terminology and the concepts that terminology entails. We'll also introduce notions related to Bayesian statistics and inference, with the latter being a prime motivator for, and beneficiary of, the development of sampling algorithms.

Terminology

If we roll a standard die, we'll get one of six possible outputs, each occurring with equal probability. We express this distribution as a bar graph with bars labeled 1 through 6, each of equal height corresponding to the fraction $1/6$. The sum of the fractions represented by each bar is 1 because the graph is a discrete probability distribution.

The *probability mass function (PMF)* tells us the probability of any discrete outcome. For a standard die, the PMF is $1/6$ for each outcome. For a binomial distribution, the PMF depends on the number of trials (n) and the probability of an event happening (p) per trial according to the formula $\binom{n}{k}p^k(1-p)^{n-k}$. Here k, $k = 0, \ldots, n$, is the number of events occurring during the n trials. Think of the continuous case as moving the discrete case to more and more possible outcomes; for example, a bar graph where the bars become increasingly narrow until they have an infinitesimal width. Talk of infinitesimals often implies calculus, as is the case here. The discrete distribution morphs into a continuous one so that:

$$\sum_i P(X = i) = 1 \quad \rightarrow \quad \int f(x)\, dx = 1$$

Here, $f(x)$ is a *probability density function (PDF)*, the continuous analog of a probability mass function. The notation $P(X = i)$ stands for the probability of a random variable, X, taking on the value i.

The $\int$ symbol is simply an old-fashioned script "S" for "sum." The thing summed is an infinite number of areas formed by rectangles of width dx (a single entity, not the product of d and x) and height $f(x)$, that is, the PDF function value at x. If limits are given, $\int_a^b$, the sum is for $x = a$ through $x = b$. No limits given implies "sum over all x that matter, even if from $-\infty$ to $+\infty$."

Sample a discrete distribution, and the probability of returning x is $P(X = x)$, or the probability associated with the bar labeled x. The probability of sampling a continuous distribution and getting a specific x is, counterintuitively, $P(x) = 0$ for any real number x. We can't talk of the probability of returning x as a sample, but instead, the probability of the sample lying in some range, $[a, b]$. That probability is:

$$P(a \leq x \leq b) = \int_a^b f(t)dt$$

This is nothing more than the area under the PDF from a to b. The variables in the integral are dummy variables. I switched from x to t to avoid confusion with the x we're asking about on the left-hand side of the equation.

We must talk about the probability over a range in the continuous case because not all infinities are created equal, a fact first realized by Georg Cantor in the 19th century. Because there are so many more real numbers than integers, the probability of selecting any one from a continuous distribution becomes identically zero. While the algorithms of this chapter return samples that appear to come from the desired continuous distribution, don't be fooled. As with all computation, we never use real numbers, but rational numbers in one form or another. Ultimately, our samples approximate the desired continuous probability distribution. However, as they say, if it walks like a duck and quacks like a duck, it's a duck—or a reasonably useful facsimile of one.

The PMF and PDF relate to the probability of sampling a particular value from a distribution. A related concept is the *cumulative distribution function (CDF)*, which we use for both discrete and continuous distributions. The CDF at x is the sum of the area under the PMF or PDF from its lowest value to x. If we sum over all the bars of the discrete distribution or integrate over all of the PDF, we get an area of 1, so the CDF is a function running from 0 on the left to 1 on the right as x increases.

For example, Figure 12-1 shows the PMF (top left) and CDF (top right) for a binomial distribution with $n = 10$ trials, each with probability $p = 0.7$. On the bottom are the PDF and CDF for a standard normal distribution. In both cases, the CDF approaches 1 from the left.

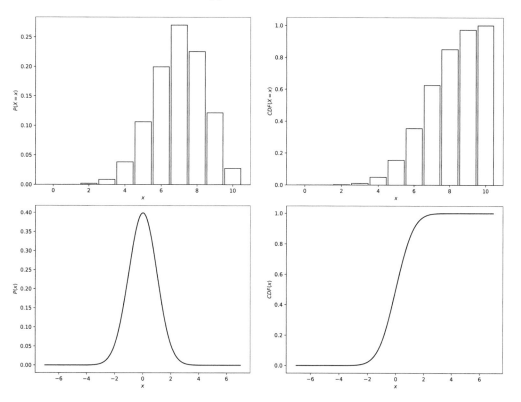

Figure 12-1: The PMF and CDF for a binomial distribution (top) and a standard normal distribution (bottom)

The code to generate Figure 12-1 is in *cdf.py*, which demonstrates how to realize pure math as code, at least with regard to probability functions. I'll skip the plotting portion of the code; see Listing 12-1 for the remaining relevant bits.

```
np.random.seed(8675309)
z = np.random.binomial(10,0.7,size=10000)
h = np.bincount(z, minlength=11)
h = h / h.sum()
cdf = np.cumsum(h)

np.random.seed(73939133)
x = np.linspace(-7,7,10000)
y = (1/np.sqrt(2*np.pi))*np.exp(-x**2/2)
cdf = np.cumsum(y*(x[1]-x[0]))
```

Listing 12-1: Generating CDFs from PMFs and PDFs

The first code paragraph generates samples from the binomial distribution. To save effort, I'm using NumPy's function to draw 10,000 random samples, meaning z is a vector of 10,000 values, each a randomly selected sample from the binomial distribution. To get the distribution itself, we need a histogram. The samples, in z, are integers, so the most efficient way to get the counts is with the bincount method. There were n = 10 trials, but there are 11 possible outcomes from 0 events up to 10, hence minlength=11 in the call to bincount.

Let's turn to the h = h / h.sum() line. The bincount method returns the per-outcome counts. We want a discrete probability distribution, which must sum to 1, so we divide each count by the total to transform them into fractions adding to 1. Therefore, h is now an *estimate* of the discrete binomial probability distribution for n = 10 and p = 0.7. For a better estimate of the true binomial distribution, increase the number of samples to 20,000 or more.

A discrete distribution's CDF is the running sum of the per-outcome probabilities. In other words

$$\text{CDF}(X = 0) = P(X = 0)$$
$$\text{CDF}(X = 1) = P(X = 0) + P(X = 1)$$
$$\text{CDF}(X = 2) = P(X = 0) + P(X = 1) + P(X = 2)$$
$$\dots \text{and so on} \dots$$

NumPy's cumsum calculates this cumulative sum for us to generate the entire CDF in a single function call.

The continuous case is similar, though we don't need to draw samples from it because the PDF of the normal distribution has a closed-form representation. For the standard normal (μ = 0, σ = 1), the PDF is:

$$f(x) = \frac{1}{\sqrt{2\pi}} e^{-x^2/2}$$

In code, we estimate this function (y) using 10,000 values of *x* equally spaced from −7 to 7 (`linspace`).

To estimate the CDF, however, we can't simply sum the values in y as in the discrete case. The area under the PDF must sum to 1, but because it's an area, we multiply each *y* value by the width of the rectangle it makes with the *x*-axis. The rectangle's width is the difference between successive x values, so we multiply y by `x[1]` - `x[0]` before summing. There's no need to scale y, as the values in y are the actual PDF values, not counts.

This chapter's goal is to sample from arbitrary distributions, where "sampling" means that we ask a distribution to give us a number according to the probabilities assigned to possible outputs. The higher the *y*-axis value of a distribution in a PMF or PDF plot, the higher the oracle's probability of selecting that number (discrete) or a number in a very narrow range about that position (continuous).

For example, the code in *nselect.py* generates the plot in Figure 12-2, in which the black dots on the *x*-axis signify 30 samples from the normal distribution.

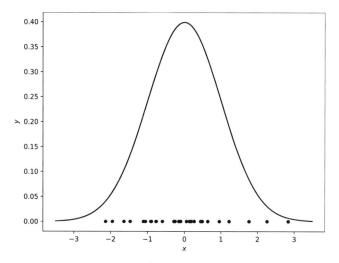

Figure 12-2: Thirty samples from a standard normal distribution

The samples are concentrated near the center of the PDF—the most likely region to be sampled. As we draw more samples, their density along the *x*-axis increases in proportion to the probability of being selected. Transforming the density into counts via a histogram approximates the PDF itself.

Now, let's discuss Bayesian inference, as its use depends on efficient sampling from complicated probability distributions.

Bayesian Inference

English minister Thomas Bayes (1701–1761) originated a seemingly simple equation that has recently turned the world of statistics on its head. Deriving the equation is an exercise in basic probability theory.

We write the probability of event B happening, given that event A has already happened, as $P(B|A)$. This is the *conditional probability* of B given A. The probability of A happening is $P(A)$, and the probability of both B and A happening is their *joint probability*, $P(A, B)$.

Probability theory states that $P(B, A) = P(B|A)P(A)$, which, switching the order of events, gives us $P(A, B) = P(A|B)P(B)$. Joint probabilities represent the probability of all combinations of the events, meaning $P(B, A) = P(A, B)$. Putting these observations together tells us that:

$$P(B, A) = P(A, B)$$
$$P(B|A)P(A) = P(A|B)P(B)$$
$$P(B|A) = \frac{P(A|B)P(B)}{P(A)}$$

The final equation is *Bayes' theorem*, which relates the *posterior probability*, $P(B|A)$, to the product of the *likelihood*, $P(A|B)$, and the *prior probability*, $P(B)$. The denominator on the right, $P(A)$, is the *evidence*. It's a normalizing value to ensure that the posterior probability is a probability—that the sum over the PDF of the posterior is 1. In practice, $P(A)$ becomes an integral that typically can't be expressed in closed form. In Bayesian modeling, the likelihood and prior probabilities are selected and known functional forms, but the evidence becomes intractable. This is where the sampling methods we'll discuss soon come into play. Drawing samples from the posterior is *Bayesian inference*. Without advanced sampling methods, Bayesian inference is all but impossible; with them, Bayesian inference becomes a paradigm shift, as Sharon Bertsch McGrayne notes in her book *The Theory That Would Not Die* (Yale University Press, 2011):

> The combination of Bayes and Markov Chain Monte Carlo has been called "arguably the most powerful mechanism ever created for processing data and knowledge."

Markov Chain Monte Carlo is one of the sampling algorithms we'll explore in this chapter. While we recognize Monte Carlo from Chapter 11, we'll discuss the Markov chain aspect in time.

Let's start sampling from arbitrary distributions.

Discrete Distributions

An arbitrary, one-dimensional discrete probability mass function might look like:

$$p_X(x) = [1, 1, 3, 4, 5, 1, 7, 4, 3]$$

While perhaps unexpected, as far as we're concerned this is a perfectly valid PMF represented as a Python list. It illustrates a distribution returning samples in the range 0 through 8 (the list has nine elements), and each sample returns an index into the PMF. As it stands, the PMF isn't *normalized*, so the sum of the "probabilities" isn't 1; it's $1 + 1 + 3 + 4 + 5 + 1 + 7 + 4 + 3 = 29$. To get probabilities, we divide each number by this sum.

The PMF tells us the proportion with which each value—0 through 8—appears, on average, after a large number of samples. For example, 6 appears seven times as often as 0 because the ratio between 6 and 0 is 7 : 1. Likewise, the ratio between 6 and 7 is 7 : 4, and so on.

Let's calculate the ratio between the probability of sampling a 6 (7/29) and the probability of getting a 0 (1/29):

$$\frac{7/29}{1/29} = \left(\frac{7}{29}\right)\left(\frac{29}{1}\right) = \frac{7}{1} = 7 : 1$$

The result is as expected.

This section explores three approaches to sampling from arbitrary discrete distributions. Two approaches expect the PMF to be normalized (sums to 1), while the third uses a PMF expressed as integer ratios between the elements. This may seem to be a drawback, but we often sample from distributions approximated by histograms, and the bins of a histogram are integer counts.

Sequential Search

In Chapter 7, we generated fractals using IFS by applying maps selected according to a given probability. In other words, we sampled from a distribution over maps. The code we used is in Listing 12-2.

```
def ChooseMap(self):
    r = self.rng.random()
    a = 0.0
    k = 0
    for i in range(self.nmaps):
        if (r > a):
            k = i
        else:
            return k
        a += self.probs[i]
    return k
```

Listing 12-2: Choosing a map

Listing 12-2 implements a version of *inversion sampling by sequential search*; at least, that's what Luc Devroye calls it in his book *Non-Uniform Random Variate Generation* (Springer, 1986). You'll find Devroye's book on his website, *http://luc.devroye.org/books-luc.html*. He's giving it away. I recommend grabbing a copy.

The implementation in Listing 12-2 isn't exactly terse. We can do better, as in Listing 12-3.

```
def Sequential(probs, rng):
    k = 0
    u = rng.random()
    while u > 0:
```

```
        u -= probs[k]
        k += 1
    return k-1
```

Listing 12-3: A more parsimonious implementation of inversion by sequential search

First, we hand the function the PMF as probs, which we expect to be normalized. The second argument is our old friend, an instance of RE configured to return floats in $[0, 1)$.

The code picks a uniformly distributed value, u, and subtracts the probabilities in probs, in order, until the result is zero or negative. We then return the number of times we subtract, k, as the sampled index into probs after adjusting for indexing from zero.

We can visualize the sampling process as in Figure 12-3 using the unnormalized PMF we discussed earlier.

$$p_X(x) = [1, 1, 3, 4, 5, 1, 7, 4, 3]$$

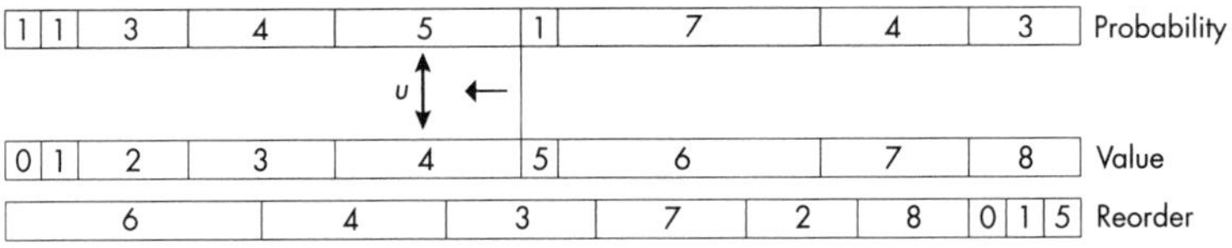

Figure 12-3: Sequential sampling of a discrete distribution

The Probability row reflects this PMF where the length of each box is in proportion to the other boxes so that the 7 box is seven times longer than the 1 box. To get the probability passed in probs, divide each box label by the sum, 29.

Figure 12-3 shows a selected u as a double arrow. This u is about 0.4 because it's a bit less than half the distance across the entire PMF, which must sum to 1. The vertical line after the box labeled 5 marks the full set of probabilities subtracted from u until $u < 0$. We subtract five boxes in this order: 1, 1, 3, 4, and 5. Adjust for indexing from 0, and the returned sample is 4—the label on the matched box below u in the Value row. Ignore the Reorder row for the time being.

This process maps $[0, 1)$ to $[0, 8]$ using the width of the boxes to transform the uniform input to a nonuniform output that will match the desired PMF if we draw enough samples. Try selecting other samples by closing your eyes and placing your finger somewhere on Figure 12-3's Probability row. Then, slide your finger down to the Value row and read off the output, which is the number of boxes covered from the left. After repeating this a few times, you should see that 6 will be chosen most often, followed by 4. Inversion sampling by sequential search is our first discrete sampling algorithm.

Let's make one minor improvement. The top row of Figure 12-3 presents the PMF in order, meaning the probability assigned to 0 is $1/29$, while the probability assigned to 6 is $7/29$. This makes sense if we use Listing 12-3 to sample from the distribution. However, to get 6 (the most frequent value) as a sample, we need to search from the beginning of the probability vector each time. If we list the probabilities in descending order, we'll select the most likely outcome after one pass through the `while` loop of Listing 12-3. The next most likely outcome requires only two passes, and so on. This notion leads to the Reorder row of Figure 12-3.

The bars of the Reorder row run from left to right in decreasing size order, with the label on each bar listing the value to return should that bar be selected. Notice, the algorithm in Listing 12-3 doesn't change; it's still subtracting probabilities from u, but they're now ordered from greatest to least. Therefore, we must map the index returned by Listing 12-3 to identify the true value selected. For example, if Listing 12-3 returns index 1, the mapping knows that $1 \rightarrow 4$ to return 4 as the selected value. This adjustment should speed things up, depending on the arrangement of probabilities in `probs`. The reordering tweak is our second discrete sampling algorithm.

Fast-Loaded Dice Roller

Our final discrete sampling algorithm is relatively new: the *Fast Loaded Dice Roller (FLDR)*, presented by Saad et al. in their 2020 paper, "The Fast Loaded Dice Roller: A Near-Optimal Exact Sampler for Discrete Probability Distributions." You'll find the code and paper on their GitHub site at *https://github.com/probcomp/fast-loaded-dice-roller*. We need only the *fldr.py* file from the *src/python* directory. Either copy that file from the GitHub repo via your browser or install the full package with `pip`:

```
> pip3 install fldr
```

If you copy *fldr.py* from GitHub, place it in the folder for this chapter.

The FLDR paper describes the algorithm and its genesis. It also refers to the Devroye book mentioned earlier, which motivated the algorithm's design. We won't discuss the details, as they're rather involved and mathematical. However, it's interesting to learn that there are more sophisticated ways of thinking about sampling from a discrete distribution.

The version of FLDR we'll use wants PMFs as vectors of integers, precisely as I've presented them so far. Using the FLDR requires two steps; the first conditions the algorithm based on the PMF, and the second draws individual samples on demand. We need the `fldr_preprocess_int` function to configure the sampler and the function `fldr_sample` to draw a sample. The FLDR code is not NumPy aware, but we can live with that.

Now that we have our algorithms, we'll pit them against each other.

Runtime Performance

Let's find out whether our algorithms work, and how they compare to each other in terms of runtime performance.

First, run *discrete_test.py* without arguments:

```
> python3 discrete_test.py
discrete_test <N> [<kind> | <kind> <seed>]

  <N>    - number of samples
  <kind> - randomness source
  <seed> - seed
```

The command line's form is familiar. The only required argument is the number of samples to draw from the nine-element distribution presented at start of "Discrete Distributions" on page 328: $[1, 1, 3, 4, 5, 1, 7, 4, 3]$.

Let's select some samples:

```
> python3 discrete_test.py 5000 minstd 476
[157 187 504 722 813 155 1251 693 518] (0.033361 s, sequential)
[164 171 526 674 904 162 1198 702 499] (0.029343 s, reordered)
[165 178 510 702 870 166 1225 683 501] (0.005494 s, FLDR)
[172 172 517 690 862 172 1207 690 517] expected
```

The command line requests 5,000 samples and displays the number of times each possible output was selected by algorithm type. For example, 1 was selected 187 times by the sequential algorithm. As we expect, 6 is the most frequent output. The final line contains the expected number of samples, found by multiplying the probability by the number of samples, rounded to the nearest integer.

The three algorithms appear to work as expected, with the results close to the expected output. Looking at the runtime on the right, the sequential algorithm is the slowest, the reordered algorithm is slightly faster, and FLDR is nearly an order of magnitude faster still.

If we ask for 50 samples instead of 5,000

```
> python3 discrete_test.py 50 minstd 476
[2  0 10  5 13  0 12  4  4] (0.000363 s, sequential)
[2  1  5  7 13  1  8  8  5] (0.000451 s, reordered)
[3  2  6  5  4  2 12  8  8] (0.000101 s, FLDR)
[2  2  5  7  9  2 12  7  5] expected
```

the output is noisy, as the expected frequency and the sampled frequencies are farther apart; for example, the sequential algorithm picked 2 in 10 instances while the expected frequency is only 5.

Figure 12-4 demonstrates this effect visually using FLDR to select 50 versus 5,000 samples.

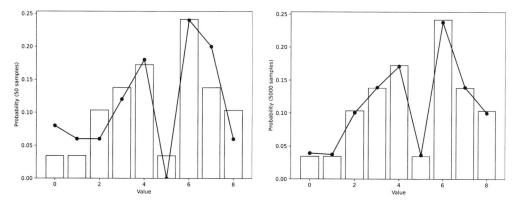

Figure 12-4: Sampling from a discrete distribution using 50 samples (left) and 5,000 (right)

The bars are the true distribution and the dots are the samples, both expressed as probabilities.

Figure 12-5 displays the runtime performance of our samplers for $\lfloor N^\alpha \rfloor$ samples with α running from 1 to 6 in 25 steps. Note that the *x*-axis is in units of 1 million.

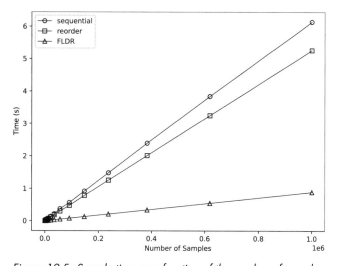

Figure 12-5: Sample time as a function of the number of samples

We can safely say that all three sampling algorithms run in $\mathcal{O}(n)$ time. However, Figure 12-5 is a good practical example for us. Big O notation ignores multiplicative factors, so while all three algorithms run in linear time, in practice we'll want to use FLDR.

Let's walk through *discrete_test.py*, starting with the setup (Listing 12-4).

```
from fldr import fldr_preprocess_int, fldr_sample
N = int(sys.argv[1])

if (len(sys.argv) == 4):
    rng = RE(kind=sys.argv[2], seed=int(sys.argv[3]))
elif (len(sys.argv) == 3):
    rng = RE(kind=sys.argv[2])
else:
    rng = RE()

probabilities = [1,1,3,4,5,1,7,4,3]
prob = np.array(probabilities)
prob = prob / prob.sum()
M = len(prob)
```

Listing 12-4: Setting up discrete_test.py

We'll focus on the beginning, where we introduce functions from fldr, and on the end, where we define probabilities.

FLDR wants integer counts, so we use probabilities in that case. When working with sequential sampling algorithms, which need true probabilities, we use prob.

Listing 12-5 uses each algorithm to draw the requested number of samples. Each case creates a vector of the samples, z, timing how long it takes. A list comprehension draws the samples.

```
s = time.time()
z = np.array([Sequential(prob,rng) for i in range(N)])
e = time.time() - s
h = np.bincount(z, minlength=M)
print(h, ("(%0.6f s, sequential)" % e))

idx = np.argsort(prob)[::-1]
p = prob[idx]
s = time.time()
z = np.array([Sequential(p,rng) for i in range(N)])
e = time.time() - s
h = np.bincount(idx[z], minlength=M)
print(h, ("(%0.6f s, reordered)" % e))

s = time.time()
x = fldr_preprocess_int(probabilities)
z = np.array([fldr_sample(x) for i in range(N)])
e = time.time() - s
h = np.bincount(z, minlength=M)
```

```
print(h, ("(%0.6f s, FLDR)" % e))

print(np.round(prob*N).astype("uint32"), "expected")
```

Listing 12-5: Sampling with each algorithm

The first code paragraph calls Sequential, which we saw in Listing 12-3, and then creates the histogram with bincount before displaying the counts and generation time.

The second paragraph ultimately calls Sequential, but first rearranges prob to be in decreasing sort order. NumPy's argsort returns the indices that sort prob in ascending order. The [::-1] idiom reverses the list to put idx in decreasing order.

We then call Sequential with p instead of prob. This means that the values in z are not the proper indices, but indices into idx, which holds the proper indices. In other words, idx is the mapping to get the proper sample values. A call to bincount using idx indexed by z generates the correct sample frequencies. Consider taking a moment to convince yourself that idx[z] makes sense.

Symmetry tells us that the final code paragraph in Listing 12-5 draws samples using FLDR by repeated calls to fldr_sample. But first, we have to pass the probabilities to fldr_preprocess_int to create the structure fldr_sample uses.

The final line of Listing 12-5 displays the expected per-value counts by rounding the product of the probability and the number of samples, N.

What if the distribution we want to sample from is two dimensional? What does that even mean? Let's find out.

Two Dimensions

We can store a one-dimensional discrete distribution in a vector. By extension, we might imagine storing a two-dimensional distribution in a matrix. But how do we interpret the distribution?

Since a one-dimensional distribution tells us how often we should expect to sample each value, then a two-dimensional distribution refers to a *pair* of values, namely the indices of each dimension.

Consider this two-dimensional distribution, for example:

$$p_X(\boldsymbol{x}) = \begin{bmatrix} 0.1 & 0. & 0.1 & 0.2 \\ 0. & 0. & 0.1 & 0.1 \\ 0.2 & 0. & 0. & 0.2 \end{bmatrix}$$

The sum of all elements is 1, so $p_X(\boldsymbol{x})$ is a PMF. Note that I have replaced x with $\boldsymbol{x}$, a vector. We can also write $p_X(x, y)$ to emphasize that we have two dimensions.

The distribution says that $P(X = 0, Y = 0) = 0.1$ while $P(X = 2, Y = 3) = 0.2$, that is, the value of the variables are the indices of the rows and columns of the matrix representing the distribution. Two-dimensional probability distributions show up when considering joint distributions—how often a pair of random variables appear together with some combination of values.

We'll use our existing sampling techniques to draw samples from a two-dimensional distribution by unraveling the distribution, sampling, and converting the samples back to two-dimensional pairs. For example, unraveling the previous distribution gives us:

$$p_X(x) = [0.1, 0.0, 0.1, 0.2, 0.0, 0.0, 0.1, 0.1, 0.2, 0.0, 0.0, 0.2]$$

If we sample from this distribution using one of the aforementioned algorithms, we'll get samples in the range $[0, 11]$. To convert the samples back to pairs, we must undo the raveling, meaning we need to know the number of rows and columns in the original two-dimensional distribution.

Let's walk through an example. Run *discrete_ravel.py* with this command line:

```
> python3 discrete_ravel.py 1000 mt19937 10101
```

The output consists of four sections. The code itself samples from the previous $p_X(x)$ by unraveling it, sampling, then mapping the samples back to (x, y) pairs that represent the frequency with which combinations of x and y appear. If all goes well, one- and two-dimensional histograms of these frequencies should approximate $p_X(x)$.

The first output line gives us:

```
[0.091 0.   0.1   0.204 0.   0.   0.091 0.104 0.205 0.   0.   0.205]
```

This is a PMF generated from the samples drawn using the unraveled histogram. The values are all around 0.1 and 0.2, which is encouraging.

The second output block is the same estimated PMF remapped to two dimensions:

```
[[0.091 0.    0.1   0.204]
 [0.    0.    0.091 0.104]
 [0.205 0.    0.    0.205]]
```

This looks very much like $p_X(x)$.

The estimated PMF looks right. As for the sampled values, here are the first eight drawn from the unraveled PMF:

```
[ 3 11 11  7  3  0  7  8]
```

Mapped back to pairs, these samples become:

```
[(0,3), (2,3), (2,3), (1,3), (0,3), (0,0), (1,3), (2,0)]
```

The conversion from one-dimensional sample to two-dimensional pair is

$$(x, y) = (z \div 4, \ z \bmod 4)$$

where z is the 1D sample value and $\div$ means integer division. The 4 comes from the number of columns in the two-dimensional PMF.

Listing 12-6 shows the unravel, sample, remap process.

```
prob2 = np.array([[0.1, 0.0, 0.1, 0.2],
                  [0.0, 0.0, 0.1, 0.1],
                  [0.2, 0.0, 0.0, 0.2]])
prob = prob2.ravel()

z = np.array([Sequential(prob,rng) for i in range(N)])
h = np.bincount(z, minlength=len(prob))
h = h / h.sum()

print(h)
print(h.reshape((3,4)))
print(z[:8])

x,y = np.unravel_index(z[:8], prob2.shape)
print([i for i in zip(x,y)])
```

Listing 12-6: Sampling a two-dimensional PMF by unraveling

The two-dimensional PMF is in prob2, which unravels to become the one-dimensional PMF prob. The second paragraph samples from prob as we did earlier. Notice rng, an instance of our RE class. I'm ignoring some of the code head of *discrete_ravel.py*, so make sure to review the file itself. As before, the sample counts come from bincout, which we then turn back into a one-dimensional PMF by dividing h by the sum of the counts.

Three of the four outputs come next as h, h reshaped as a 3×4 matrix, and the first eight samples from z.

When mapping samples to pairs, we save time by calling unravel_index, which needs the one-dimensional indices along with the shape of the source array—here 3×4 from prob2. NumPy returns the *x*- and *y*-coordinates, so a pair is (x_0, y_0) and so on, as given by the list comprehension using zip.

We can also use this unraveling approach for distributions of more than two dimensions. If we have three random variables—*X*, *Y*, and *Z*—then samples from a three-dimensional PMF, $p_{XYZ}(x, y, z)$, are triplets, (x, y, z), according to the probability with which a particular combination of values appears. We unravel, sample in one dimension, and use unravel_index to map back to triplets. Bear in mind that as the dimensionality increases, the number of samples necessary to reasonably approximate the distribution goes up dramatically.

Suppose every random variable in our system takes on one of 10 possible values. If we have only one variable, we must sample from a probability distribution representable as a vector of 10 elements. With two random variables, we need a matrix to represent the joint distribution, a $10 \times 10 = 100$-element vector when unraveled. If we have three random variables, we unravel to a vector of $10 \times 10 \times 10 = 1{,}000$ elements; for four random variables, we need 10,000 elements.

Fixing the number of values at 10, an *n*-dimensional PMF unravels into a vector of 10^n elements—the distribution size scales exponentially with dimensionality. Therefore, this trick works best for only two or three dimensions.

Images

Now for a bit of fun. The code in *discrete_2d.py* knows how to use grayscale versions of images as discrete two-dimensional probability distributions, so we can sample from them. A grayscale image is a matrix of integer values from 0 to 255, making an unraveled grayscale image immediately useful to FLDR as a distribution.

Samples become pixel locations. The higher the image intensity at a pixel, the more likely it is to be sampled. Therefore, if we draw enough samples, scale them to [0, 1], and multiply by 255, we can transform the estimated distribution back into an image and compare it with the original. That's a lot of words; let's look at some code.

The four paragraphs in Listing 12-7 present the essential code, minus imports and command line processing.

```
image = Image.open(iname).convert("L")
row, col = image.size
row //= 2
col //= 2
image = np.array(image.resize((row,col),Image.BILINEAR))
p = image.ravel()
probabilities = [int(t) for t in p]

x = fldr_preprocess_int(probabilities)
z = np.array([fldr_sample(x) for i in range(N)])

x,y = np.unravel_index(z, (col,row))
im = np.zeros((col,row))
for i in range(len(x)):
    im[x[i],y[i]] += 1
im = im / im.max()

os.system("rm -rf %s; mkdir %s" % (oname,oname))
Image.fromarray(image).save(oname+"/"+os.path.basename(iname))
Image.fromarray((255*im).astype("uint8")).save(oname+"/histogram2d.png")
```

Listing 12-7: Treating images as two-dimensional distributions and sampling from them

To turn the input image into a one-dimensional distribution, we first load the image, resize it to half its original dimensionality, and then unravel it into a list of pixel intensities, [0, 255]. Using a list comprehension with int is necessary because FLDR doesn't work with NumPy arrays.

The next paragraph configures FLDR (x) and then uses it to draw N samples (z), with N given on the command line.

Samples in hand, unravel_index turns the one-dimensional samples into (x, y) pairs, or pixel locations. We then use the pixel locations to populate im, a two-dimensional histogram counting the number of times FLDR selected each pixel. To convert im into an image, we must scale it so that the most often sampled pixel has a value of 1, which we get by dividing by the maximum.

The final few lines of code create an output directory and dump the original and sampled images into it. We must multiply im, now [0, 1], by 255 and make it an unsigned int before writing it to disk as an image.

Run *discrete_2d.py* without arguments to learn the command line options. Try experimenting with the images in *test_images* and those in *images*. The latter contains high-contrast images that might make it easier to see where samples are coming from, especially the inverted images (those with "_inv" in their filename). The ramp images present a linear, quadratic, and cubic ramp, from left to right. We'll sample brighter regions first as they're more probable.

Figure 12-6 shows one of the high-contrast images where white areas are most likely to be sampled. This version of the image prints well. The inverse version requires far fewer samples in general.

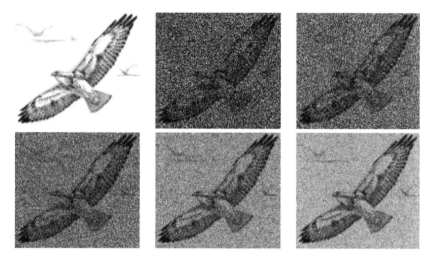

Figure 12-6: Original hawk image (top left) and sampled images with an increasing number of samples

In Figure 12-6, the original image is in the upper left, followed by sampled images using 60,000, 120,000, 240,000, 480,000, and 960,000 samples, left to right and top to bottom. I used this command line

```
> python3 discrete_2d.py images/hawk.png 120_000 tmp mt19937 19937
```

varying the number of samples as needed.

This experiment concludes our investigation of sampling from discrete distributions. Let's move on to the more mathematically relevant case of sampling from continuous distributions and learn some new techniques, culminating in our introduction to the world of Markov Chain Monte Carlo.

Continuous Distributions

Let's change focus to consider continuous distributions, represented by PDFs that admit any real number input over their range. The techniques of the previous section no longer work in this case—at least not without alteration—but other methods exist, three of which we'll explore: inverse transform sampling, rejection sampling, and Markov Chain Monte Carlo.

Inverse Transform

We represent continuous distributions by PDFs. Note the word *function*, which tells us there's a mathematical relationship describing the shape of the PDF. The CDF for a given PDF is an integral:

$$F(x) = \int_{-\infty}^{x} f(t)\, dt$$

The integral is the continuous version of summing discrete probabilities. It represents the area under the PDF from $-\infty$ to some x. Replace $-\infty$ with any value below which the PDF is always zero.

The CDF runs from 0 up to 1, meaning that a plot of the CDF produces y-axis values that begin with 0 and end with 1; review Figure 12-1's CDFs. If we pick a random value on the y-axis of the CDF plot, slide horizontally from there to the curve, and move down to the x-axis, we'll have a sample from the PDF. Pick another y-axis starting point and repeat the process to get a new x and yet another sample from the PDF. Uniformly sampling y-axis values in $[0, 1]$ produces x values that, when histogrammed, follow the form of the PDF.

To express this process mathematically, flip the graph of a function, $F(x)$, along the line $y = x$ (which runs 45 degrees up from the x-axis in the first quadrant), and you have a graph of the inverse of the function, $F^{-1}(x)$, if it exists. The inverse flips x and y values, meaning inputs to the inverse function act like y values for the function, and the output of the inverse function is the x producing that input for the function itself.

Therefore, if we know the functional form of the inverse of a function representing a CDF, we can sample from the PDF by selecting random values in $[0, 1]$ as inputs and keeping the outputs of the inverse CDF as the desired samples. This process is *inverse transform sampling*.

Let's work through an example. Suppose we want to draw samples from an *exponential distribution*, whose PDF is

$$f(x) = \lambda e^{-\lambda x}$$

with λ (lambda) being a constant that decides how quickly the PDF decays from a maximum of λ at $x = 0$. The most probable samples from this PDF are close to zero, with samples farther from zero less likely.

The CDF for this PDF is an integral:

$$F(x) = \int_{0}^{x} \lambda e^{-\lambda t}\, dt = \lambda \left[-\frac{1}{\lambda} e^{-\lambda t} \Big|_{0}^{x} \right] = - \left[e^{-\lambda x} - 1 \right] = 1 - e^{-\lambda x}$$

Therefore, $F(x) = 1 - e^{-\lambda x}$. If we find the inverse of this function, we can generate exponentially distributed samples from uniformly distributed inputs. To find the inverse, set the CDF equal to u (for *uniform sample*) and solve for x:

$$u = 1 - e^{-\lambda x}$$
$$1 - u = e^{-\lambda x}$$
$$\log(1 - u) = -\lambda x$$
$$\frac{-\log(1 - u)}{\lambda} = x = F^{-1}(u)$$

We now have $F^{-1}(u)$, a mapping from uniform inputs u in $[0, 1]$, the range of the CDF, to x, a value selected based on the shape of the exponential distribution PDF. Before putting the inverse function to work, let's make one more observation.

We don't care, specifically, about the exact pairing of this u to that x in terms of a sequence of u's. We plan on picking u values at random. Because u is in $[0, 1]$, $1 - u$ is also in $[0, 1]$, but "flipped" along the u-axis because it's the complement of u, giving us two values that sum to 1. So, we can replace $1 - u$ in $F^{-1}(u)$ with u, and our samples will still be from the exponential distribution. This step isn't necessary, but it makes the plot of the inverse function look less strange to us, as we're used to curves that decay from a high point as x increases to the right.

Figure 12-7 shows a plot of $F^{-1}(u) = (-\log u)/\lambda$, where specific u values have been mapped to their respective x outputs.

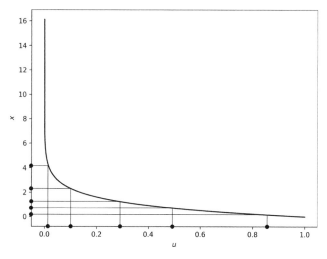

Figure 12-7: Inverse transform sampling from $e^{-\lambda x}$ using $-\log(u)/\lambda$

The distribution of the x values—or a properly scaled histogram of many x values from many u inputs—will become a better approximation of $\lambda e^{-\lambda x}$ as the number of samples increases.

The file *inverse.py* samples from functions supplied as the inverse of their CDF. In other words, we give the code $F^{-1}(u)$ and the corresponding PDF, $f(x)$, along with the desired number of samples, and it gives us the samples, along with a plot of the PDF and the histogram of the samples.

Let's use the code to draw samples from $f(x) = 2e^{-2x}$. The inverse CDF for this PDF is $F^{-1}(u) = (-\log u)/2$. To draw 1,000 samples, use a command line like so:

```
> python3 inverse.py 1000 "-np.log(u)/2" "2*np.exp(-2*x)" tmp minstd 90210
```

The first argument is the desired number of samples. The second, enclosed in double quotation marks, is a NumPy-aware version of the code implementing $F^{-1}(u)$ that uses NumPy functions and u as the independent variable. The next argument, also enclosed in double quotes, is $f(x)$, the PDF. Note that it's a function of x, not u. The remaining arguments are the output directory and the usual randomness source with an optional seed.

Figure 12 shows the output of *inverse.py* for 1,000 (left) and 10,000 (right) samples.

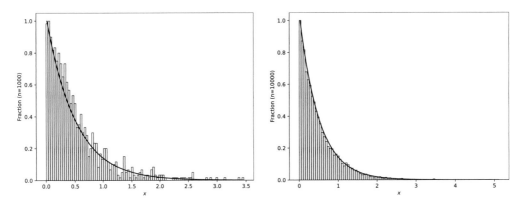

Figure 12-8: 1,000 (left) and 10,000 (right) samples from $2e^{-2x}$

The code scales the output so the curve and histogram match. The samples follow the desired distribution, with more samples better representing the PDF. The 10,000 samples are beginning to select x values farther from zero.

Let's walk through another example. The Kumaraswamy distribution is like the beta distribution, but the functional forms of the PDF and CDF are conducive to inverse sampling. Specifically:

$$f(x) = abx^{a-1}(1-x^a)^{b-1}$$
$$F(x) = 1 - (1-x^a)^b$$
$$F^{-1}(u) = (1-(1-u)^{1/b})^{1/a}$$

Here, a and b are constants that define the shape of the distribution, much like the a and b of the beta distribution. I leave it as an exercise for you to show that $F^{-1}(u)$ comes from $F(x)$.

Let's draw samples from this distribution for $a = 2$ and $b = 5$. The command line we need is:

```
> python3 inverse.py 10000 "(1-(1-u)**(1/5))**(1/2)"
                       "10*x**1*(1-x**2)**4" kumaraswamy pcg64 42
```

The resulting plot is in Figure 12-9. As expected, the samples follow the shape of the distribution.

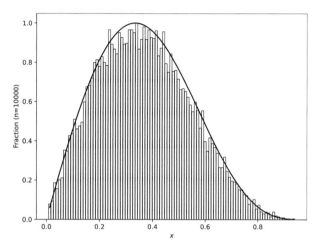

Figure 12-9: Sampling Kumaraswamy (2,5)

The primary code in *inverse.py* is straightforward because we supply the PDF and inverse CDF on the command line in a form that lets us use Python's eval function:

```
samples = np.zeros(N)

for i in range(N):
    u = rng.random()
    samples[i] = eval(ifunc)
```

That's all there is. We create samples to hold the N requested samples, and then loop to generate samples[i] from u by evaluating the inverse CDF function passed from the command line as the ifunc string. The remainder of *inverse.py* creates the output plot.

Inverse transform sampling is direct and works from closed-form functions, but it's of limited applicability because of the two conditions that must be met to allow its use. The PDF must produce a closed-form CDF, and that CDF must be invertible to get $F^{-1}(u)$. This doesn't happen too often, especially for arbitrary continuous PDFs. In "Exercises" on page 358, I suggest another PDF/CDF combination that works with *inverse.py*, but only if you restrict u to something other than $[0, 1)$.

Let's explore the next continuous PDF sampling algorithm, rejection sampling. Unlike inverse transform sampling, rejection sampling works with arbitrary PDFs.

Rejection

We want to draw samples from a function $q(x)$ so that a histogram of many samples from the function converges on the shape of the function itself. While we don't know how to sample directly from $q(x)$, we can sample from a *proposal function* that we'll call $p(x)$. If we find a constant, c, such that

$$q(x) \leq cp(x), \ \forall x$$

then we can use samples from $p(x)$ to draw samples from $q(x)$. Recall that $\forall$ means "for all."

First, we draw a sample from the proposal function, $x \sim p(x)$, where $\sim$ means "draw a sample from." This gives us a candidate x position.

Next, we pick a y value that's some fraction of the way up from x but still less than $cp(x)$. In other words, we pick a uniform value in the range $[0, cp(x)]$, or $y = ucp(x)$ for some u in $[0, 1]$.

If $y \leq q(x)$, we keep x as a sample from $q(x)$; otherwise, we reject x and repeat with another sample from $p(x)$. We stop when the desired number of samples from $q(x)$ have been kept.

Figure 12-10 shows the situation for two candidate x positions.

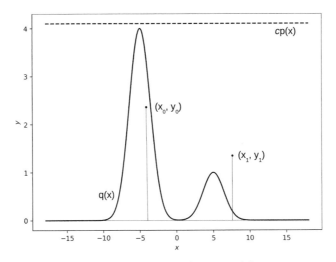

Figure 12-10: Rejection sampling with two candidate x positions

The solid curve is $q(x)$, the function we want to sample from. The dashed curve, here a uniform value over the range of $q(x)$, is $cp(x)$.

Let's look at $x = x_0$ first. The algorithm says to sample from $p(x)$, which gives us x_0. Next, we pick a y some fraction of the way up from x_0 to $cp(x_0)$. We can write this as $y_0 = u_0 cp(x_0)$. While we know that y_0 will always be less than or equal to $cp(x_0)$, we're wondering whether y_0 is less than $q(x_0)$. For x_0, this is the case, so we accept x_0 as a valid sample from $q(x)$.

In x_1, $y_1 = u_1 cp(x_1)$ is greater than $q(x_1)$, so we reject x_1 as a valid sample from $q(x)$ and the process repeats.

Algorithmically, the process boils down to the following:

1. $x \sim p(x)$.
2. $u \sim U[0, 1)$.
3. If $ucp(x) \leq q(x)$, accept x as a sample; otherwise, reject x.
4. Repeat from step 1 until we've accepted N samples.

You might see the condition of step 3 written as

$$u \leq \frac{q(x)}{cp(x)}$$

which we get by dividing by $cp(x)$. In that form, we're asking whether u is less than the fraction of the way to $cp(x)$ covered by $q(x)$ for the selected x. If not, reject x and try again.

Think of rejection sampling as randomly throwing darts at the xy-plane. If the y-coordinate of the dart is less than both $cp(x)$ and $q(x)$, we accept the dart's x-coordinate as a sample from $q(x)$. In effect, we're keeping all the x-coordinate values for darts that land under the $q(x)$ curve. We did the same in Chapter 3 to estimate π.

Let's put this process into practice with *rejection.py*:

```
> python3 rejection.py
rejection <N> <proposal> <c> <func> <limits> <outdir> [<kind> | <kind> <seed>]

    <N>         - number of samples
    <proposal>  - uniform|normal_mu_sigma (e.g. normal_0_1)
    <c>         - proposal multiplier (e.g. 1)
    <func>      - function to sample from (e.g. 2*x**2+3)
    <limits>    - lo_hi limit on sampling range (e.g. -3_8.8)
    <outdir>    - output directory name (overwritten)
    <kind>      - randomness source
    <seed>      - seed
```

Rejection sampling works for any proposal function, $p(x)$, so long as we can draw samples from it, but *rejection.py* restricts us to two options: a uniform distribution, represented by the dashed line in Figure 12-10, and a normal distribution with a given mean (μ) and standard deviation (σ). The Box-Muller transform lets us sample from a normal distribution (see Chapter 1).

Let's reproduce the example in Figure 12-10. The proposal function is a uniform distribution multiplied by 4.1, as this puts the proposal function just above the highest part of the sampling function, $q(x)$

$$q(x) = e^{-\left(\frac{x-5}{2}\right)^2} + 4e^{-\left(\frac{x+5}{2}\right)^2}$$

which is the sum of two normal curves centered at ± 5, with one being four times higher than the other.

Now, we sample:

```
> python3 rejection.py 100000 uniform 4.1
    "np.exp(-((x-5)/2)**2)+4*np.exp(-((x+5)/2)**2)" -18_18 reject0 pcg64 1313
832745 trials to get 100000 samples (30.7419 s)
```

The output tells us we need over 830,000 candidate samples to get the requested 100,000. That's a conversion rate of 12 percent, meaning we rejected 88 percent of the candidates. The efficiency of rejection sampling depends critically on the closeness between the proposal function, $cp(x)$, and the sampling function, $q(x)$. The closer the proposal function is to the sampling function, the better.

Figure 12-11 shows a histogram of the samples from $q(x)$. The proposal function is in Figure 12-10. Note that rejection sampling doesn't care whether $p(x)$ and $q(x)$ are normalized (in which the area under the curves is 1). So long as $cp(x)$ is above $q(x)$, all will be well (if perhaps slow).

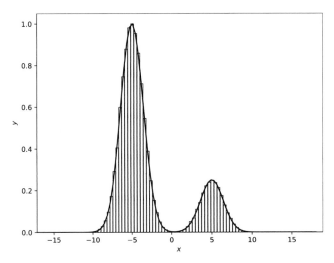

Figure 12-11: Sampling with a uniform proposal function

Let's use a normal curve for the proposal function:

```
> python3 rejection.py 100000 normal_0_1 4
    "np.exp(-((x-5)/2)**2)+4*np.exp(-((x+5)/2)**2)" -18_18 reject1 pcg64 1313
1511344 trials to get 100000 samples (62.7675 s)
```

The proposal function is now a normal curve with a mean of 0 and a standard deviation of 1. The multiplier is 4.

The output is in *reject1*. We're told that we need 1.5 million candidates to get 100,000 samples for a conversion rate of 6.6 percent. Figure 12-12 shows the histogram (left) and a plot of the proposal and sampling functions (right). The plots might seem strange to you, for good reason.

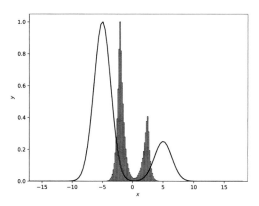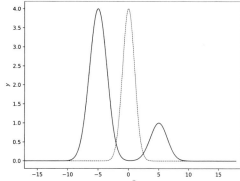

Figure 12-12: Using N(0,1) as the proposal function

The proposal function is the dashed curve on the right in Figure 12-12. It's centered between the normal curves making up $q(x)$, and is larger than $q(x)$ over only a small region around zero. The algorithm can't select samples in areas where $p(x) < q(x)$; therefore, it selects only in the region covering the overlap between the right part of the leftmost normal curve and the left part of the rightmost normal curve.

In the histogram on the left, the code scales $q(x)$ and the histogram of the samples, so both have 1 as their maximum y value. The histogram peaks at two locations, the left and right maxima of the overlap region. While not what we're after, the output from this run is correct, given the constraints.

Let's try a few more examples. The file *run_rejection_examples* contains Listing 12-8.

```
python3 rejection.py 100000 normal_0_20 4.2
    "np.exp(-((x-5)/2)**2)+4*np.exp(-((x+5)/2)**2)" -18_18 reject2 pcg64 1313
python3 rejection.py 100000 normal_-5_2.4 4
    "np.exp(-((x-5)/2)**2)+4*np.exp(-((x+5)/2)**2)" -18_18 reject3 pcg64 1313
python3 rejection.py 100000 uniform 4
    "np.exp(-((x-5)/2)**2)+4*np.exp(-((x+5)/2)**2)" -11_4 reject4 pcg64 1313
python3 rejection.py 100000 uniform 158
    "2*x**2+3" -3_8.8 reject5 pcg64 1313
```

Listing 12-8: Additional examples

Figure 12-13 shows the generated plots for *reject2* through *reject5* from top to bottom.

The topmost plot covers the entire $q(x)$ range, with samples drawn from each peak. The next row shows samples from only the left peak, as the proposal function covers it and a tiny part of the right peak. The third row limits selection to $x \in [-11, 4]$, restricting the range of samples while still mirroring the shape of $q(x)$. The final row of Listing 12-8 switches to a new function, $q(x) = 2x^2 + 3$, and a uniform proposal function.

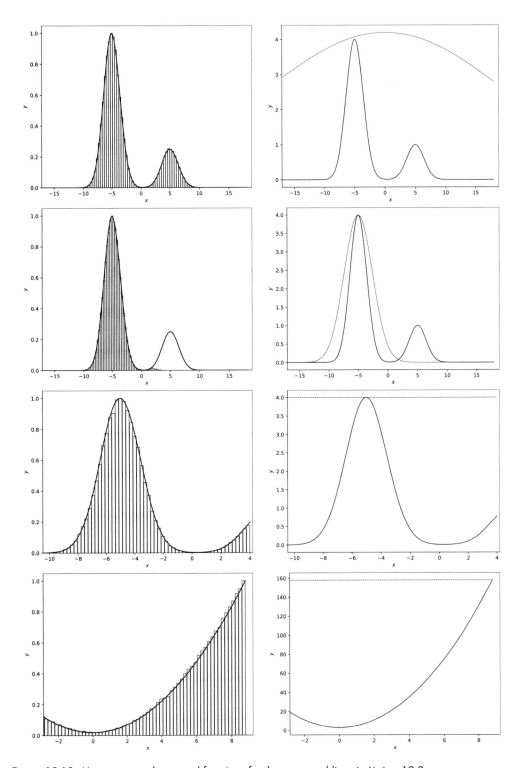

Figure 12-13: Histograms and proposal functions for the command lines in Listing 12-8

Experiment with *rejection.py* with $q(x)$ functions and proposal functions that either completely or partially cover $q(x)$. Remember to pick a c such that $q(x) < cp(x)$ for the regions from which you want to sample.

Rejection sampling isn't restricted to one dimension. For example, if we have $q(x, y)$, we can draw samples as long as we can sample from $p(x, y)$. The uniform function, which is 1 for all (x, y) points, is an easy $p(x, y)$ to use. A multivariate normal distribution will also work, though it's harder to code and visualize. The algorithm remains the same, but instead of drawing x from $p(x)$, we draw (x, y) from $p(x, y)$—a random point in 2D space. The test is still $ucp(x, y) \leq q(x, y)$, or:

$$u \leq \frac{q(x, y)}{cp(x, y)}$$

If the condition holds, the point (x, y) is a sample from $q(x, y)$.

The extension to arbitrary dimensions, $\boldsymbol{x} = (x_0, x_1, x_2, \dots)$, follows as long as we can sample from $p(\boldsymbol{x})$. However, as the dimensionality increases, the number of samples rejected tends to increase exponentially (for each new dimension) unless $p(\boldsymbol{x})$ follows $q(\boldsymbol{x})$ very closely, which is quite difficult to do while maintaining easy sampling from $p(\boldsymbol{x})$. This effect, where the utility of rejection sampling decreases with the problem's dimensionality, is a form of the *curse of dimensionality* that frequently plagues machine learning models as well.

The inefficiency of rejection sampling as the dimensionality of the problem increases leads to our final sampling algorithm, which handles high-dimensional problems: Markov Chain Monte Carlo.

Markov Chain Monte Carlo

Our final sampling algorithm is the most powerful: *Markov Chain Monte Carlo (MCMC)*. We learned about Monte Carlo algorithms in Chapter 10. The phrase *Markov chain*, named after Russian mathematician Andrey Markov (1856–1922), refers to a process where the *transition probability* of moving from a current state to a new state depends only on the current state and nothing that came before. Markov chains are helpful in simulations because what happens next depends solely on what the system is currently doing, regardless of its history.

MCMC uses Markov chains to approximate sampling from a complex probability distribution. A *stationary distribution* in Markov chains, typically denoted as π, is a vector in the discrete case or a PDF in the continuous case. Regardless of the initial distribution, walking the Markov chain eventually reaches the stationary distribution governed solely by the transition probabilities if specific criteria are met.

We'll first delve into stationary distributions; then, we'll explore the Metropolis-Hastings algorithm and use it to walk a continuous Markov chain—only to realize that its stationary distribution is the very PDF we want to sample from.

Walking a Markov Chain

Let's work through an example. Fezzes are all the rage this year. Everyone's wearing them, and there are three colors: red, green, or blue. The probability of a person changing their fez color next year depends on the color they wear this year. The probabilities are fixed from year to year. What happens to an initial distribution of fez colors as time passes? Will the distribution of colors change continuously or eventually settle down to a specific, perhaps stationary, distribution?

We encode transition probabilities in a matrix where the row shows a current state (a fez color) and the columns of that row represent the probability of a transition from the current state to a new state—a new fez color, which may be the same as the current.

Consider this transition matrix:

$$P = \begin{bmatrix} 0.53 & 0.05 & 0.42 \\ 0.13 & 0.83 & 0.04 \\ 0.14 & 0.29 & 0.57 \end{bmatrix}$$

The rows are red, green, and blue fezzes, as are the columns. Therefore, if someone is wearing a red fez this year, they have a 53 percent chance of wearing a red fez again next year, a 5 percent chance of changing to a green fez, and a 42 percent chance of donning a blue fez. The row sums to 1, as it must. Similarly, a green fez aficionado has an 83 percent likelihood of continuing to wear a green fez next year, though 13 percent will switch to a red fez, and a rogue 4 percent will go all in on a blue fez.

To use the transition matrix, we need an initial distribution of fezzes:

$$\pi = \begin{bmatrix} 0.70 & 0.24 & 0.06 \end{bmatrix}$$

The vector tells us 70 percent of the population owns a red fez, 24 percent a green one, and only 6 percent a blue one.

To find out what the distribution looks like next year, we need to see what happens to the proportion of the population wearing each fez color when acted on by the transition matrix. Those wearing a red fez transition to new colors according to the transition matrix's first row, [0.53, 0.05, 0.42]. We multiply the red-fez wearers by the transition probabilities to get the fraction of next year's fez colors from those currently wearing a red fez:

$$\begin{bmatrix} 0.7(0.53) & 0.7(0.05) & 0.7(0.42) \end{bmatrix} = \begin{bmatrix} 0.371 & 0.035 & 0.294 \end{bmatrix}$$

For the greens, we multiply the second row of the transition matrix by 0.24, and for the blues we multiply the last row by 0.06. Finally, we sum across to get the new distribution of fez colors.

In the end, these steps involve nothing more than multiplying the current distribution as a row vector by the transition matrix:

$$\pi \leftarrow \pi P = \begin{bmatrix} 0.70 & 0.24 & 0.06 \end{bmatrix} \begin{bmatrix} 0.53 & 0.05 & 0.42 \\ 0.13 & 0.83 & 0.04 \\ 0.14 & 0.29 & 0.57 \end{bmatrix}$$

$$= \begin{bmatrix} 0.4106 & 0.2516 & 0.3378 \end{bmatrix}$$

Next year, 41 percent of the population will wear red fezzes, 25 percent will wear green, and nearly 34 percent blue. The Markov property tells us that the following distribution is this distribution multiplied again on the right by the transition matrix, and so on.

The file *markov_chain.py* accepts an initial distribution of fez colors (π) and a transition matrix (P) and generates the Markov chain until the distribution becomes stationary. To make things more interesting, the distribution of red, green, and blue fezzes is treated as an RGB color so that the output file, *markov_chain.png*, shows the transition from initial to stationary distribution as a color bar running from left to right.

Run the code with the previous values:

```
> python3 markov_chain.py 70 24 6 [[53,5,42],[13,83,4],[14,29,57]]
```

The first three values are the initial distribution: red, green, and blue fezzes. The final argument, which must not contain spaces, is a Python list representing the transition matrix. The values are not precisely the same as before, but the inputs are scaled by their respective sums so that

$$\begin{bmatrix} 70 & 24 & 6 \end{bmatrix} \to \begin{bmatrix} 0.70 & 0.24 & 0.06 \end{bmatrix}$$

and likewise for the individual rows of the transition matrix.

We print the Markov chain

```
[0.7  0.24 0.06]
[0.4106 0.2516 0.3378]
[0.297618 0.32732  0.375062]
[0.25279782 0.39532448 0.3518777 ]
[0.23463791 0.44280374 0.32255835]
[0.22708075 0.47280092 0.30011833]
[0.22383348 0.49081312 0.2853534 ]
[0.22238693 0.50131905 0.27629402]
[0.22171771 0.50733942 0.27094286]
[0.22139651 0.51075104 0.26785245]
[0.22123713 0.5126704  0.26609247]
[0.22115578 0.5137451  0.26509912]
[0.2211133  0.51434497 0.26454173]
[0.22109074 0.51467909 0.26423017]
[0.2210786  0.51486493 0.26405647]
[0.221072   0.5149682 0.2639598]
[0.2210684  0.51502555 0.26390605]
[0.22106642 0.51505738 0.2638762 ]
[0.22106533 0.51507504 0.26385963]
[0.22106473 0.51508484 0.26385043]
[0.2210644  0.51509028 0.26384532]
```

which tells us that the stationary distribution is 22 percent red, 51.5 percent green, and 26.4 percent blue fezzes.

Try experimenting with the code, changing the input distribution to 1 0 0 (100 percent red) or 0 0 1 (100 percent blue). You'll always end up at the stationary distribution, though the number of links in the chain might differ. Then, alter the transition matrix and see what happens.

The only portion of the code worth discussing builds the chain:

```
eps = 1e-5
last = np.array([10,10,10])
chain = []

while (np.abs(d-last).sum() > eps):
    print(d)
    chain.append(d)
    last = d
    d = d @ transition
```

The while loop runs until the difference between the new distribution and the last distribution is less than or equal to eps. The chain list holds the sequence of distributions. We use @ to perform the matrix multiplication, $\pi \leftarrow \pi P$.

The power behind MCMC comes from the fact that the Metropolis-Hastings algorithm, to which we now turn, runs the Markov chain without directly generating it, but as a proxy returns samples from π once the chain is long enough to reach the stationary distribution.

Exploring Metropolis-Hastings

A paper titled "Equation of State Calculations by Fast Computing Machines" appeared in the June 1953 edition of *The Journal of Chemical Physics*. The authors were Nicholas Metropolis, Arianna W. Rosenbluth, Marshall N. Rosenbluth, Augusta H. Teller, and Edward Teller. The paper introduced the *Metropolis algorithm*, named as such because Metropolis's name is first on the paper. However, as so often happens in the thoroughly human enterprise of science, the process that led to the algorithm is disputed. It appears more likely now that the actual inventors were Marshall and Arianna Rosenbluth, not Metropolis. All five authors have since passed away, so it's doubtful we'll ever know the whole story. We'll refer to the algorithm by its modern name, the Metropolis algorithm, knowing full well that credit may belong elsewhere.

In 1970, Wilfred Hastings extended the algorithm to the case where the proposal distribution is not symmetric; hence the algorithm is now known as *Metropolis-Hastings (MH)*. We'll restrict ourselves to a symmetric normal distribution as the proposal distribution, so, technically, we're using only the Metropolis part.

MH generates samples using a proposal distribution in much the same way as rejection sampling; however, in this case, the proposal distribution walks around randomly (we'll learn what that means soon) and, as a consequence, alters a Markov chain distribution. Run the random walk with proper rejection and acceptance of moves for long enough, and eventually, we'll reach the Markov chain's stationary distribution. At that point, the samples MH returns are from the stationary distribution, which is the distribution we want to draw samples from in the first place. How convenient!

NOTE *A full mathematical treatment of MH and what it's doing under the hood is beyond what we tackle here. Those interested will find a good summary on Gregory Gundersen's blog:* https://gregorygundersen.com/blog/2019/11/02/metropolis-hastings.

For our purposes, we'll accept MH's claims and instead look at the random walk version of the algorithm. MH requires a function to sample from, the functional form we want our samples to follow when we make a histogram of their distribution. This is the stationary distribution for a Markov chain, so we'll call this function $\pi(x)$ (not to be confused with the number, π). MH works well in the multidimensional case, but we'll limit ourselves to one dimension, so it's $\pi(x)$ and not $\boldsymbol{\pi}(x)$.

MH also requires a proposal distribution, $Q(x)$. We'll use a normal distribution because it's symmetric, and we know how to sample from it efficiently, $x' \sim N(x, \sigma)$ for a user-supplied σ and mean x.

With $\pi(x)$ and $Q(x)$ on hand, random walk MH is straightforward:

1. Pick an initial sample, x; for example, $x = 1$.
2. Propose a new sample based on the current: $x' \sim N(x, \sigma)$.
3. Define $A = \dfrac{\pi(x')}{\pi(x)}$.
4. Define $\rho = \min(1, A)$.
5. Define $u \sim \text{uniform}(0, 1)$.
6. If $u < \rho$, accept x', $x \leftarrow x'$.
7. Otherwise, keep x.
8. Output x as a sample.
9. Repeat steps 2–8 until all desired samples are collected.

The acceptance value, A, evaluates the function we want to sample from using the current sample position, x, and the proposed sample position, x'. If this value, passed through ρ (rho) to limit it to a maximum of 1, is less than a random uniform sample, accept the proposal (x') as a new sample; otherwise, stick with the current sample, x. Before looping, output whatever x is as a sample from the distribution, $\pi(x)$.

This algorithm is the simplest way to implement MH. In practice, we can make it even simpler because there's no need to define ρ; we can use A as it is because u is always in $[0, 1)$.

In step 2, the new proposal position, x', comes from the proposal function, a normal distribution centered on x, the current sample. This is the random walk part. In step 6, if x' is ultimately accepted, it becomes the new x we use to pick the next proposal position. In other words, the normal curve jumps to a new position on the x-axis when a proposal is accepted. We'll soon generate animations showing this behavior.

We base acceptance or rejection on the value of $\pi(x')$ and $\pi(x)$, that is, the ratio of π's y value at the proposed new sample position and the current. If π has a high value at the current position, the fraction, A, will be small, meaning the comparison in step 6 is less likely to succeed. If the proposal is rejected, x is output again as a sample from π. We want this because π has a high y value at that x. If $\pi(x)$ is tiny, $\pi(x')$ is more likely greater, meaning A is greater than 1. If $A > 1$, $\rho = 1$ and the proposal position will always be selected because $u < 1$. Therefore, the parts of π less likely to be selected when viewing π as a PDF will be less often sampled.

Given this behavior, we can imagine that, in time, the random walk based on samples from the normal distribution will wander over π in proportion to π's value at each position, thereby generating samples in the desired proportions. I haven't commented on σ, the user-supplied parameter to MH, yet. We'll experiment with it shortly and understand its effect then.

As for the definition of A, I've written it as the ratio of $\pi(x)$ evaluated at the current and proposed x positions. This is the Metropolis version of the algorithm, which works with symmetric proposal distributions. If the proposal distribution isn't symmetric, the numerator and denominator each have an additional multiplicative factor. In the symmetric case, this factor is the same for the numerator and denominator, so it cancels.

A is a ratio, and Bayes' theorem writes the posterior as the product of the likelihood and the prior, all divided by a normalizing factor that, in practice, is usually an intractable integral. Since it works with the ratio, MH cancels this integral, so we don't need to compute it in the first place. MH makes it possible to sample from posterior distributions using only the likelihoods and priors. This makes Bayesians very happy and leads to the dramatic quote earlier in the chapter about Bayes and MCMC.

Run the algorithm to see that your samples *don't* follow $\pi(x)$. We've neglected a key statement about the MH algorithm: it doesn't claim to immediately generate samples from $\pi(x)$, but only in the limit, after some period of time. How long a time, and how many samples do we generate before we trust that the samples are coming from $\pi(x)$? There is no foolproof answer to that question. Our experiment with fezzes generally converged to the stationary distribution after a dozen or fewer iterations. That might be the case with MH, but it's generally accepted that complex $\pi(x)$ functions require many thousands of samples or more before they come from $\pi(x)$. Therefore, when we implement MH in code, we'll specify a number of *burn-in* samples, which we'll throw away, and keep only those that come after. We did something similar in Chapter 7 when playing the chaos game to generate points on the attractor of an iterated function system.

This is a random walk algorithm and a Markov algorithm because we randomly draw the next candidate sample, x', from a distribution with a mean value based on the current sample, x. In a random walk, the next position is relative to the current position. It's a Markov algorithm because history doesn't matter; only the current sample position, x, influences any possible new position. Finally, it's a Monte Carlo algorithm because it depends on randomness and isn't guaranteed to generate accurate samples from $\pi(x)$, at least initially.

Let's dive into some code and contemplate the *mcmc.py* file. You'll find code to parse the command line, sample from a normal distribution, and generate a series of plots using the samples—all of which we've seen several times before.

The heart of *mcmc.py* is the MH function (Listing 12-9).

```
def MH(func, nsamples, sigma=1, q=1, burn=1000, limits=None):
    samples = [q]
    while (len(samples) < (burn+nsamples)):
        p = normal(q, sigma)
        if (limits is not None):
            lo,hi = limits
            if (p <= lo) or (p >= hi):
                p = q
        x = p; num = eval(func)
        x = q; den = eval(func)
        if (rng.random() < num/den):
            q = p
        samples.append(q)

    samples = np.array(samples)
    return samples[burn:], samples[:burn]
```

Listing 12-9: A random walk Metropolis-Hastings sampler

As with rejection sampling, func is a string defining $\pi(x)$. The rules for its composition are the same as with *rejection.py*. We ultimately want nsamples' worth of samples, excluding the first burn's worth, which we discard. This explains the while loop condition knowing that the list samples holds all the generated samples.

The body of the while loop is a direct implementation of the MH algorithm, ignoring the explicit definition of A and ρ. First, we sample a proposal position, p, from a normal curve centered on the current sample position, q. Then, if we've given MH limits, they restrict the portion of $\pi(x)$ we sample from in the end. We did the same with rejection sampling.

We define func with x as the independent variable, so we need to call eval and assign to x to get the numerator (num) and denominator (den). Finally, if u is less than num/den, accept p as the new q before appending q to samples.

Once we've acquired all samples—including those marked as burn-in, for plotting purposes—return samples as a NumPy vector after excluding the burn-in samples.

Run *mcmc.py* without arguments to see the command line arguments it expects:

```
> python3 mcmc.py
mcmc <N> <func> <limits> <q> <sigma> <burn> <outdir> yes|no [<kind> | <kind> <seed>]


    <N>         - number of samples
    <func>      - function to sample from (e.g. 2*x**2+3)
    <limits>    - limits for samples (lo_hi, -18_18) or 'none'
    <q>         - initial sample (e.g. 0)
    <sigma>     - proposal distribution sigma (e.g. 1)
    <burn>      - initial samples to throw away (e.g. N//4)
    <outdir>    - output directory name (overwritten)
    yes|no      - show or don't show the initial proposal distribution
    <kind>      - randomness source
    <seed>      - seed
```

There are a lot of arguments here, but we know what most of them do. We want N samples after ignoring the first burn samples. We know that func is a string defining $\pi(x)$. If limits isn't none, it restricts the *x*-axis range sampled.

We use q to supply an initial sample position. Finally, the shape of the proposal function, the normal distribution from which x' is drawn, depends on sigma. If sigma is too small, the normal distribution is narrow, and it's harder to jump to other parts of $\pi(x)$. On the other hand, if sigma is larger, it's easier to sample from all of $\pi(x)$, to a point.

We understand outdir, kind, and seed. The final argument is the required string, either yes or no. If yes, the output plot showing $\pi(x)$ and the histogram of samples will also show the normal distribution centered on the initial q with standard deviation sigma. Read through *mcmc.py* to understand how the output plots are made. Let's run the code to understand what it produces. We begin with this command line:

```
> python3 mcmc.py 100000 "np.exp(-((x-5)/2)**2)+4*np.exp(-((x+5)/2)**2)" none 0 3 10000 tmp
            yes pcg64 2256
100000 samples in 6.7056 s
```

We're asking for 100,000 samples after 10,000 were thrown away as burn-in. We use the same sum-of-two-normal-curves function for $\pi(x)$ as with rejection sampling. The none option opens all of the *x*-axis to sampling, though we'll end up sampling only where $\pi(x)$ is nonzero. The initial sample position is 0 and sigma is 3. Finally, we want to see the initial distribution function in the output plot; we're fixing the pseudorandom generator and seed and dumping all output in tmp.

Figure 12-14 shows the plots *mcmc.py* creates.

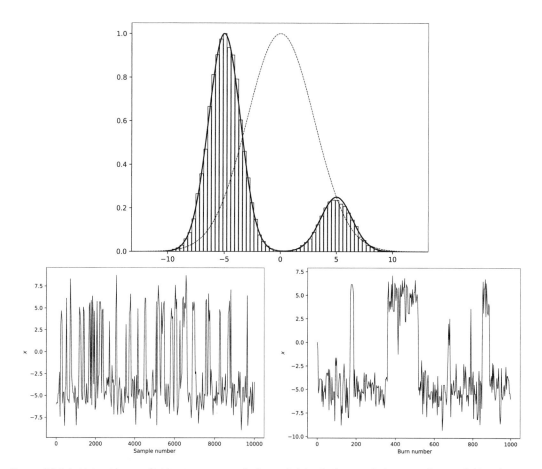

Figure 12-14: Using Metropolis-Hastings to sample from π(x) (top) along with the trace (bottom left) and burn-in plots (bottom right)

The top plot is $\pi(x)$ along with the histogram of the samples MH produced. Also included, because we said yes on the command line, is the initial proposal distribution, a normal curve centered at $x = 0$ with $\sigma = 3$. Note that the curves are scaled to have 1.0 as their maximum value. As with rejection sampling, we're looking for the shape of $\pi(x)$ and the histogram to match.

The graphs on the bottom of Figure 12-14 are *trace plots* that show the sampled x as a function of the sample number. Think of "sample number" as time, so the graphs show how x changes over time. The plot on the left follows samples generated after the burn-in period, while the plot on the right shows the burn-in samples.

The plots were created by the same command line as the top plot, but the total number of samples was set to 10,000, with 1,000 for burn-in. On the left, most samples are near $x = -5$, the peak of the larger normal curve from which $\pi(x)$ is made. The remaining samples center on $x = 5$, the smaller peak. The burn-in plot on the right, however, jumps around near the respective peaks.

There are many things to explore with *mcmc.py*. I'll offer two suggestions as starting points. I recommend running these command lines, then contemplating the output to see if you fully understand it. Remember to look at the trace plots as well, especially for $\pi(x) = 2x^2 + 3$:

```
> python3 mcmc.py 200000 "np.exp(-((x-5)/2)**2)+4*np.exp(-((x+5)/2)**2)"
    none 3 0.1 100000 tmp yes pcg64 1313
> python3 mcmc.py 10000 "2*x**2+3" -3_8.8 0 1 1000 tmp yes pcg64 2233
```

Earlier, I promised that we'd create a movie showing the random walk inherent in our implementation of MH; I'll make good on that promise now. Run this command:

```
> python3 mcmc_movie.py 10000 "np.exp(-((x-5)/2)**2)+4*np.exp(-((x+5)/2)**2)"
    -18_18 0 3 1000 tmp yes 900 pcg64 66
10000 samples in 181.5245 s
```

The file *mcmc_movie.py* is very similar to *mcmc.py*. The output directory, *tmp*, contains a new directory, *frames*. This directory contains files running from *frame_0000.png* to *frame_0899.png* showing each proposed sample (thin vertical line) along with each accepted sample (thick vertical line) as MH moves through its random walk. Use an image viewer that can page through a directory of files in alphabetical order to view the walk as a movie, or download *mcmc_movie.mp4* from the book's GitHub page.

Exercises

Here are some things you may wish to try:

- Update ChooseMap in *ifs.py* to use Sequential (Listing 12-3).
- Use *inverse.py* with $F^{-1}(u) = \frac{-\log(1-u)}{\lambda}$ instead of just u. Does anything change about the samples? What does this $F^{-1}(u)$ look like?
- The Cauchy distribution is characterized by μ and γ. The PDF is

$$p_X(x) = \frac{1}{\pi\gamma\left[1 + \left(\frac{x-\mu}{\gamma}\right)^2\right]}$$

with corresponding CDF:

$$CDF(x) = \frac{1}{\pi}\tan^{-1}\left(\frac{x-\mu}{\gamma}\right) + \frac{1}{2}$$

Try to sample from this function with *inverse.py*. Set $\mu = -2$ and $\gamma = 1$. For example:

```
> python3 inverse.py 30000 "-2+1*np.tan(np.pi*(u-0.5))"
    "1/(np.pi*1*(1+((x+2)/1)**2))" cauchy pcg64 42
```

What do you see? Now, replace *inverse.py* with *inverse_cauchy.py*. What's the difference between the two programs? There are times when algorithms need to be tweaked to succeed.

- Execute the shell script *run_rejection_c* and explain the output in terms of the rejection test for cases when $cp(x) \gg q(x)$ versus just barely exceeding $q(x)$. Hint: consider the likelihood of picking a y value for a given x when $q(x) \approx cp(x)$.

- Experiment with *mcmc.py* using functions that are always positive over some given range. What happens as you make the burn-in larger or smaller? Try changing σ. Do large or small σ values work better? Here's a function to try:

$$p_X(x) = \sin^3(x) + 1$$

I suggest a command line like this one:

```
> python3 mcmc.py 100000 "np.sin(x)**3+1" -9.4248_9.4248 0 3 100000 tmp no pcg64 2256
```

Can you explain the plot and histogram? The limits are, roughly, -3π to 3π.

Summary

In this chapter, we sampled from arbitrary distributions. First, we introduced terminology and concepts from Bayesian inference, a primary user of sampling techniques. After that, we sampled from discrete distributions, which often appear when working with histograms. We learned about sequential sampling and the FLDR, both of which run in $\mathcal{O}(n)$ time—though, practically, the dice roller is some five to seven times faster.

We then experimented with sampling from a two-dimensional discrete distribution by unraveling the two-dimensional distribution to manipulate it as a one-dimensional distribution. As grayscale images are, in effect, two-dimensional discrete distributions, we sampled from them and observed that more intense pixels were sampled most often.

Continuous distributions, characterized by PDFs, came next. In certain cases, sampling becomes a simple process if the cumulative distribution function is invertible. When such is not the case, we explored two approaches: rejection sampling and MCMC with the MH algorithm.

Rejection sampling works well in one dimension, but suffers as the dimensionality of the samples increases. We explored how the algorithm behaves for two proposal distributions, the uniform and the normal, to realize that the closer the proposal function is to the actual PDF, the fewer samples are rejected and the more efficient the algorithm becomes.

When the distribution we want to sample becomes complex or is of high dimensionality, rejection sampling is best replaced by MCMC. We learned about MH in one dimension using a symmetric normal distribution as the proposal distribution. Animated plots showed the progress of the sampling algorithm over time.

RESOURCES

Random Processes

This section includes books that discuss notions of randomness at multiple levels. Expect to be challenged.

Exploring Randomness **by Gregory Chaitin (Springer, 2012)**

This book explores algorithmic information theory, Chaitin's theory about randomness. It's related to Chaitin's earlier books, *The Unknowable* (Springer, 1999) and *The Limits of Mathematics* (Springer, 2002). I recommend brushing up on your LISP before diving in.

Algorithmic Randomness and Complexity **by Rodney G. Downey and Denis R. Hirschfeldt (Springer, 2010)**

A tour de force in regard to theoretical computer science and notions of randomness. Pick and choose topics, or consider the introduction alone.

"Chance versus Randomness" by Antony Eagle (*Stanford Encyclopedia of Philosophy***, 2018)**

This article (*https://plato.stanford.edu/entries/chance-randomness*) is a deeply philosophical exploration of chance, randomness, and their relationship. Check out the "Other Internet Resources" section at the end.

"Stochastic Processes" Lecture Notes by Amir Dembo (Stanford, 2021)

Found at *http://adembo.su.domains/math-136/nnotes.pdf*, these are lecture notes for a graduate course on stochastic (that is, random) processes. Start with the aforementioned *Stanford Encyclopedia* article.

The Drunkard's Walk **by Leonard Mlodinow (Pantheon Books, 2008)**

A popular treatment of randomness in everyday life.

Steganography

Steganography is a relatively obscure topic, especially when compared to its big sister, cryptography.

"Hide and Seek: An Introduction to Steganography" by Niels Provos and Peter Honeyman (IEEE, 2003)

Available from IEEE and other sites on the web (for example, *http://niels.xtdnet.nl/papers/practical.pdf*).

***Steganography Techniques for Digital Images* by Abid Yahya (Springer, 2018)**

An academic presentation of steganography for digital images. Chapters 1 and 2 are most likely to be of interest.

Steganography Lecture Notes by Andrew D. Ker (Oxford, 2016)

This collection of detailed lecture notes can be found at *https://www.cs.ox.ac.uk/andrew.ker/docs/informationhiding-lecture-notes-ht2016.pdf*.

Simulation and Modeling

There are many resources to choose from in this category, given the long history of using computers to simulate and model the real world. I'm offering three to get you started.

***Modeling and Simulation in Python* by Allen B. Downey (No Starch Press, 2023)**

This is an excellent place to continue the explorations of Chapter 3.

***Foundations and Methods of Stochastic Simulation* by Barry L. Nelson and Linda Pei (Springer, 2021)**

This book uses Python to introduce simulation methods. The companion website has code and datasets: *https://users.iems.northwestern.edu/~nelsonb/IEMS435*.

***Computer Simulation Techniques: The Definitive Introduction!* by Harry G. Perros (NC State, 2009)**

A free simulation book courtesy of Harry Perros at North Carolina State University: *https://repository.lib.ncsu.edu/handle/1840.2/2542*.

Metaheuristics: Swarm Intelligence and Evolutionary Algorithms

There are plenty of academic books on swarm intelligence and evolutionary algorithms techniques, but most are quite mathematical and very expensive. I've listed a few, plus some less technical sites and introductions.

***Essentials of Metaheuristics* by Sean Luke (Lulu, 2012)**

An affordable and readable introduction available from the author's website: *https://cs.gmu.edu/~sean/book/metaheuristics*. I recommend starting with this one.

Particle Swarm Central

> This website, *http://www.particleswarm.info*, is a hub for all things related to particle swarm optimization.

Applied Evolutionary Algorithms for Engineers Using Python **by Leonardo Azevedo Scardua (CRC Press, 2021)**

> This book seems right up our alley but is rather expensive.

Hands-On Genetic Algorithms with Python **by Eyal Wirsansky (Packt Publishing, 2020)**

> This affordable book covers a wide range of topics, including applying genetic algorithms to tuning machine learning models.

Swarm Intelligence: Principles, Advances, and Applications **by Aboul Ella Hassanien and Eid Emary (CRC Press, 2015)**

> This rather expensive book runs the gamut of nature-inspired swarm intelligence applications. Regardless of the inspiration, many of these algorithms work well enough in practice.

Swarm Intelligence **(Springer)**

> The link at *https://www.springer.com/journal/11721* goes to the home page for a leading journal related to swarm intelligence applications.

Machine Learning

While a thorough resource list of machine learning is out of the question, the resources here will point you in helpful directions.

Deep Learning: A Visual Approach **by Andrew Glassner (No Starch Press, 2021)**

> This book is a great place to start if you want an overview of modern deep learning without a mountain of math.

The Hundred-Page Machine Learning Book **by Andriy Burkov (2019)**

> This is another popular overview book.

Hands-On Machine Learning with Scikit-Learn, Keras, and TensorFlow **by Aurélien Géron (O'Reilly Media, 2019)**

> This book is for when you're ready to dive into code.

Machine Learning Specialization **by Andrew Ng and others**

> This is a Coursera specialization (found at *https://www.coursera.org/specializations/machine-learning-introduction*) for those who want to learn machine learning from video lectures. Ng founded Coursera along with Daphne Koller.

Neural Networks and Deep Learning **by Michael A. Nielsen (Determination Press, 2015)**

> This free online book is an excellent place to start learning more about the details of neural networks, the heart of deep learning; it is found at *http://neuralnetworksanddeeplearning.com*.

Generative Art and Music

There is much to choose from in this area. Here are a few that caught my eye.

Computational Music Synthesis by Sean Luke (2021)

A free text available from the author's website: *https://cs.gmu.edu/~sean/book/synthesis*.

Chromata by Michael Bromley (2015)

Chromata is a generative digital art tool. Experience it at *https://www.michaelbromley.co.uk/experiments/chromata*.

AI Generative Art Links

A collection of links to many AI-based generative art tools: *https://pharmapsychotic.com/tools.html*.

OpenProcessing

OpenProcessing teaches you how to code in Processing, a language designed for creating generative art; it is found at *https://openprocessing.org*.

SuperCollider

SuperCollider is an open source tool for generative sound and music: *https://supercollider.github.io*.

"How Generative Music Works" by Tero Parviainen

This website, *https://teropa.info/loop/#/title*, contains an excellent presentation on generative music.

Compressed Sensing

Like swarm intelligence and evolutionary algorithms, books on compressed sensing are generally heavy on math.

A Mathematical Introduction to Compressive Sensing by Simon Foucart and Holger Rauhut (Springer, 2013)

This is an academic book, but Chapters 1, 2, 3, and 15 might be worth your consideration.

An Introduction to Compressed Sensing by M. Vidyasagar (SIAM, 2019)

A book similar to the previous but somewhat newer.

Compressed Sensing Magnetic Resonance Image Reconstruction Algorithms by Bhabesh Deka and Sumit Datta (Springer, 2019)

Magnetic resonance imaging is a fascinating practical use of compressed sensing. This book is a current summary of compressed sensing in magnetic resonance imaging.

Compressive Sensing Resources

A long list of compressive sensing resources covering all aspects and application domains: *http://dsp.rice.edu/cs*.

Experimental Design

Proper experimental design is critical to conducting solid research that produces trustworthy results. If you're designing a research study, I strongly recommend finding a statistician or biostatistician to work with.

Design and Analysis of Experiments by Angela Dean, Daniel Voss, and Danel Draguljić (Springer, 2017)

This book is a one-stop resource for the design of experiments, from structure to collection and data analysis. Chapters 1, 2, and 3 are suitable adjuncts to our discussion in Chapter 10.

Design and Analysis of Experiments with R by John Lawson (CRC Press, 2014)

This book is similar to the preceding one but uses R, a popular choice among researchers and stats people.

A First Course in Design and Analysis of Experiments by Gary W. Oehlert (WH Freeman, 2000)

This book is older but freely available. You'll find the text and associated files at *http://users.stat.umn.edu/~gary/Book.html*.

pyDOE

Found at *https://pythonhosted.org/pyDOE*, pyDOE is a Python package for experimental design. I installed it using `pip3` but haven't evaluated it. The website includes documentation. Caveat emptor.

Experimental Design Courses on Coursera

Of the courses on *https://www.coursera.org*, I recommend "Experimental Design Basics" and "Factorial and Fractional Factorial Designs."

Randomized Algorithms

Randomized algorithms are of substantial interest to computer scientists. The resources here are only a taste of the many texts on the subject.

Probability and Computing by Michael Mitzenmacher and Eli Upfal (Cambridge, 2005)

This popular text addresses the same kinds of topics we discussed in Chapter 11 but from a more formal perspective.

The Design of Approximation Algorithms by David P. Williamson and David B. Shmoys (Cambridge, 2011)

Another book, which is older but freely available: *https://www.designof approxalgs.com*.

Notes on Randomized Algorithms by James Aspnes (Yale, 2023)

A long set of lecture notes on randomized algorithms for a spring 2023 Yale course: *http://www.cs.yale.edu/homes/aspnes/classes/469/notes.pdf*.

"Randomized Algorithms" by David R. Karger (MIT, 2002)

In this MIT Open Courseware class, you'll find lecture notes and problem sets with solutions: *https://ocw.mit.edu/courses/6-856j-randomized-algorithms-fall-2002*.

Sampling

These references address MCMC and other advanced techniques, while the Devroye book mentioned in Chapter 12 discusses more straightforward methods.

"Introduction to Markov Chain Monte Carlo" by Charles J. Geyer (Institute of Mathematical Statistics, 1992)

You'll find the PDF for Chapter 1 of the book *Handbook of Markov Chain Monte Carlo* at *http://www.mcmchandbook.net/HandbookChapter1.pdf*.

"Bayesian Inference via Markov Chain Monte Carlo (MCMC)" by Charles J. Geyer (UMN, 2021)

These notes discuss MCMC along with Bayesian inference. You'll find them at *https://www.stat.umn.edu/geyer/3701/notes/mcmc-bayes.pdf*.

***MCMC from Scratch* by Masanori Hanada and So Matsuura (Springer, 2022)**

This book is a step beyond what we discussed in Chapter 12.

Videos

YouTube has many channels that touch on the math behind the topics addressed in this book.

PBS Infinite

Found at *https://www.youtube.com/c/pbsinfiniteseries*, this series discusses many topics, including randomness and pseudorandom generators.

3Blue1Brown

This masterful channel (*https://www.youtube.com/c/3blue1brown*) presents many topics in mathematics. The instruction is clear, with visuals that set the stage for how mathematical concepts should be presented.

Numberphile

Numberphile's videos, found at *https://www.youtube.com/c/numberphile*, typically include a conversation with an expert and drawings on old-school printer paper.

Computerphile

While Numberphile deals with mathematics, Computerphile (*https://www.youtube.com/user/Computerphile*) focuses on computer topics, typically related to mathematics, theoretical computer science, or machine learning.

Mathologer

Mathologer (*https://www.youtube.com/c/Mathologer*) presents longer videos about topics from college-level math courses, particularly calculus. The discussion and presentation are top-notch.

INDEX

RDRAND instruction, 26
    seed, 21
pseudorandom process, 1, 21
p-value, 9

Q

quasirandom generator, 24
Halton sequence, 24
quasirandom process, 1, 21
Quicksort, 315–318

R

radioactive decay, 20
random forest, 173, 198
    bagging, 201
    bootstrapping, 201
    ensembling, 203
    random feature selection, 205
randomization
    block, 276
    combining, 292
    in experiments, 272
    simple, 275
    stratified, 277
randomized algorithm, 295
    Las Vegas algorithm, 296
    Monte Carlo algorithm, 296
randomness engine (RE), 29
random noise, 271.
        See also experimental design
Random Numbers and Computers
        (Kneusel), 29
random process
    atmospheric noise, 17
    coin flipping, 8
        de-biasing, 10
    dice rolls, 11
    hybrid, 26
    physical, 16
    pseudorandom, 1, 21
    quasirandom, 1, 21
    radioactive decay, 20
    roulette wheel, 12
    testing, 6
    truly random, 8
random variable, 1
random walk, 215

RDRAND, 26
RE class, 30, 35–36
Regan, Kenneth W., 305
Reinhart, Alex, 5
reverse Polish notation (RPN), 117
Riddle, Larry, 237
Rosenblatt, Frank, 183

S

sampling, 324
    Fast Loaded Dice Roller, 331
    inverse transform, 340
    Markov chain, 349
        stationary distribution, 349
        transition probability, 349
    Markov Chain Monte Carlo, 349
    Metropolis-Hastings, 352
        algorithm, 353
        burn-in, 354
        trace plots, 357
    rejection, 344
        algorithm, 345
        proposal function, 344
    sequential search inversion, 329
    sound, 240
scientific method, 271
scikit-learn, xxiii, 187.
        See also Glorot initialization
sequential search inversion, 329
Shannon, Claude, 12
Sierpiński, Wacław, 225
simple randomization, 275
simulation, 73–74, 278
    birthday paradox, 80–84
    estimating π, 74–80
    evolution, 84–99
        catastrophic, 93
        genetic drift, 93
        static, 89
    experimental, 278–291
    sanity check, 91
software
    Audacity, 12, 240
    ent, 6
    GenJam, 253
    IFS Construction Kit, 237
    lame, 249

The fonts used in *The Art of Randomness* are New Baskerville, Futura, The Sans Mono Condensed, and Dogma. The book was typeset with $\LaTeX 2_\varepsilon$ package `nostarch` by Boris Veytsman *(2008/06/06 v1.3 Typesetting books for No Starch Press)*.

RESOURCES

Visit *https://nostarch.com/art-randomness* for errata and more information.

More no-nonsense books from **NO STARCH PRESS**

MATH FOR DEEP LEARNING
What You Need to Know to Understand Neural Networks
BY RONALD T. KNEUSEL
344 PP., $49.99
ISBN 978-1-7185-0190-4

MODELING AND SIMULATION IN PYTHON
An Introduction for Scientists and Engineers
BY ALLEN B. DOWNEY
280 PP., $39.99
ISBN 978-1-7185-0216-1

PRACTICAL DEEP LEARNING
A Python-Based Introduction
BY RONALD T. KNEUSEL
464 PP., $59.95
ISBN 978-1-7185-0074-7

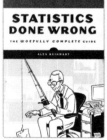

STATISTICS DONE WRONG
The Woefully Complete Guide
BY ALEX REINHART
176 PP., $24.95
ISBN 978-1-59327-620-1

DEEP LEARNING
A Visual Approach
BY ANDREW GLASSNER
768 PP., $99.99
ISBN 978-1-7185-0072-3
full color

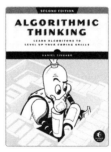

ALGORITHMIC THINKING, 2ND EDITION
Learn Algorithms to Level Up Your Coding Skills
BY DANIEL ZINGARO
480 PP., $49.99
ISBN 978-1-7185-0322-9

PHONE:
800.420.7240 OR
415.863.9900

EMAIL:
SALES@NOSTARCH.COM
WEB:
WWW.NOSTARCH.COM